T0245166

CAMBRIDGE LIBRARY COLLECTION

Books of enduring scholarly value

Physical Sciences

From ancient times, humans have tried to understand the workings of
the world around them. The roots of modern physical science go back to
the very earliest mechanical devices such as levers and rollers, the mixing
of paints and dyes, and the importance of the heavenly bodies in early
religious observance and navigation. The physical sciences as we know them
today began to emerge as independent academic subjects during the early
modern period, in the work of Newton and other 'natural philosophers',
and numerous sub-disciplines developed during the centuries that followed.
This part of the Cambridge Library Collection is devoted to landmark
publications in this area which will be of interest to historians of science
concerned with individual scientists, particular discoveries, and advances in
scientific method, or with the establishment and development of scientific
institutions around the world.

Memoir and Scientific Correspondence of the Late Sir George Gabriel Stokes, Bart.

Lucasian Professor of Mathematics at Cambridge and President of the
Royal Society, Sir George Gabriel Stokes (1819–1904) made substantial
contributions to the fields of fluid dynamics, optics, physics, and geodesy,
in which numerous discoveries still bear his name. The *Memoir and
Scientific Correspondence of the Late Sir George Gabriel Stokes*, edited by
Joseph Larmor, offers rare insight into this capacious scientific mind, with
letters attesting to the careful, engaged experimentation that earned him
international acclaim. Volume I (1907) includes a memoir – culled from
the reminiscences of family, friends, and colleagues – and letters, including
early correspondence with Lady Stokes during the time of their engagement
and early marriage. Professional correspondence covers Stokes' discoveries
in the areas of spectroscopy, fluorescence, and colour vision. The result is an
intimate portrait of a brilliant mathematician – both in the early stages of his
career and at the height of his intellectual powers.

Cambridge University Press has long been a pioneer in the reissuing of out-of-print titles from its own backlist, producing digital reprints of books that are still sought after by scholars and students but could not be reprinted economically using traditional technology. The Cambridge Library Collection extends this activity to a wider range of books which are still of importance to researchers and professionals, either for the source material they contain, or as landmarks in the history of their academic discipline.

Drawing from the world-renowned collections in the Cambridge University Library, and guided by the advice of experts in each subject area, Cambridge University Press is using state-of-the-art scanning machines in its own Printing House to capture the content of each book selected for inclusion. The files are processed to give a consistently clear, crisp image, and the books finished to the high quality standard for which the Press is recognised around the world. The latest print-on-demand technology ensures that the books will remain available indefinitely, and that orders for single or multiple copies can quickly be supplied.

The Cambridge Library Collection will bring back to life books of enduring scholarly value (including out-of-copyright works originally issued by other publishers) across a wide range of disciplines in the humanities and social sciences and in science and technology.

Memoir and Scientific Correspondence of the Late Sir George Gabriel Stokes, Bart.

Selected and Arranged by Joseph Larmor

VOLUME 1

GEORGE GABRIEL STOKES
EDITED BY JOSEPH LARMOR

CAMBRIDGE
UNIVERSITY PRESS

CAMBRIDGE UNIVERSITY PRESS

Cambridge, New York, Melbourne, Madrid, Cape Town, Singapore,
São Paolo, Delhi, Dubai, Tokyo

Published in the United States of America by Cambridge University Press, New York

www.cambridge.org
Information on this title: www.cambridge.org/9781108008914

This edition first published 1907
This digitally printed version 2010

ISBN 978-1-108-00891-4 Paperback

SIR GEORGE GABRIEL STOKES

MEMOIR

AND

SCIENTIFIC CORRESPONDENCE

George G. Stokes

1857.

MEMOIR

AND

SCIENTIFIC CORRESPONDENCE

OF THE LATE

Sir GEORGE GABRIEL STOKES, Bart.,

Sc.D., LL.D., D.C.L., Past Pres. R.S.,

KT PRUSSIAN ORDER *POUR LE MÉRITE*, FOR. ASSOC. INSTITUTE OF FRANCE, *ETC.*
MASTER OF PEMBROKE COLLEGE AND LUCASIAN PROFESSOR OF MATHEMATICS
IN THE UNIVERSITY OF CAMBRIDGE.

SELECTED AND ARRANGED BY

JOSEPH LARMOR, D.Sc., LL.D., Sec.R.S.,

FELLOW OF ST JOHN'S COLLEGE AND LUCASIAN PROFESSOR OF MATHEMATICS.

VOLUME I.

CAMBRIDGE:
AT THE UNIVERSITY PRESS.
1907

Clear mind, strong heart, true servant of the light,
True to that light within the soul, whose ray
Pure and serene, hath brightened on thy way,
Honour and praise now crown thee on the height
Of tranquil years. Forgetfulness and night
Shall spare thy fame, when in some larger day
Of knowledge yet undream'd, Time makes a prey
Of many a deed and name that once were bright.

Thou, without haste or pause, from youth to age,
Hast moved with sure steps to thy goal. And thine
That sure renown which sage confirms to sage,
Borne from afar. Yet wisdom shows a sign
Greater, through all thy life, than glory's wage;
Thy strength hath rested on the Love Divine.

R. C. JEBB (1899).

PREFACE

SOON after the death of Sir George Stokes, representations were made from various authoritative sources, including Lord Kelvin and Lord Rayleigh, that his papers should be carefully examined; as the experience of his friends and correspondents had shown that he was in possession of valuable improvements and advances in scientific subjects, which had not been adequately published to the world.

Fortunately it had been his custom to preserve all his papers; but for many years they had not been sorted, and their great bulk, as they appeared in the numerous packing cases to which they had been consigned from time to time, demanded an organised plan of attack.

They were in the first place sorted and arranged, after ephemeral printed matter had been rejected, by Mr S. Matthews, the Librarian of the Cambridge Philosophical Society, during the summer of 1902.

It then appeared that the bulk of formal manuscript material that was at all suitable for publication was small. What was found consisted largely of rough sheets containing jottings of arithmetical reductions and calculations, of which the net results had been either published by himself or communicated by letter to other workers. By far the greater part of the material was made up of scientific and official correspondence, amounting probably to more than ten thousand letters and memoranda, in many cases containing matter of high scientific value.

These papers, arranged according to dates and the names of his correspondents, were then further examined and a preliminary selection was made from them. The collection thus obtained was further weeded out by a second scrutiny, and again as it was sent to press. It is possible that in some respects the gleanings might have been still further restricted with advantage. But it had been impressed on the Editor that he must err by excess rather than omission: and indeed, as it was hardly likely that the material would be examined again, it seemed reasonable in cases of doubt to lean to the side of inclusion.

While the material for a volume of Scientific Correspondence was being prepared, the question of an authoritative Memoir of the life of Sir George Stokes arose. On ascertaining that his daughter, Mrs Laurence Humphry, would undertake, on behalf of her brother, Sir Arthur Stokes, and the surviving relatives, to prepare materials for the personal part of a Memoir, it was arranged that the two projects should be combined. Accordingly the first section of the present volumes has been prepared by Mrs Humphry: and for detailed advice in difficult questions relating to the choice of material she desires to express grateful acknowledgment to many helpers, including Mr and Mrs Horace Darwin, Mr and Mrs T. F. C. Huddleston, Mr Francis Darwin, and Dr G. W. Prothero, in addition to those friends of Sir George Stokes who have contributed personal appreciations and to many of those mentioned below as having promoted the undertaking as a whole.

The sonnet by the late Sir Richard Jebb, which is printed on the back of the title-page, was published in the *Cambridge Review* at the time of the Jubilee Celebration: it is specially appropriate here, as Sir Richard, just before his final illness, was looking forward to writing out his recollections of his friend, who was also his colleague for several years in the Parliamentary representation of the University.

In the general arrangement of the volumes there is much that could now be improved. No attempt has been made to provide a continuous narrative: indeed the gradual accumulation of fresh material, in the form of groups of letters kindly sent by their owners as the printing went on, would have rendered that course impossible in any case. The printing has been spread over a considerable time, and has claimed the attention of the Editor in the midst of very varied occupations. Considerable care has, however, been taken in the choice of head-lines for the pages, in the expectation that by showing the main contents at a glance they would to some extent take the place of a narrative. Care has also been expended on the index, which is of paramount importance in a work of this kind, if the hope is to be fulfilled

that, in addition to its biographical interest, it will serve as a
book of reference for the scientific history of the half-century in
which Sir George Stokes played so influential a part. With
this object in view, some series of letters to Sir George Stokes,
notably the remarkable one from Prof. Clerk Maxwell inserted
at the beginning of the second volume, have been included,
although the replies are not available.

In order to appreciate to what extent the scientific activities
of a past time may be elucidated in this manner, reference may
be made to the numerous and increasing Continental collections,
and especially to the correspondence of Huygens, recently published
verbatim in ten sumptuous volumes by the Academy of Haarlem.

The same aim for historical perspective supplies a reason for
the inclusion in the volumes of many explanations that are matters
of common knowledge now, but of which even on that account it
is desirable to exhibit the early development. As regards the
considerable proportion of the letters that are expository or didactic,
it has been felt that they may to some extent take the place of
the projected Treatise on Optics, the abandonment of which, owing
to pressure of other scientific occupations, gave rise to so much
regret twenty-five years ago.

A considerable part of the contents of the second volume
relates to Prof. Stokes' activity in the organisation of the scientific
side of the British Meteorological Service. Advantage has here
been taken of the opportunity to collect together various notes from
the Reports of the Meteorological Council and of the Solar Physics
Committee, which had not been received in time to be included in
the *Mathematical and Physical Papers.* Even where these notes
and letters are wanting in finality they embody the fruits of
prolonged experience and thought, and cannot fail in their
collected form to be of value to others essaying to unravel the
complex problems of practical meteorology. The very interesting
series of letters on waves and rollers at sea and their meteorological
relations opens up new developments of a subject eminently his
own, and to which at a later period Helmholtz applied all his
powers.

Another large section of the second volume is concerned with
difficult questions relating to pendulum affairs, in connexion with
the original Indian Gravity Survey, and with subsequent surveys
in other countries which profited by the experience there gained,

in the direction of evading intricate corrections as well as uncertainties arising from climate. This section has been supplemented by extracts from Col. Burrard's recent memoir, describing the substantial corrections which have been found necessary in a revision of the earlier Indian results.

Throughout the volumes the footnotes are introduced mainly by the Editor; it has not been thought necessary to mark them specially, except in the few cases where the subject-matter had footnotes of its own, in which cases the new ones are enclosed in brackets. Considerable condensation has been made in the formal beginnings and endings of letters: in a succession by the same writer these have usually been entirely omitted except in the first letter of the series.

A circumstance that will at once attract attention is the absence of correspondence with Lord Kelvin. This omission arises from wealth rather than deficiency of material. The existing letters from Lord Kelvin, with the complementary letters from Sir George Stokes which Lord Kelvin has promptly supplied, suffice to form a collection by themselves, and a memorial of the lifelong friendship and collaboration of the writers; and it has been decided, with Lord Kelvin's consent, in view of the length to which the present collection has run, that this mode of treatment would be preferable.

A rather full account of the proceedings in connection with the Jubilee Celebration of 1899 has been inserted. It had been felt desirable to place on record for future reference a description of this interesting and unusual event; while the contents of the various formal addresses received from public bodies on the occasion have been printed in an Appendix, in acknowledgment of the courtesy shown by the senders to the University of Cambridge and to Sir George Stokes.

A work of this nature could not have proceeded without very substantial assistance. The Editor desires to acknowledge help from M. Henri Becquerel, who advised regarding the French part of the correspondence, Prof. G. Chrystal, Prof. Sir G. H. Darwin, Prof. F. W. Dyson, who examined and made extracts from the Greenwich letter books, Prof. A. R. Forsyth, Sir Howard Grubb, Dr F. G. Hopkins, Prof. H. Kayser and Prof. G. Quincke, who advised regarding the German letters, Mr E. H. Minns, Librarian of Pembroke College, Sir H. E. Roscoe, Dr E. J. Routh and

Mr Wilfrid Airy, Dr W. N. Shaw, who examined the records of the Meteorological Office, Prof. A. Smithells, Prof. J. J. Thomson, the Master of Peterhouse, Prof. E. Wiedemann, Mr W. E. Wilson, and others, in addition to those already mentioned and to the contributors of sets of letters as set forth in the Tables of Contents. For assistance in linguistic matters he has to thank Mr W. A. Cox, Fellow of St John's College.

Finally, it is a pleasure to acknowledge the efficient assistance, always very cordially rendered, of the officials of the Cambridge University Press: they undertook to set up in type difficult and often imperfect documents direct from the original manuscripts, and in many other ways they have lightened the task of preparation of these volumes.

J. L.

December, 1906

ADDENDA AND CORRIGENDA.

Vol. i. p. 398. The Solar Physics Committee was not constituted until 1879, when Prof. Stokes became Chairman. Sir Norman Lockyer was Secretary of the Royal Commission on Scientific Instruction and the Advancement of Science of 1870–74, which consisted of the Duke of Devonshire, Lord Lansdowne, Sir John Lubbock, Sir J. P. Kay-Shuttleworth, B. Samuelson, W. Sharpey, T. H. Huxley, W. Allen Miller, and G. G. Stokes, with H. J. S. Smith added subsequently, and which published Reports and evidence in two large volumes.

Vol. ii. p. 362, line 5, *for* ramified *read* rarefied; p. 375 headline, *read* sensitive.

CONTENTS OF VOL. I.

SECTION I.

PERSONAL AND BIOGRAPHICAL.

NOTES AND RECOLLECTIONS.

By Mrs Laurence Humphry.

My father's descent can be traced back to Gabriel or Gaberill Stokes, son of John Stokes, born 1680, who appears to have inherited good brains. He was a well-known engineer of Dublin, where he lived in Essex Street. He suggested a plan for supplying that city with water without the use of pumps, wrote a Treatise on Hydrostatics, and designed the Pigeon-House Wall in the Harbour. He was Deputy Surveyor General for Ireland, and maps exist in the Record Office of Dublin which are countersigned by him.

The history of the Stokes family, previous to 1618, is involved in obscurity. There was a very ancient family of that name who lived in Gloucestershire and owned a good deal of land in that County. Two representatives of that family were living in 1876, Dr Thomas Stokes, of Mailsworth, and his nephew Adrian Stokes, of Southport, both advanced in life and both without issue. Dr T. Stokes possessed an old parchment pedigree, by which his descent could be traced back to the year 1312. It gives the name in different forms as de Stokke, afterwards Stokys. On the tomb of Adam de Stokke in the parish church of Great Bedwyn, Wilts., is the figure of a Crusader with the legs crossed. Dr T. Stokes expressed his belief that the family about whom this memoir is written was a younger branch of his own family. One of the grounds of his opinion was that the Irish branch possessed a seal bearing bezants and a crescent used by Dr John Stokes, Scholar and Fellow of Trinity College, Dublin, not a man likely to use a seal to which he was not entitled.

The Irish branch of the family never rose to high rank or wealth, but, what is of greater importance, they have been distinguished by a more than ordinary degree of intellectual power. Gabriel Stokes married Elizabeth King in 1711; in the muster-roll of their descendants during three generations there are five professors, five fellows and eight scholars of Trinity College, Dublin, and some of them have earned a name likely to endure. In more recent times Margaret Stokes, the Irish antiquary, and the Celtic and Indian scholar Whitley Stokes, children of the eminent physician, Dr William Stokes of Dublin, have been celebrated.

Sir George Gabriel Stokes' father was Gabriel Stokes, Rector of Skreen, a village twelve miles south of Sligo, and Vicar-General of Killala; he married Elizabeth, daughter of the Rev. John Haughton, Rector of Kilrea, County Derry, while she was still very young. They had eight children, of whom two died in infancy. Those who lived were

John Whitley, born 1800, Archdeacon of Armagh, Rector of Aughnacloy; who obtained mathematical honours during every year of his course at Trinity College, Dublin; married Caroline Elrington, daughter of the Bishop of Ferns.

William Haughton, born 1802, died 1884; 16th Wrangler, Fellow of Gonville and Caius College, Cambridge, examiner in the Natural Science Tripos at Cambridge in 1865 and 1866; Rector of Denver, Norfolk.

Henry George, born 1804; Rector of Ardcolm, County Wexford, musical, who did well at Trinity College, Dublin, and wrote *The Secret of Life* in blank verse, privately printed; married Ann Maria, daughter of the Rev. W. Hickey.

Elizabeth Mary, the maiden sister, who died in her ninety-fourth year in 1904; it was to her care that George Gabriel, eight years her junior, the youngest member of the family, was specially committed, and a very close friendship existed between them throughout their lives.

Sarah Ellen, born 1813; who married her cousin, Hudleston

Stokes, of the Madras Civil Service, and lived for many years in India.

George Gabriel, born August 13th, 1819, died Feb. 1st, 1903, the subject of these recollections.

He was thus one of a large family living in a country village near the sea. The home-life in the Rectory at Skreen was very happy, and the children grew up in the fresh sea-air with well-knit frames and active minds. Great economy was required to meet the educational needs of the large family; but in the rustic simplicity of a place where chickens cost sixpence each and eggs were to be bought at the rate of five or six a penny, they were easily provided with food. No coach or vessel came near Skreen, so that it was an out-of-the-way, quiet place.

All the children were much influenced by their mother, a beautiful and somewhat stern woman, of whom they stood in awe while they were deeply attached to her. Her husband, her senior by many years, was a man of rather taciturn nature, who had been a scholar of Trinity College, Dublin.

His sister Elizabeth Stokes wrote, some years before her death, that she was away when her brother George was born, having been sent on a visit to England; she first saw him when he was a few months old:

"Such a pleasant-faced, plump, very fair, rosy baby. His cap, for babies in those days always wore caps, had half fallen off his head, showing his pretty hair. He grew into a dear good child, a little passionate, but with a strict sense of honour and truthfulness, I do not think he ever told a lie.

"We had a very large, handsome Newfoundland dog, called Brontë; he and George were great friends. They used to run about, his arm in Brontë's mouth. The child had an enquiring mind and asked many questions, as, for instance, 'But, William, is the hawk a bird, and does he really eat mice?' One day my mother discovered him on the floor measuring a dead bird, and saying to himself, 'That's right,' probably comparing it with some description which he had read. My mother found him very slow in learning to read: she was teaching him out of *Cobwebs to Catch Flies* in words of one syllable, a book which was probably too childish for his mind. One day he asked her if he might read the Psalms instead, and she

thought that she would try him; he got on at once and she had no
further difficulty. I believe that he was very well grounded in
Latin grammar by my father before he went to school.

"When a very little boy he had small-pox. We other children
were never kept from him, and I used to play draughts and cat's
cradle with him to amuse him. George thought that I was playing
the latter so that he might win, which did not suit his strict
sense of honour. One day he pretended to be a bird, and William
said that he ought to eat a worm; he naturally objected, but
when William told him it was the right thing he opened his
mouth. It was a trial on William's part of George's steadfastness
to his principles.

"You ask about his dress. Very little boys were not dressed
like little men in those days, as they are now. They wore jackets
and kilts, a much prettier dress, I think. George had a very
pretty nankeen kilt worked all round in green shamrocks by my
aunt Truelock. Another dress was of green plaid. The disloyal
people were called 'White Boys': George did not like to go out
with his pinafore on, for fear he should be taken for a 'white boy.'

"Walking on the road one day, in passing the smith's forge,
a bar of iron just turning from red to black attracted him. He
clasped his hand round it and got a bad burn; he shivered with
pain, but never cried. He was so brave that my uncle George
called him 'Wellington.'

"He was exact even when a child; one day on being sent to
see what o'clock it was he said, 'I can't tell you what o'clock
it is now, but when I was at the clock it was such an hour.'

"He was very absent-minded sometimes. I remember one
day when sent on a message he said, 'I forget who sent me and
what it was about.'

"He had a tender heart; on hearing a little poem read out
from *Rhymes for the Nursery*, by Ann and Jane Taylor,

> 'What! go and see the kittens drowned
> On purpose, in the yard!
> I did not think there could be found
> A little heart so hard,'

he burst into tears. One day when reading the twenty-second
chapter of Genesis, he was so amused by the names Huz, Buz,
and Pildash, that to keep himself from laughing he had to twist
about; for he had a high sense of reverence. I do not think that

he ever enjoyed riding; but he owed much to my brother William, who trained him to be hardy and brave and to have good walking powers, as William took him with him when he went out shooting. When quite a little boy he used to amuse us by dancing like a dancing dog, his arms up and his hands hanging like paws, and also by repeating the 'Three Blind Mice' in character, and he did both very well.

"He learnt arithmetic from the Parish Clerk, George Coulter, who used to say with delight that Master George had soon found out new ways of doing sums far better than those given in Voster's *Arithmetic*; and clever people were surprised by the questions which he used to solve by the arithmetical rule of False Position. There is a tradition that he did many of the propositions of Euclid as problems without having looked at the book."

In 1832, at 13 years of age, he was sent to the Rev. R. H. Wall's school in Hume Street, Dublin. There he remained for three years, attending school as a day boy, and living during term time with his uncle, John Stokes, a course probably necessitated by the many calls on the limited incomings at Skreen. He pursued the usual school studies, and attracted the attention of the mathematical master by his solutions of geometrical problems. While there the news reached him that his father had suddenly died from heart failure. This made a great impression upon him. I remember that in 1901, when I was wearing a pair of small and unremarkable silver buckles he, usually most unobservant of details in dress, looking earnestly at them requested me not to wear them again, as they were knee-buckles worn by his father on the morning when he was found lying dead in 1834.

While at Dr Wall's school he took lessons in riding; the horse tripped, threw him, and he broke his arm; he had the good sense to go straight to his cousin Gabriel Stokes in Harcourt St. to get it set, instead of returning home and frightening his mother, who was then residing with his Uncle John. He was much interested when the Röntgen Rays shewed the fracture years afterwards. His sister Elizabeth found him caterpillars, and he had the pleasure of seeing "thirteen turn into butterflies" during his convalescence.

In 1835, he went to Bristol College, a school that no longer exists, of which Dr Jerrard, his brother William's friend and a mathematician of some note, was principal. In making his first crossing from Waterford to Bristol the packet "Killarney" nearly sank, and William gave a great account of his coolness in face of danger, telling how the boy took off his great coat so as to be ready to swim.

He remained for two years at Bristol College; he considered that when there he owed much to the teaching of Francis Newman, brother of the Cardinal, a man of charming character and great attainments, afterwards made manifest in many ways, who was then Lecturer in Elementary Mathematics, and subsequently cor responded with him on mathematical subjects when both had become famous. Distinction awaited him at this school, as at the previous one, for proficiency in mathematics, although he said that he had not read so far as some of the senior boys. I have in my possession Roscoe's *Life of Lorenzo de' Medici*, a "Prize presented to George Gabriel Stokes for eminent proficiency in Mathematics at an examination held at the Bristol College in January, 1837, by J. H. Jerrard, D.C.L., Principal." In a letter to him of date June 30, 1837, expressed in frank and intimate terms, Dr Jerrard writes, "I have strongly advised your brother to enter you at Trinity, as I feel convinced that you will in all human probability succeed in obtaining a Fellowship at that College."

In his early days he is related to have been somewhat rash on foot and in the water; he had various narrow escapes amongst the mountains of Westmorland and Cumberland, and he was a venturesome swimmer on the North Coast of Ireland.

He is said to have been even more silent in early life than in later years. This characteristic is supposed to have originated when his brothers laughed at him before he went to Bristol and warned him not to give "long Paddy answers"; in consequence he formed the habit of simply answering yes or no. As a little boy he was subject to violent though transient fits of rage; but this tendency was so completely overcome that in later years, though eager, he was almost uniformly calm and even in temper. He had a real love of Botany and a practical knowledge of it: in youth and early manhood this formed a distraction, and he was in the habit of searching for flowers during his walks. In later years he became more and more engrossed by work and thought, and usually walked with his eyes upon the ground, even passing people he knew well, close enough to brush against them, without perceiving them.

Between the ages of sixteen and seventeen he was keen in the study of butterflies and caterpillars. One day while he was at Cambridge, returning from a walk, he failed to respond to the salutation of some ladies of his acquaintance; when asked the reason of such odd behaviour he answered that he could not bow, as his hat was full of beetles.

He entered the University of Cambridge in 1837, at the age of eighteen years. In his undergraduate days sports were not the fashion with reading men, who took "grinds" or country walks instead. This habit he maintained in youth, and until long past middle life long walks were the custom, both summer and winter, at a pace of nearly four miles an hour. At eighty-three years of age he still went the Grantchester "Grind," of three or four miles, and other equally long walks as his afternoon exercise. When he was an undergraduate everybody dined very early, and he maintained the practice, doubtless conducive to work, for many years afterwards.

My father told me that he never read more than eight hours a day, even before an examination, and held that this was enough for anyone reading mathematics: he added laughingly that he had never been reduced to binding his head up in a wet towel.

Mr Glazebrook, who had undertaken to write an article in *Good Words* for May, 1901, asked me to obtain some information for him; my father in response wrote the following short paper.

"I entered Pembroke College, Cambridge, in 1837. In those days boys coming to the University had not in general read so far in mathematics as is the custom at present; and I had not begun the Differential Calculus when I entered College, and had only recently read Analytical Sections. In my second year I began to read with a private tutor, Mr Hopkins, who was celebrated for the very large number of his pupils who obtained high places in the University examinations for Mathematical Honours. In 1841, I obtained the first place among the men of my year as Senior Wrangler and Smith's Prizeman, and was rewarded by immediate election to a Fellowship in my College. After taking my degree

I continued to reside in College and took private pupils. I thought I would try my hand at original research; and, following a suggestion made to me by Mr Hopkins while reading for my degree, I took up the subject of Hydrodynamics, then at rather a low ebb in the general reading of the place, notwithstanding that George Green, who had done such admirable work in this and other departments, was resident in the University till he died. My earlier papers are mostly on Hydrodynamics, and have for the most part been printed in the *Cambridge Philosophical Transactions*. Perhaps the most important of these papers is one in which the equations of a viscous fluid are applied to the determination of the resistance of the air to pendulums. In 1849, when thirty years of age, I was elected to the Lucasian Professorship of Mathematics, and ceased to take pupils. The direction of the Observatory was at that time attached to the Plumian Professorship; and the holder of that office, Professor Challis, used to give lectures in Hydrodynamics, Hydrostatics, and Optics, as had been done by his predecessor Airy. On my election Challis, who was desirous of being relieved of his lectures on Hydrostatics and Optics, so as to be free to take up the subject of Astronomy, which was more germane to his office at the Observatory, agreed with me that I should undertake the course of lectures which he himself had previously given. This turned my attention to the subject of Optics. Having casually heard in 1851 that there was something peculiar about the solution of quinine, I procured some and examined Sir John Herschel's papers on the subject in the *Philosophical Transactions* for 1845. The peculiarity of that phenomenon, taken in connection with the remarkable analysis of light which Herschel had discovered, in relation to the epipolic dispersion of light, but had left unexplained, intensely interested me. At first I took for granted that the blue light coming from the solution could only have arisen from light of the same refrangibility in the incident beam. But in following out the necessary consequences of this assumption, I was compelled to make suppositions as to the behaviour of the fluid towards the incident light, which were of so complex and artificial a character as to bear no resemblance of truth. In thinking over the matter it occurred to me that if we may suppose the blue light given out by the solution to be an effect of the rays of higher refrangibility incident upon it, everything would fall into its place, and the whole of the phenomena be explained in the simplest manner. Such an idea, when once it had occurred to the mind, can be confronted with observation without any difficulty; and experiment showed that the suggested explanation was the true one. A new field of research was thus thrown open, on account of the facility with which the presence of invisible rays of high refrangibility could be made evident. The first of my papers on this subject is published in the *Philosophical Transactions* for 1852, and was rewarded by the Royal Society by the award of the Rumford Medal. In 1849, and for many years after, most of the Professorships in the University, the Lucasian among them, were very inadequately endowed. Accordingly I did not deem it to be inconsistent with the retention of the Professorship at Cambridge to accept a few years later a Lectureship in the Royal School of Mines, though it obliged me to live in London for some part of the year. I resigned this on being appointed as additional Secretary to the Cambridge University Commissioners of 1856."

His writings were chiefly memoirs on mathematical and physical subjects. He specially mentions another publication belonging to pure mathematics; it related to a determination "which was believed to be novel at the time *, of the curious way in which the coefficient of a Bessel's function of a mixed imaginary changes with the argument of the variable, when the function is expressed in a form convenient for calculation for large values of the variable, by means of a semi-convergent series."

This account of the detection of the change of refrangibility of light may be compared with a contemporary memorandum, endorsed "A Discovery," and evidently written as a provisional record, which has been found by Prof. Larmor among his papers.

"I discovered on Monday, April 28th, 1852, that in the phenomenon of interior dispersion a ray of light actually *changes its refrangibility*. In sulphate of quinine (a solution of about $\frac{1}{100}$ part in which is dilute sulphuric acid) the violet rays of a certain refrangibility produce the interior dispersion noticed by Sir D. Brewster, while the invisible, or at any rate barely visible, rays beyond the extreme violet produce the narrow band of light described by Sir J. Herschel (*Phil. Trans.* 1845). The dispersed

[* It was so. See *Math. and Phys. Papers*, Vol. IV. pp. 80, 298.]

ray is compound, although dispersed from a ray of definite refrangibility. In alcoholic tincture of laurel leaves, rendered
almost colourless by exposure to the light, a dispersed blood-red
ray is produced not only by the red rays but by the indigo; that
is to say, *a part* of the compound dispersed ray is blood red. The
red component of the dispersed ray appears to be of definite
refrangibility, the refrangibility, however, being a function of the
refrangibility of the ray from which it was dispersed.

G. G. Stokes."

This fundamental and entirely unexpected discovery fixed the
attention of the scientific world and brought him the award of the
Rumford Medal of the Royal Society in the same year. At the
meeting of the British Association at Belfast in September, 1852,
he delivered one of the two evening lectures, on that subject:
the occasion is thus described by his sister Elizabeth.

  " We fixed upon Friday for our journey to Belfast that
I might go to George's Lecture, and I had that great pleasure,
the pleasure of seeing the very high estimation in which he is
held. He spoke for about two hours and was listened to with the
deepest attention, and such stillness, except a burst of applause
now and again. He was most perfectly at his ease, and spoke so
distinctly that I did not lose a word. When he had concluded
Col. Sabine rose and said he was sure he was only fulfilling the
wishes of the ladies and gentlemen present in conveying to him
their sincere thanks for his kindness in coming forward, and for
the very clear explanation he had given of his discovery; that he
felt sure there were but few present who could follow the subject
in all its details; yet he was also sure there were none who had
not derived pleasure and profit, and that many would look back
with delight to their presence there that evening, as they watched
the onward progress of him whose present discovery was but a
first step, of him who, if God is pleased to spare his life, promises
to be one of the first scientific men of his age or of any other; that
his countrymen have good reason to be proud of him, and so on.
Had Col. Sabine been his father I think he could not have taken
more deep interest in him than he appeared to do."

He considered that many of the men who worked under
him in later years had been much overtrained, and that this

tended to weaken their minds and diminish their power of originality. He had heard of a machine for stuffing live fowls to which he likened the process, and sometimes said, "They are stuffed, they won't do anything more, the thinking has been done for them." He considered this to be the reason why some of those who took very high places in the Mathematical Tripos did so little afterwards. In crossing the fen to Newnham one day I asked him if he thought that there were many "mute inglorious" mathematicians in the world, for instance Newnham millers, who might have done great things, but, for want of opportunity, only counted their sacks and did their accounts. He replied that that kind of genius wanted a good deal of help and cultivation to bring it out; he thought that in the past men of great ability in these directions might easily have been buried and lost, but that now a clever National schoolmaster would discover boys who had ability. But he considered that this sort of power was slow in its development.

He was of course acquainted with a number of foreign languages, that is, he knew them sufficiently to follow the meaning of scientific literature with mathematical signs to guide him; but it was very difficult to him to speak in any foreign tongue. It was interesting to see him sit down with a pamphlet in an unknown language; he took a dictionary and hammered away, "finding the Latin by the sense, not the sense by the Latin."

The days of the Smith's Prize and Bell Scholarship Examinations were always marked days with us, as the house was turned upside-down; we lunched in the drawing-room, and the dining-room mahogany supported the elbows of those who were examined. If they were in awe of my father during their papers, he was quite afraid of them at lunch. He considered it a part of his duty to help to relax their mental strain, and used to lament that he found it so difficult to entertain them and did not know what to say. It once happened that during the recreation interval in the garden after lunch two candidates ran away. It was particularly unfortunate, as one of them had done rather well. The event was long spoken of in the family with bated breath, and afterwards the garden gate was kept locked on these occasions.

He often regretted his habit of silence; sometimes when asked why he had not taken pity on an uncomfortably shy person, he would answer that he wished to speak, but that he could think of nothing worth saying. Though silent he was very sympathetic, and almost everyone felt keenly attracted to him; even children, whom it might have been thought that his abstractedness would have daunted, went to him at once. His little grand-daughter was quite devoted to him, and would coax him to take her to see the ducks, tortoises, and gold-fish in the Botanical Gardens. Had he not married, he might have had a lonely old age; for people did not often visit him socially, even his old pupils. He used to be disappointed on Ray Club evenings when few men came. He would prepare things to show with an eager expression and be ready and anxious to talk about them; and then members did not come, or if they came were more interested in other subjects.

There was something intensely attractive in his personality, in the mixture of calm and activity in his body and mind, in his strength and gentleness, dignity and humility of character, wisdom and absolute simplicity, firm faithful nature and simple courtesy. He was constant in his willingness to give his time and his services, and had a deep regard for the lives and interests of other people. His appearance was a true index of the mind within, which in moments of intense feeling or thought seemed to shine through his bodily frame. His smile was most attractive, coming suddenly upon a gently grave face in a glow of kind sweetness, which gave the impression of light. In stature he was rather short than tall, firmly and strongly knit, strong rather than graceful in person, his head very large but so well moulded and proportioned that it did not seem too large for his height, very finely shaped over the forehead, with a splendid sweep of brow; the eyes rather small than large, of a grey blue tone and bright; his complexion clear and slightly ruddy, giving an impression of health. He used to fluff out his hair, always curly and in later years like a silvered halo round his head, when his problems were too distracting; "making his fur fly," we said. Perhaps the Lowes Dickinson portrait in the Combination Room at Pembroke College* is most like his expression when in deep thought. But he was apt to look bored when being painted, and

* This portrait has been reproduced as a frontispiece to Vol. IV. of the *Mathematical and Physical Papers*.

to draw down the corners of his mouth. Thus the portrait by
Herkomer at the Royal Society is not satisfactory to those who
knew him best; but the busts by Hamo Thorneycroft in the
Hall of Pembroke College and the Fitzwilliam Museum, and the
medallion by G. W. de Saulles, all done at the time of his
Jubilee, are beautiful and true portraits. The medallion by Hamo
Thorneycroft in Westminster Abbey is also an excellent memorial.
Possibly few men have been photographed so often, and the idea
of sun-pictures was always attractive to him; some of the results
were very successful, for instance, Mrs Frederic Myers' photograph,
reproduced, by permission, in Vol. v. of the *Mathematical and
Physical Papers*.

Perhaps his silence was most painful to him when some
foreigner came from a distance on purpose to meet and talk with
him. It might have been imagined that polite and sociable
Frenchmen, interested in the same subjects, and coming with the
express object of conversing with him, would have vanquished
any degree of reserve; but they were not always successful. Yet
he very much enjoyed talking, and hearing conversation. It
seemed on the whole easier to him to converse with women than
with men—that is among unscientific people—and during the last
few years of his life, when less busy, he showed much pleasure in
the society of ladies. He would often sit at tea-time amused by
the chat, and then would suddenly launch into the conversation.
The absence of obligation to talk inclined him to do so.

But anyone evincing a real desire for information or advice
was always sure of attention and kindness. He had none of the
impatience of ignorance and stupidity that is often shown by
clever people, nor did he seem surprised by absence of the most
ordinary knowledge, provided there was no pretence and no
humbug of any kind; these however he did not expect, and he
was never on the watch for them. I do not remember having
heard him speak of any one as stupid, though if questioned he
might say 'Well, perhaps not very bright.' Nor indeed did he
often refer to men as having great intellectual power; but of one
of the younger scientific men he once remarked, 'The breadth of
his knowledge amazed me, but he is so modest that he will never
get credit for all he knows.' The differences in the quality
of his silence were very interesting, also the way in which his
strong character could be displayed through silence and be keenly

felt. If his silence were one of disapprobation it was known at once; and few people have exerted a more potent influence by well-chosen words than he wielded by this power of silence. It was rarely that unkind and unjust things were said before him; inaccurate people became more exact in his presence.

He seldom gave opinions upon the characters of persons, and only when asked to do so, when he would give his unvarnished opinion, but with pain if it were of an unfavourable nature; beginning, " I am disposed to think," and springing slightly on his chair, just hitching it a little, his two hands holding the seat in a way he had when something was drawn from him which he must say because he thought it true, but did not wish to say lest it might be lacking in charity or justice.

It was curious that with this fixed habit of silence he could speak with great ease when formal speech was needful, not only on scientific matters and business, but quite unexpectedly on subjects of a far lighter nature. For instance, he had taken me as his lady to a City dinner, and after it had begun a message came asking him to make one of the speeches of the evening, as the person chosen to do it was prevented from being present. It was reassuring to see him eating his dinner very calmly, and not appearing in the least dismayed. He made a very good after-dinner speech, clearly spoken in well-chosen language, with no harping on previous jokes but with jokes of his own, courteous and easy, neither too long nor too short, and without any of the hesitation which is not unknown on such occasions. For although shy he was not a nervous man.

His marriage was a singularly happy one. He first met Miss Robinson, daughter of the Rev. Thomas Romney Robinson, Astronomer of the Armagh Observatory, at a Meeting of the British Association, and had some difficulty in re-discovering her, as he imagined her to be a daughter of Sir David Brewster. It was but a cursory first interview, but he was so much charmed by her appearance and her manner that he cherished the hope that this might be the lady of his affection. They next stayed together at Lord Rosse's; report said that he proposed to her in the tube of the Great Telescope, but this is absolute fiction.

Dr Thomas Romney Robinson was the son of Thomas Robinson, portrait painter, pupil and friend of Romney, to whose life by

Hayley he considerably contributed. Bishop Percy, who wrote *Reliques of Ancient Poetry*, took a great interest in him as a boy, and helped in his education. He was elected in 1814 a Fellow of Trinity College, Dublin, and for several years lectured in Dublin University as Deputy Professor of Natural Philosophy. He became an eminent and active scientific man and a prominent Fellow of the Royal Society, and in various ways was a very remarkable man.

In connection with his work as a teacher he published a volume entitled *A System of Mechanics*, in 1820. After nine years' residence in Dublin University, and some sojourn in two country parishes, he was appointed head of the Armagh Observatory, where he produced *The Armagh Catalogue of Stars*, in a large volume of more than 900 octavo pages. In recognition of the excellence of this work he received the Copley Medal from the Royal Society in 1862. He worked upon the improvement of the Mural Circle, and had a great share in the construction of the Great Melbourne Reflector, and in the invention of the Robinson or Cup Anemometer. When the Board of Trade established seven first-class meteorological stations, Armagh was selected as one of the seven. But the work of the Observatory was much crippled by the disestablishment of the Irish Church in 1868, which reduced the income from £216 to £60, besides preventing for the future the liberality of the Archbishops of Armagh, whose incomes also were much reduced. He held many other medals and was a knight of the Prussian Order *Pour le Mérite*. He was President of the British Association.

In addition to his scientific knowledge Dr Robinson was a linguist, with a good knowledge of Greek and Latin authors; besides reading Spanish, French and Italian, he had great love and admiration of the old Icelandic Sagas in the original. He was one of the early and devoted admirers of E. FitzGerald's work. He was a most charming man and a delightful talker, for joined to an extraordinary memory he had a most eloquent tongue. He married first Miss Rambaut, the mother of his children, and secondly Lucy Edgeworth, half-sister of Maria Edgeworth, and the original of "Lucy" in her novel *Harry and Lucy*. There is an interesting account in Maria Edgeworth's *Memoirs* of a visit to the Observatory.

Dr Robinson died in February 1882, having nearly reached the age of ninety. My father was to have written an obituary

notice of him for the Royal Society, but the years which followed
were very busy ones, and he never did this piece of work.

It may be imagined that my father found this bright and
talkative Irish home-circle very pleasant after his quiet and lonely
college life, and that he gained in his future father-in-law a most
affectionate and congenial companion.

His suit was successful; but on one occasion a letter of fifty-
five pages about his scientific preoccupations gave room for
misunderstanding, and it seemed that the engagement might
terminate. The passages printed below are chosen from many
letters that were written at this period. At first sight it seemed
as though even these were of too intimate a nature for publication;
but on thinking the matter over and taking advice from friends
the letters were so unlike ordinary love letters, so dignified and
impersonal in their expression, that, written, as he said, to explain
his character, they must be of legitimate interest to others as containing the only self-revelation that he apparently ever consciously
made. They are remarkable also from the curious place which he
assigned to his original investigations; it almost seems as if he
considered them the height of dissipation, and everything else a
duty. He evidently thought that his correspondent had not been
unjust in thinking his nature deficient in warmth at this period,
and that he was conscious of a too overwhelming absorption in his
investigations and experiments. Nor is this surprising, when we
consider that this period coincided with the development of some
of his most striking discoveries. As she felt this anxiety when
about to sever herself from her old home, she was wise and true in
expressing it at the risk of pain to them both. He never afterwards heard of a broken engagement without pain, holding that
if not two, anyhow one person usually suffered acutely. Even if
he hardly knew people, he grieved at such news.

But his patience, sweetness, and good sense convinced her that
she had been right in her first decision, and she stood firm. They
were married in 1857, and went to Switzerland for their honey-
moon. The following entries from my mother's journal are
characteristic: 'George is so fond of lightning'; later, 'he puts
his head under all the waterspouts he can find'; then, 'he flew
about, now up, now down, trying to find a better path; he quite
enjoys dangerous places and looks so happy where his neck might
be broken.' Their first Cambridge residence, in 1858, was in tiny

G. G. Stokes in 1839. Aged 19 Years.

rooms over a nursery gardener's, now 'Willers',' some way out, on the Trumpington Road.

Then after a brief sojourn in London lodgings, while he lectured at the School of Mines, they took a small isolated house in Cambridge, Lensfield Cottage (recently re-named Stokes Lea), in which the remainder of her life and most of his were passed. It was gradually enlarged to meet the needs of a larger family party, but always retained its simple cottage character.

His fellowship at Pembroke was vacated by marriage, but twelve years later, in 1869, his College was enabled to re-elect him under a new Statute.

My mother, though not educated in the modern manner, was cultivated in her tastes, a great reader of good English prose and poetry, as well as of the *Belles Lettres* of France and Italy. My father never wrote anything not purely scientific without consulting her, reading the proof-sheets to her and accepting many of her suggestions. He read over his Gifford Lectures on Natural Theology in this way to an audience of two. These lectures gave him great difficulty, and he felt much hampered and handicapped by the terms of the bequest, which prevented him from treating the subject from the standpoint of revealed religion.

He took great pleasure in his wife's music; though no executant, she played with taste and feeling and with sweetness of touch. He especially delighted in Handel, always his favourite composer; before deciding whether he would go to a concert he often asked whether there would be any Handel in the programme.

In later years she went very little into society with him, never having recovered from the shock of their younger son's sudden death. But although he greatly missed her company abroad, perhaps the quiet of home was more restful to him on that account, and he always returned to it as a peaceful haven, sure that she would be there and ready with loving sympathy for all his work.

This is perhaps the place to insert the letter in which long afterwards he communicated to his wife the news that he had been recommended for a baronetcy of the United Kingdom.

LENSFIELD COTTAGE, CAMBRIDGE.
24 *May*, 1889.

DEAREST MARY,

......On arriving I found a letter which surprised me. It was marked private, "Salisbury" was in the corner, in the same hand as the direction, and it was sealed with the seal of the Foreign Office.

Here is a copy:—

May 22, 1889.

DEAR PROFESSOR STOKES,

I am glad to be able to inform you that in recognition of your remarkable distinction as a man of science, as well as of the high official position that you occupy, the Queen has been pleased to confer upon you a Baronetcy of the United Kingdom. It gives me much pleasure to be the medium of making this communication.

Believe me, yours very truly,

SALISBURY.

I had a mixed feeling about it. It is of course a very high honour, but there is something of a white elephant about it, more especially looking to the time when I am gone....At the same time I felt as if on public grounds I ought to accept it, or more strictly not to decline it; for from the words of the letter it would seem to be a *fait accompli*....

My own feeling was that there was an awkwardness for myself, and still more for Arthur, in having a Baronetcy without means to support it. On the other hand I felt it a sort of public duty not to decline, even if it were open to me to do so, which from the words of the letter seemed very doubtful. For it was through me an honour to science, of which I am *pro tempore* the representative, as being President of the Royal Society, and an honour to the University; and I felt that there might also be a political motive in it as a set off against the recent attack on the representation of the Universities, which was defeated in the Commons by a good majority, 91. I thought however I would consult Sir George Paget confidentially. He felt the awkwardness as to money matters, but thought I ought not to decline. I was thinking of further consulting Professor Browne, but Isabella suggested Sir Thomas Wade, which at once commended itself to me. I had a long talk with him. He did not think the money objection was at all serious; he thought I ought not to decline without some very strong reason; and supposed that it would appear in the *Gazette* to be published to-day—the Queen's birthday. I don't think that you had much idea on July 4, 1857, that you were going to be a Baronet's wife.

Probably few people have been more completely absorbed in their experiments. In fact his devotion to them sometimes gave rise to misapprehension. In his college days, when convalescent from a long fever, he began to experiment on chlorophyll. A friend calling to enquire found him absent and asked his bed-maker if he were better. ' Yes,' said she, ' he is getting strong again, but (tapping her forehead significantly) ' the poor young gentleman is always playing with green leaves.' When sitting with his family he might be seen trying the different way in which he saw colour with his two eyes,—engaged at his 'blinkings,' as his children called it. He would eagerly seize upon any brilliant piece of knitting, etc., which he found lying about, and he became devoted to a shawl of his wife's, on which he founded a series of experiments. Many of his most interesting experiments were made in a narrow passage room behind the pantry, which was his study until Lensfield Cottage was enlarged. He had a shutter fitted into the small window, and a bracket was fixed before a slit cut in the shutter on which to place crystals and prisms; in these simple and narrow surroundings he carried on his work, so that it was aptly said, "that if you gave Stokes the Sun there was no experiment he could not do for two-pence."

As a child I loved to watch him working at experiments in his study; I can still see the Rembrandt effect of the strong light and shade cast upon his face, when he opened the shutter from time to time to alter the position of the things resting on the bracket, and the absorbed and delighted expression of his countenance.

He rejoiced in silent companionship, often taking one of us on his long quick walks, perhaps not saying a single word for miles. Then some day his interest would be aroused by something heard or seen, and he would have a sudden fit of eloquence. One day it would be caused by the humming of the telegraph posts*, on

[* See the fundamental memoir " On the Communication of Vibration from a Vibrating Body to the Surrounding Gas," *Phil. Trans.* 1868, *Math. and Phys. Papers*, Vol. ii. pp. 299—324.]

several occasions by the beauty and interest of rainbows, especially by one near Bray Head when we were sitting close to the cliff's edge and saw a far larger part of the arc than is usually seen. He then told me that he once had to speak at one of the Royal Academy dinners, and took as his text the help which Science might be to Art, criticising a fine landscape on the walls, much spoilt to him by a rainbow with its colours in the wrong order. He said that Millais stood up and owned to the rainbow, which he said he would amend, and made a very amusing speech.

Another time when we were at Malahide I went with a cousin to post letters late in the evening, and we noticed something curious in the sky; she said that it was only sea-fog arched by the wind, but it appeared to be something more, though even if it were only sea-fog I was certain that my father would want to know why it arched like that. He came out at once and said that it was a lunar rainbow, and that he had seen one only once before in his life; he was delighted at not having missed it, and walked and talked until quite late.

After taking his degree, he was once observed at a party given at the Cambridge Observatory employing himself by making waves with the spoon in his tea-cup, watching their formation in abstracted silence*; he acquired a habit which he never lost of stirring his tea an immense time.

A favourite amusement of his was making a musical note on the edge of his finger-bowl at dessert; his mind seemed divided by interest in the number of different notes which could be produced, and how they were made, and by the peculiarities of the shivers in the water to which the vibrations gave rise. The different bowls gave different results, so that at one period he was constantly tuning; he was always much interested in the subject.

He never worked in the garden, but used to stroll in it, picking off the dead flowers from the geraniums, one of his favourite plants. He loved brilliant colours intensely; and my mother and I used to be amused at the vehement colours he wished us to select when we had patterns of dress stuffs. He

[* The illustration mentioned by Helmholtz, in his great memoir of 1858, on vortex motion, namely, the formation of half vortex-rings by drawing a semi-immersed spoon through liquid, will come to mind.]

would say of some portentously powerful specimen, "Now, that is a nice quiet colour, *don't* wear dunduckety-mud-colour," an expression which he often used, and applied to Morris' wall-papers. I used to say that he would like us to be dressed in the colours of the spectrum, a different one each time. He delighted in stained glass, and was very much interested in the optical reasons which caused the beauty of fine old glass. He talked quite a long time about it once in King's College Chapel. I begged that he would write it down, but fear that he never found time to do so. It was about the juxtaposition of certain colours and in certain quantities giving satisfaction, and the immense help which was given by the thickness of the leads in old glass, because they prevented the rays from the different colours from blending before they reached the eye, so that they arrived fresh at the brain and gave full sense of pleasure. He thought that the leads ought therefore to be thicker the further the windows were from the eye, and that the want of leads was one reason why Munich-glass is so unsatisfactory. I dabbled a little in stained glass, and he used to prowl round my painting table, looking at the light through the little glass samples which are the pallet of a glass painter. He was pleased at having bits of his favourite colours cut off, and carried them away to his study as precious possessions. When we walked home from Pembroke Chapel on Sunday evenings in summer he would lament that the Chapel had no colour, and wish that the windows were full of rich glass, and the panels on the walls were filled with fine painted colours. He took the greatest delight in bright sunsets; he used to watch to see if there were going to be any glow, and often walked with me up to the turn, and past the nursery garden where he lodged when he was first married, as we got a fine clear view of the west from that place. The Krakatoa sunsets interested him very much, and also the splendid colour prints of them in the *Report* published by the Royal Society, which were made by Mr Dew-Smith at Cambridge; he said that they were the finest colour prints that he had ever seen.

He was much interested too in some colour photography shown at a Soiree of the Royal Society. One print was of a cockatoo with its yellow crest; one of a horse all dappled like a ripe chestnut. Much consideration was given to a photograph sent him in which the chimneys of some houses had come out red.

One day he talked a long time about colours and Chevreul's work on colour, because he found me admiring Chevreul's fine head by Roty on a medal of the Académie des Sciences. It was a pity no one was there who understood; but when those who understood were there he often said nothing.

One Hallow E'en as we children sat burning nuts and salt he came in and threw our precious salt upon the fire by pinches till it was all gone, saying as he watched it burn, 'how beautiful'; then suddenly noticing our disappointment he went to his study for materials for finer flames. When we blew soap bubbles in the garden, if he were passing he almost always joined our game, he was so much interested in the colours and the varying duration of the bubbles. He was interested in iridescent glass too and bought vases of that sort.

I am inclined to think that one of the exciting causes of his experiments was a natural love of pure and beautiful colour; and in looking through the fourth volume of his *Collected Papers* for reminiscences of past years it was interesting to find the following passage at p. 264, "I had no sooner looked at the spectrum than the extreme sharpness and beauty of the absorption bands of blood excited a lively interest in my mind, and I proceeded to try the effect of various reagents." In another place he speaks of the beautiful celestial blue given by quinine,—which gave the clue to one of his greatest discoveries. He never got tired of that particular blue, and always kept a large bottle of quinine in his dressing-room as a family tonic, often calling us when he was doling it out to sympathise with him over the lovely mysterious colour. The exquisite crystals of carbolic acid also attracted his admiration.

He was very keen about the Academy and the Winter Exhibition of Old Masters, and always went to see the pictures, sometimes several times. His favourite time for seeing them was before breakfast, as soon as the Exhibition opened*; he looked at them for an hour or two, took a late breakfast and looked at them again, until people began to arrive, when he departed to his work. It was serious to play games with him, for in backgammon he would pause to calculate chances and in bowls to consider the mysteries of bias. He not only worked but played strenuously, thoroughly, and with enthusiasm.

* He had then a room at the Royal Society, next door to the Academy.

He would be quite interested in little things; and when I wondered why the hats of an acquaintance were so distractingly lovely, he asked to have the lady pointed out, and after watching her narrowly, told me that fundamental lines of beauty were involved and that the hats illustrated the principle of double curves.

He was always very measured in his words, and this habit came out strongly when he wrote testimonials. Moderation in speech and conduct was one of the touchstones of his character, as also that outcome of moderation, tolerance for the opinions and views of others; but he had one or two strong expressions, such as 'I own it amazes me,' and would speak of 'astounding notions.'

Perhaps because Skreen had been so quiet, and economy in his early days so urgent, he had the freshest pleasure in shows and exhibitions of all kinds, and often went on the pretext of taking us children. One of my early recollections is his taking me to see the Chinese giant Chang, and the terrible grief it was to me to find that my father was not far taller, and how amused he was to find the cause of my woe. I remember, too, our going to see Japanese tight-rope dancers. He was greatly interested in watching their progress across the Guildhall, and the wonderful way in which they gracefully balanced themselves with paper fans and umbrellas. He was evidently quite excited at seeing principles of equilibrium so daringly demonstrated, and was surrounding the tiny figures with imaginary angles, which they ought not to be able to transgress.

Many years afterwards we had a very entertaining afternoon at the American "Wild West" show: he might have been a schoolboy, aged twelve, out for a holiday, he so enjoyed himself. I gently tried several times to dislodge him in order that we might visit some Irish cousins, great friends who lived near Earl's Court, but at each effort he said, "I think the next scene will be interesting. It would be a pity for you to miss it." So we stayed on till the end of the performance. He confessed afterwards that he had spent the whole of another afternoon there by himself; this too at a very busy time in his life. I used to tell him that it was in vain to try and persuade me that he spent all those long days in the august shades of the House of Commons and Burlington

House; then playful questioning would elicit that he had been to see the oddest things. He used to say, "I was passing and I thought I would just go in for a few minutes and see what it was like," and further enquiry would perhaps reveal that he had sat out several hours all by himself. He did not go to the theatre, probably from habit and bringing up; but he never objected to other people liking to go. His objections were without prejudice, for he once took me to see moving photographs at one of the large Music Halls in London; but when some of the rest of the entertainment seemed decidedly unexpected, on looking round at him I saw that he was quite uncomfortable, and he said that, to make sure, he had been that morning all the way to this Music Hall, and had asked the Hall Porter if the afternoon performance would be the sort of thing to take a lady to see, and that the porter had said that the most particular lady could not object.

It was striking that, having lived so long in the world, he still always expected all sorts of people to adhere to the truth, even to their own disadvantage. This was so much the case that regular scamps used to get money from him by their plausible tales. He had certain cottages to sublet which were leased to him with his house; but he permitted such obviously unsuitable people to have them, that at last my mother persuaded him to put the matter into the hands of an agent. One man, afterwards discovered to have been in jail for getting money on false pretences, used to waste a great deal of his time by talking for hours upon religious doubt; but he did not at all enjoy this sort of person being unmasked.

He was occasionally very dilatory about business, and would take quite a dislike to something he had to do. Having received Sir Isaac Newton's manuscript papers (mainly those on optical subjects) to arrange and catalogue *, he kept these precious documents so long that there was some anxiety as to whether they had been overlooked, and after letters had been written to him on the subject in vain, other members of his family had to be approached; but this was during the busiest time in his life. It was noticeable how he would occasionally put off some big

[* The Portsmouth MSS., now deposited in the Cambridge University Library. A catalogue with descriptive preface was published in 1888 by a University Syndicate consisting of Luard, Stokes, Adams, and Liveing.]

piece of work which he knew he ought to grapple with, even sometimes making other tasks in order to excuse himself. If he had only been given a few of Newton's papers at a time he would probably have got through them all by degrees: but, with numerous other matters pressing for attention, he sometimes got daunted by a large piece of work and quite began to hate it, naturally more so the longer it was deferred.

It was the same with the editing of his own *Collected Papers*: he felt it most distinctly a great bother. Sometimes we asked him where he had arrived in them and coaxed him to go on with them, but he would complain that it was very hard work, as so much new work had been done since he first wrote his papers*; yet all the time he was writing ardently on religious questions.

He occasionally developed marked preferences for people. He scarcely ever saw one lady friend without saying kind things of her afterwards. When one of my husband's friends came from town to vote, and had lunched with us, my father pronounced emphatically that he had a fine presence, looked as if he had good ability and seemed a very nice man. No one would have guessed that all these kindly observations were being made. He did not often take these keen fancies for people, seldomer against them ; but if there were anything droll in manner, if they poured out torrents of words, or gave an impression of special vanity or egoism, he would be very much entertained, as appeared in little amused twitches of his mouth, and reined-in smiles, with a certain twinkling of the eyes. We were often amused by the details he noticed in people staying in the house, a keen observation of which they were quite unaware because of his silence and apparent absence of mind.

In spite of his resolution and decision of character, he would occasionally find it most difficult to make up his mind about some trifle, as for instance, on which day to make a journey, or some similar matter in which the *pros* and *cons* seemed to balance about equally; he would hesitate, and rehearse his reasons again and again.

He would occasionally take very unexpected things much to heart; once he was quite annoyed with me, in connexion with

---

* The delay was occasioned largely by his scruples about completing them and bringing them into line with more recent advances: thus the volumes contained a considerable amount of very valuable new work.

the proposal to legalize marriage with a "Deceased Wife's Sister,"
because, walking with him to the station when he was going up
for a House of Commons debate on that question, I asked him
why there was to be no Bill for "Deceased Husband's Brother."
He was quite shocked and quoted texts from the Old Testament
and Jewish customs, and was not pleased at my quoting that "the
seven had her." But he was a little ruffled that day, for all his
feelings were against people intermarrying in this manner. He
had felt obliged to read up the subject, and had come to the
conclusion, against his wish, that the Law needed alteration, and
that he must not vote against the change, and might even feel
obliged to vote in favour of it, dead against his inclination and
bias.

He never could have done all he managed to do as Member of
Parliament, Lucasian Professor, Secretary or President of the
Royal Society, and doing them all so hard, if it had not been for
his wonderful power of napping soundly on every likely or unlikely
occasion, in the House, in the train, at concerts or evening
meetings; yet he always managed to be awake when he ought
to be. He was never heavy after sleep, but wakened from
these short naps quite refreshed, and fully alive to all that was
going on. At night, too, he slept most excellently, never being
kept awake by business, importunate letters, or such things.

Another cause of his enduring hard work, as well as much
dining out, was his excellent digestion. Even when travelling
and happening upon strange or tough fare, it never seemed to
make the slightest difference to his health. He would get bad
chills, wet feet, rush to catch trains after dinner, and do anything
or everything that usually bowls over the average or even the
strong man, with perfect impunity.

He hardly ever ate luncheon until he was eighty, and then
under protest. He usually breakfasted at nine o'clock, and had
nothing more until dinner at five o'clock. Then, as dinner was
gradually pushed to a later hour till it reached half-past seven, on
pressure being brought to bear he would accept a cup of tea and
a biscuit or piece of dry bread if it were taken to him.

After dinner he drank strong tea—two large cups at least at
about nine o'clock, and if he meant to sit up until half-past two
or three o'clock, as frequently happened, he liked to have a
"Brown Jenny" and a kettle left with him, and would brew

himself terribly strong tea late at night to keep himself awake ; it never seemed to affect his nerves. As a rule he did not go to bed before half-past twelve or one, except in Vacation if he happened not to be very busy. It should be added that when he was at home his breakfasts and dinners were not fancy, light meals, but real and solid ; though when he travelled backwards and forwards, questioning would bring out that he had dined at the Athenaeum on two poached eggs, cheese and beer, or on sausage rolls at the railway station. He drank two glasses of beer with his dinner at home, besides a glass of port at dessert.

When we were children we used to make him sing a drinking song in praise of beer. He had only one other song, an imitation of a very highly sentimental one, which a sister used to sing when he was young. They were sung to an accompaniment of strumming on the table and were performed with great expression and fun. When past eighty he performed them both for the edification of his little grand-daughter.

In the year 1877 an Irish cousin came over to spend a long visit with us. On hearing her express a wish to study Euclid my father suddenly announced that he was going to try her paces, and would take her for an hour every evening when he was at home, and that I might come too. He read through the first book of Euclid with her during the month of her sojourn with us. It was apparent during the first lesson that I was keeping them back, and that it was better to withdraw ; but I felt most wretched and abased at losing the chance of learning from him. That night when bidding him good-night he kept my hand in his and said he wished to talk to me. He first spoke of things not mathematical which he wished me to study. He then gave me the most beautiful account of the growth of knowledge, and said that even the wisest people knew very little. He spoke of himself as only apprehending slightly in advance of others, as standing on the edge and looking into the unknown, and said that people were then only born who would perhaps know far more than anyone yet dreamt of. Then after speaking of human knowledge as it had been and as it was, he passed on to imagine it in an infinite degree, and from that to Divine Wisdom as the root of all things which are or can be, and yet as willing to dwell in every creature who in humility desired true wisdom.

When a baby sister died he took me into the room where she lay dead,—for he thought it well to accustom people to the notion of death early in life in order that they might cease to dread it,— and spoke of the immortality of the soul. He always inspired a feeling of calm and security. One knew that his mind and heart had some resting-place far from the troubles and changes of the world.

When it was becoming customary for the University Professors to admit lady students to their lectures someone wrote to him asking for permission for the ladies to attend his lectures also. He said that he had almost decided to refuse; but I begged that he would take them, and asked him how he would feel if a Mrs Somerville had asked him to teach her, and he would not. At last he promised to admit them, and he became much interested in his lady students, and always knew how much they understood. He was much amused by one. After the first lecture he said, " She frowns "—after the second, " She is frowning horribly !"—after the third, " Her forehead is one mass of corrugations; she won't be there next time !" and she was not. But some of the ladies got on splendidly, and he was much pleased when a Newnham lady who had attended his lectures brought him some original work which he approved.

I wonder if those who worked with him had any idea how merry he could be. His humour was somewhat peculiar to himself, and perhaps not quite developed in proportion to the rest of his mind; and the things which amused him did not always amuse others, though they laughed with him because of the sympathy which he always inspired. His fun belonged to a different order from that usual to a grown-up person. Of course this does not apply to his scientific jokes. It was interesting to see him with Professor Clerk Maxwell, whose sense of humour was of a very high order, and very varied. They were greatly attached to each other, but it is doubtful whether they had much meeting-ground upon the humorous side of their characters.

My father delighted in playing with my little white Pomeranian, a dog called Pearl, of whom he made a great pet. This was a dog of acute instincts, who in illness used to creep to my mother; but when in high spirits he would fawn upon my father, begging for a game of romps, and then there often ensued wild races round the garden. As children we sometimes thought that it was papa who had "fits of the funnies," and sometimes that they originated with Pearl. My father and I were walking with Pearl one day when he was run over by a cart, and he carried the dying dog home most tenderly. Then we buried him in the garden, but first my father went to the house for something white to wrap the little dog in, because he thought it would be so sad to see the dark earth fall upon his pretty white fur.

I remember one delightful day when we had to walk from Port Ballintrae, near the Giant's Causeway on the coast of Antrim, to Portrush, where we were to attend a picnic and then walk back. Before going he was anxious to make some experiments on waves in a cave near Portrush, which could only be reached by boat at high water, or at low water by wading. It was such a pretty cave that he offered to carry me in on his back, and it may be imagined that the offer was accepted. He tucked up his trousers and waded in with a stick. We had a most delightful time in this lovely cave. He worked away at his waves, so that we waited rather long, and the tide was quite high when we set off back again. The rocks and stones were slippery with seaweed, and I made him laugh unguardedly. Presently, plump we went in, and being wet all over thought we might just as well amuse ourselves in the water awhile. Anyhow, with the best intentions we had safely avoided that picnic, a severe form of entertainment even in Ireland.

He was very fond of walking on the Velvet Strand near Malahide when we stayed with his sister Elizabeth, and we often walked all along it and back with bare feet. I was for burying our boots and setting up a stone to mark the spot, but he thought it more prudent to retain them, so that in case we saw correct persons approaching they might be hastily resumed.

He said that he dreamed a good deal until he went to school, but little afterwards; he thought that frequent dreams showed that the mind wanted exercise. Yet in after-life, when his mind was certainly busy enough, he sometimes had queer dreams. Once, for instance, that he arrived at the North Pole, where he found a young lady sitting on a large iceberg, very elegantly dressed, and sheltering herself from the rays of a hot sun with a pink parasol; he then quoted

> The sun's perpendicular rays
> Illumined the depths of the sea,
> And fishes cried out in amaze,
> Oh! bless us, how hot we shall be.

In spite of this dream, he maintained that the South Pole would be much more interesting to explore than the North Pole and would give a better scientific return.

He used sometimes to be rather entertained by his travelling experiences, and by the way in which persons who recognised him through having seen his photographs introduced themselves. One occasion in particular amused him, when a fellow traveller, after a long conversation, glancing at the name on his hat-box, congratulated himself on having had the honour of making the acquaintance of the illustrious author of *The Art of Memory.*

Every Long Vacation we spent about a month with his brother, William Stokes, Rector of Denver, near Downham Market in Norfolk, a very charming old man, a delightful talker and a member of the "Family," an old Cambridge Dining Club. He was a gentleman of the old school, who always wore a top hat, a very high stock collar, and Wellington boots. His tastes were scientific, and being much the elder brother he found it difficult to remember that little George had become a most eminent scientific man; and certainly he was never reminded of it by my father, who used to venture to differ from him in the most deferential and gentle manner. While there we went for long drives, drawn in the old carriage at the very slowest rate by the ancient and fat horse, to visit the fine churches and view the flat and picturesquely wooded country.

Later in the Long Vacations we always went over to stay at the Observatory at Armagh, and afterwards went with the Armagh party to some seaside place, oftenest the magnificent neighbourhood of the Giant's Causeway, or occasionally some other equally quiet locality. My father got through a good deal of work at these times, giving the morning and evening to it, spending the afternoon in long walking expeditions, and climbing, if there were anything to climb. He would scramble along the Spanish Path and other dangerous cliff paths on the face of the Causeway rocks, to my mother's considerable anxiety, showing us afterwards from the sea the little thread-like tracks he had been along.

There was a cave called the Land Cave which we always visited after storms had been ploughing up the Atlantic. It had a sort of window opening into it from the land, so that we could

see the great waves come in, making the cave dark; it was
striking to see such great masses of water fall without sound
upon a bed of foam. He made a good many wave-observations
there, not about steep sea-waves, for that was much earlier, but
I think he was trying to find out the relation of the waves to one
another and why the ninth wave was so much larger than the
others. He told me that he was nearly carried away by one of
these great waves when bathing as a boy off the coast of Sligo,
and this first attracted his attention to waves. Once after a
great storm we found a number of the shells of the Ianthena
Communis, a sort of purple snail with a nautilus-like float.
He said it was a member of the same family from which the
Tyrians made their purple dye: we tried to dye linen and wool
with them but only succeeded in dyeing our cuffs well. We often
explored the coast and looked for the fairy pools in the rocks
where the water was so clear that you could examine the sea-
anemones and the waving seaweed. He delighted in the unusual
geology of the place with its strange basaltic columns, and would
wonder if they went under the sea and came out at Staffa and
Iona. The fisher folk, fine manly fellows who sailed along that
inhospitable shore at great personal risk, would bring us their
queerest fish, like the Gar-fish with its pale green bones and the
Devil-fish. They used to come for advice as to how to race their
heavy boats and make them slip through the water, and those who
took the "Professor's" advice won at the regattas and got their
fish first to market. He visited one of these fishermen often
during one vacation; the man had lost his eye in a fight, and was
suffering great pain of body and agony of mind.

Foreign travel was not then so much the fashion as it is now;
and after his wedding tour in Switzerland he only once or twice
went abroad in order to represent the University or receive a
degree.

My mother sketched well in pencil. After an illness in 1880
she went to stay at Hastings and writes from there in April, "The
waves are so flat and undefined to-day I can get nothing sure
out of them." He was evidently working at waves that year, and
employing her to sketch them.

About the year 1882 we had much amusement over some
experiments he made while on the Madras Harbour Commission,

which sat to discover how the Harbour should be rebuilt, so that the tremendous seas coming into the Bay of Bengal should not sweep it away. He had a long wooden trough made, and got heavy little bricks—of lead, I think—with which to build the harbour, my part being to demolish his work with waves of varying magnitude, while he timed the resisting power of the uncemented harbour, building it at different angles. The work of wave-destruction was done with a rake without teeth, which fitted the width of the trough. We used this apparatus near the garden tap, which was provided with a hose, and we usually worked with bare feet, as boots soon got soaked through upon the sopping grass. Previously, in 1873, when we were staying near Portpatrick in Wigtonshire, he had studied with intense interest the action of the waves in demolishing the harbour that had been constructed at great cost on that exposed coast.

In 1874 we ascended Slieve Donard when staying at Newcastle, County Down, getting a magnificent view of shining ranks of great cumulus clouds beneath, when he talked a good deal about cloud-formation. He usually talked far more in vacation than in term time; he was less busy; then too he loved the beauty of nature, and it seemed as if the novelty of things round him kept his mind at its out-posts where things enter by the senses, and so made communication easier to him.

He was much interested, as also was Prof. Clerk Maxwell about the same time, in cat-turning, a word invented to describe the way in which a cat manages to fall upon her feet if you hold her by the four feet and drop her, back downwards, close to the floor*. The cat's eyes were made use of, too, for examination by the ophthalmoscope, as well as those of my dog Pearl: but Pearl's interest never equalled that of Professor Clerk Maxwell's dog, who seemed positively to enjoy having his eyes examined by his master.

He always took the keenest interest in thunderstorms, and used to make us anxious by the way in which he stood out in the garden to watch them, especially in the great storm of 1889, of which his house and the Roman Catholic Church seemed the centre. He would go any distance to see houses and trees that had been struck, and did his best to extract their sensations from people who had been rendered unconscious by lightning.

[* A few years ago this subject excited attention in France, and the dynamical explanations were set forth afresh in the *Comptes Rendus*.]

He was very much interested about earthquakes and earthquake waves. When Mont Pelée in Martinique erupted, and 40,000 people were killed, and St Pierre destroyed, he sat for long spells thinking about it and the conditions of the earth which caused it. He bought all the newspapers with accounts of the disaster, but considered that the account given in *Pearson's Magazine* for September, 1902, written by an eyewitness, Captain E. William Freeman, of the steamship *Roddam*, was by far the best. He bought several copies at once to send to friends.

In 1902 he was also strongly attracted by photographs sent him by Mr F. H. Neville and Mr Heycock, showing the structure of thin sections of an alloy quenched at different temperatures, and would sit looking at them and comparing them for long times. Even in his latest days he was always ready for anything new.

It was interesting to notice in my father's discussions with scientific men the horror he had of theorizing from unproved facts; he would always say, "But have you proved your facts?" and then would often show them that this had not been done. He could not bear "scientific romancing," as he called it.

He had a strong admiration for Charles Darwin's character and patient research, but could not understand the way in which, as he thought, scientific men had accepted the theory of evolution before the chain of evidence was completed: he used to say that this surprised him exceedingly, and that he knew of no similar instance in the history of scientific thought.

When we were children our great day was Sunday, because my father was usually at home on that day and less busy. He would take us for long walks on those afternoons, and read and talk with us in the evening. He was always very fond of reading aloud. He did not read poetry well, always, whatever the line, bringing out the last word strongly to make sure of the rhyme, which had a jingling effect; he read us one or two of Keble's Hymns every Sunday evening. But he read prose beautifully, and in particular interpreted *The Pilgrim's Progress* for us in the most charming way; he had an immense admiration for it, and illustrated it by his own experiences in travelling and walking amongst the mountains. When he came to the Fair and Flourishing Professor, I called out "That's you, Papa"; and we all laughed so much whenever he came to the name, and he had such

difficulty in keeping his own countenance, that the reading had to
be abandoned for that evening; for he said that it was a book
which should be read with gravity, as it was written by a
good man about holy things. So he told us instead the story of
Bunyan's life, and of how he had suffered for the right to think
as he chose, and how he had long lain in Bedford Gaol, not far
from Cambridge.

We often walked round by the Roman Road to Trumpington,
and often to the Observatory to see Professor and Mrs Adams.
He was also very fond of taking the Cherryhinton round, to hear
the larks sing in the fields, which have since been covered with
houses. Or we would walk by Madingley and Girton, when he
would sometimes repeat Gray's *Elegy* by Madingley Church,
I think the only piece of poetry not a hymn which he knew by
heart.

He had had very little time for general reading ; but he had a
very high opinion of books like Hooker's *Ecclesiastical Polity*
and Locke's *Human Understanding*, which he had chosen in early
years as Pembroke College prizes. Once when we were reading
a *History of the Inquisition* he took it up and became quite
absorbed, speaking often of it afterwards with horror. He spent
a good deal of time in perusing doctrinal books which were sent
him by the authors.

He had two really wicked characteristics, that he would never
allow anyone to help him with his work, not even permitting
invitations to be answered for him, and that he kept every single
thing he received by post, even advertisements. His study was
enough to drive any housemaid " wild." He used gradually to
acquire tables from the rest of the house, until there were as
many tables as the room would hold, with narrow passages
between, through which to squeeze if you could. On these tables
papers were piled a foot or more deep. It may be imagined
that, keeping everything, he could find nothing. One remembers
the hunts there used to be before he went to London, every
person in the house sometimes enlisted. It always began by
his hunting alone and refusing all help rather fiercely. Then
gradually, as the quest grew more desperate, the rejected suitors
fell quietly into the ranks. Sometimes it was grim earnest, and
the necessity was urgent; as for instance when one of the Gifford
Lectures was missing, and it was nearly time to catch the only

train that would connect with the mail, if he were to arrive in Edinburgh in time to deliver it. At last it was discovered, but only in the nick of time, in the round basket of an aged relative, which was called ever after " the magpie's nest," and came in for first search on subsequent occasions. When the study and the inner study had reached a state of repletion, my mother would wait for the Royal Society day, Thursday, for then he was often away for two nights or so, don apron and sleeves, and fall upon those rooms, when clothes-baskets full of unnecessary matter would be removed. Then for a long time afterwards we were considered the cause of the disappearance of every missing object or paper. We often presented him with letter-cases and other domestic inducements to tidiness, but they were usually found empty, or with things quite unimportant inside. Those home-comings after clearances always reminded us of the scene in the *Antiquary* when Oldbuck finds his study being cleaned. But we endeavoured not to be caught red-handed, and even if we had safely finished, hung the head and went softly.

I used to go with him to his lecture-room sometimes to help to look for things, or to assist in hanging strings and wires to break the echo to which it was subject.

Then of course both in his lecture-room and in his study at home he made strange infusions which smelt horribly, horse-chestnut bark and leaves and all sorts of other things, at one period bullock's and sheep's blood for his experiments, not even fresh but kept for ages in soup plates close to where he wrote. He never seemed to notice bad smells; once when we had the planks up in the dining-room for a horrible smell which he said only existed in our imaginations, we found a dead mother rat and her young immediately under his chair. After this, it seemed right to be more firm, as these things, though inoffensive to him apparently, might harm his health.

The late Queen's Jubilees were occasions which he thoroughly enjoyed, for like all Irishmen of his way of thinking, he was a very loyal subject. I accompanied him to the first in 1887, which was also the jubilee year of the electric telegraph, and he enjoyed it keenly, from the Procession to the cruise in the Admiralty launch around the fleet at Portsmouth and the illumination of the fleet in the evening: getting back to bed at three o'clock in the morning was not of the slightest consequence in his eyes.

It was the same at the Naval Review, after the King's Coronation, to which my husband accompanied him: he would not hear of going home until everything had been seen and done.

Lord Kelvin's visits were occasions of enjoyment to him, and great were the discussions between them, which anything served to begin; for instance, the eggs were always boiled in an egg-boiler on the table, and Lord Kelvin would wish to boil them by mathematical rule and economy of fuel, with preliminary measurement by the millimetre scale, and so on.

He had no love of sport of any kind, and I think the nearest he ever got to fishing was once on the coast of Antrim, when he got limpets and baited my hook, and took the fish off the hook, carrying the seventeen home in his silk pocket-handkerchief.

The absence of mind which was noted in him as a child remained in later years. Once after he had been spending several days as a guest in a country-house, when we enquired if he had liked his hostess, he said she was very kind and pleasant; but we found that he had never discovered whether she were the wife or daughter of the house.

Naturally people often came to see him on business, and it was very interesting to observe his method of getting through that kind of work. He had a way of reducing business to its simplest expression, and then sticking to the point whatever other people said. His charity was that which thinketh no evil, and he was so far above the petty jealousies of everyday life, that even when in the midst of them he would be totally unconscious of their existence. This must often have been useful in keeping the mind in a serene state of calm: still, it must occasionally be easier to decide or guide, if the conditions of other people's minds can be clearly seen, even if they prove an unpleasing prospect. He was so totally unaware of small-mindedness that it made him quite sad to have it pointed out.

Perhaps what most differentiated him from other men in his tone of mind was the blending of the active and the passive attitudes towards life; one does not often find the resolute active enthusiastic spirit blended with calm aloofness and power of cool judgment. It seemed that at any moment and in the hurry of business he could withdraw into himself to decide on the best course of action or of thought.

He naturally much disliked to find that others had been

angry with him for doing business in what he considered the right way. On one occasion in particular I remember that a gentleman came to see him about the publication of a friend's work through the Royal Society. It was interesting to hear the different lines of argument, my father's very simple, quietly and calmly repeated opinion, that the work was not good enough, the visitor's varied arguments becoming finally rather heated. When the visitor finding it useless had gone, I asked him if he had happened to notice that the man had left in a very bad temper. My father hitched his chair and got rather red, and seemed aware for the first time that this had been the case. He said, looking most uncomfortable, " I thought that he seemed a little warm, but that he could not possibly be angry with me about what was purely a matter of business." Then he became somewhat roused and said, "It is most unjust!" We then went out walking, and he talked for a long time on the subject, but naturally without any personal bias. He gave me a most eloquent lecture on the Royal Society, its foundation, growth, position in England, position in the world, and spoke of the misfortune which it would be to Science if it ever departed from its isolated position, or got entangled socially or politically, or ever published anything for any reason than its first-rate excellence. When he ended, for it lasted from before we left Lensfield Cottage until past Trumpington, he remarked that he gave up an immense amount of time to the improvement of hopeful work, but that he could not make bad work good.

The same personal feeling which some people have about others or themselves was applied by him to institutions or societies, a strong sort of intimate affection and regard. He naturally felt it particularly towards the Royal Society, with which he had so long an official connection. Elected Fellow in 1851 he was made a Member of the Council in 1853, and was appointed one of the Secretaries in 1854, an office which he held until 1885, when he was elected President. He remained President for five years, and for two years longer was a member of the Council, on which he had therefore sat for thirty-nine consecutive years when he withdrew in 1892. In the following year the Council took the first opportunity that was open to them, to confer on him the Copley Medal, which is the crowning award of the Society.

In later years, as he grew older, the continued marks of special appreciation from learned societies, such as the Arago Medal from the Institute of France, specially struck and sent over with the deputation to his Jubilee celebration, and last of all the Helmholtz Medal from the Berlin Academy, gave him great pleasure and satisfaction.

Besides the Royal Society he had this feeling of great affection and keen personal gratitude in a high degree for Pembroke College. He often told us of all he owed to Pembroke, and of how his re-election to a fellowship had helped him and given him ease and liberty for his own work and investigations. He always delighted in speaking of the great and good men who had belonged to his College: and one of the books which he was reading a short time before his death was the *Life of Bishop Ridley*. When any Pembroke man did well in the mathematical examinations, Tripos or Smith's Prize, it gave him the most lively pleasure. He delighted in showing friends over the College, and would often say, "Don't take them to Pembroke. I will show them that myself," and this even when he was very busy.

When he had business to do with people he almost always went to them instead of their coming to him. Whether it was because he was prompt and got to them while they were still meditating on coming to him I don't know, but it would happen with persons far younger than himself.

People may hold different views with respect to the amount of time which he gave to the study of the question of Conditional Immortality and cognate theological topics, on which it is well known that he held opinions of his own. To some it may seem that as nothing can ever be proved objectively about these things, however interesting, it was time taken from what was the real work of his life, the search for absolute truth regarding Natural Phenomena, truth which is capable of being demonstrated. Indeed, considering the enormous amount of time and thought which he gave to correspondence and discussion relating to religious questions, and in science to improving the work of others, it must always seem amazing how he found time for his own investigations.

His attitude to the Royal Society is brought out in the following letter to his father-in-law, which has been passed on to me by Professor Larmor.

LENSFIELD COTTAGE, CAMBRIDGE,
1st *December*, 1877.

MY DEAR DR ROBINSON,

Somehow or other time has slipped by and I have not yet sent you the plottings I contemplated. My hands have been as usual full with one thing or another and this matter I felt did not press.

And now I must ask you to excuse me if for the present I drop anemometry and pass to a subject concerning myself and my family, about which I must come to a decision before long, though there is no occasion for an immediate conclusion.

I have now held the Secretaryship of the Royal Society for a long time, having yesterday been re-elected for the 23rd time, and having therefore entered on my 24th year of office. I don't look on a post of that kind, which is continued by re-election, and which is hardly regarded as honorary (though since the raising of the stipend from £100 to £200 I think there is a very fair remuneration), like a professorship which one is expected to hold for life, or so long as one's powers last. I have always felt that I ought to anticipate rather than otherwise any wish on the part of the Society for a change in the office.

I have no reason to suppose that any such wish is felt at present in my case. I asked Mr White confidentially if he had heard any expression of a wish that I should make way for fresh blood, and he said not; that on the contrary the Fellows seemed to be very well content with me. Still I can't help thinking that for a non-life office 24 years' tenure is a very long one.

What precedes is by way of introduction to the immediate question. When the office of President was last vacant there was a division of opinion as to whether there should be an understanding that the office should be held for a limited number only of years, say 4 or 5, or whether when we had got a President we had best keep him so long as he is fit for work and does not feel the duties too onerous.

The majority of those who had to decide were in favour of the shorter tenure. Such was Hooker's view when he took office. He himself considered that 5 years was about the thing.

He has now intimated to the Officers, and expressed his intention of intimating to the new Council, his wish that that view should be acted on. As yet it has not gone beyond the Officers.

Before the meeting of the Council yesterday Sir Joseph Hooker told me that the Officers were strongly of opinion that I would be the proper person to succeed to the office. After the dinner Huxley, sitting behind me on a sofa where we were by ourselves, strongly pressed it upon me to express my willingness to accept the office.

Among other solutions there is one which will occur to everybody. For a combination of exalted social position with the highest moral and intellectual qualities the Duke of Devonshire stands pre-eminent. He is universally respected, and if we come to have a nobleman at all I think that he is *par excellence* the man.

I think, however, that many would feel that our President ought to be a man who would really work for the Society, and *that* we could hardly expect of a man in the Duke's position. Though in good health he is getting advanced in years, and I confess I much doubt whether he would *now* consent to take the office.

Then there is Spottiswoode, who was one of those talked of on the last occasion. Of the two I am probably the better known as a man of science, but he has the advantage of a fortune which would enable him to show hospitality to distinguished foreign *savants*, which his residence in London would enable him to do.

However as the office *has* been pressed upon me by my colleagues, and I think it probable that the Council might make the same suggestion, at least if they had any inkling of what had been in the minds of the Officers, the practical question is (supposing at least the Duke not available) what answer I should give.

Of course I feel that it would be a great honour, the highest or one of the highest scientific posts in the country. But then I have a wife and family depending on me, and mainly a life income, and the Presidentship would involve of course the surrender of the £200 which I should have so long as I held the Secretaryship, besides some positive outlay (though not to any great amount) attending the position. Besides I am naturally of rather a retiring character, and should feel not a little out of my element in being brought so prominently forward. On the other hand, if I am deemed the fitting person (and I am not the one to judge of that) my duty to the Society of which I have so

long held the Secretaryship would seem to require me to consent;
and having now an endowed professorship and a fellowship as
well, I ought to be able to dispense with the salary of Secretary.

I have now pretty well put before you the *pros* and *cons*, and
should be glad to know your feeling in the matter.

Yours affectionately,

G. G. STOKES.

P.S. I mean to sound Cambridge F.R.S.s. to see whether
there is anything of a general feeling here in favour of the
Duke of Devonshire.

Although of a very quiet and silent disposition, he by no
means liked being alone; he would often bring his work into
the drawing-room in the evening, and had a folding-table kept
close to the door of that room so that he could work in family
surroundings. He did not like the talk to stop on his account;
indeed his power of concentration prevented its being a worry
to him; it just seemed to reach him as a cheerful and soothing
buzz. Still it was interesting to note that one never knew
when he was listening; and the most unexpected subjects occasionally arrested his attention, when he would launch suddenly
into the conversation in the intervals of his work.

He had a great love of parties and public functions of all kinds,
and rarely refused invitations. I used laughingly to tell him that
whether it were a wedding or a funeral did not make the slightest
difference. This keen sympathy with human joys and griefs was
characteristic of him; it was felt even towards those with whom
he was but slightly acquainted, and he sometimes expressed a wish
that he had the power of conveying his sympathy in words.

When friends were ill he would often go to see them and
sit with them quite a long time. His fondness for frequent
church-going on Sunday seemed to have its root in this same
turn for companionship and good-fellowship, especially when
linked by union of action or idea. But we used to wish that
after the labours of the week he would not go three or even four
times to church the same day,—including the University Sermon,
of which playful questioning would sometimes reveal that he had
heard remarkably little. He rarely went to church on week-days
except on special occasions, Ash Wednesday, Good Friday, etc.

It was one of the greatest disappointments of my youthful life that I never went with my father to visit Skreen. He had been there once alone when I was too small to go, and he used often to talk to me of his happy child-life, and of returning to visit the dear place again, and always I begged that we might go there together. It seemed odd that we never went, as for many years a good part of every Long Vacation was spent in Ireland. At last, one Long Vacation, he said that he would take me to visit his old home. Then, just as it was all settled, they sent him piles of papers to look over from the Royal Society, and our interesting plan fell to the ground. It was very disappointing, for had we walked together in those old places always so dear to him, he would have told me much about his early years, and I should perhaps have better understood how he came to be all he was. His early life always seemed more interesting to him than his undergraduate days. What probably made it difficult to extract reminiscences from him about his past life was that he did not feel interested in it, as most people would, because it was a part of himself. He was always much more inclined to talk about things quite outside himself than of things connected with him, and this not perhaps so much from natural shyness and reticence as because he was not very much interested in the latter. Anything in the shape of natural phenomena was a different matter, and on such he would enlarge at great length, even to the uninitiated.

As the amount of writing he had to do grew steadily greater, one shoulder grew somewhat higher than the other; he had slight congenital malformation of both little fingers; they crooked inwards, and this served to make writing more difficult to him. At last he consented to try the key-board of a writing machine, and tapped away one Long Vacation until he was sure that he could acquire the art. This typing machine was a great boon to him and to his correspondents, as his handwriting had become very difficult to read.

The impersonal quality of mind already mentioned is perhaps rare, even amongst men of thought. It gave him great restfulness in active affairs, and saved him much wear and tear. He was not troubled by fear of what others might think of his actions and his work, and this gave simplicity and unity to his life. It was not solely strength of character, but arose in great measure from conscientiousness and the continual habit of feeling himself in the presence of God. It was very interesting to be quietly at home with him during the time his Jubilee was being celebrated by the University. He was so absolutely simple about it all, enjoyed it all so thoroughly and in such a perfectly un-selfconscious way. He thought it most kind of people to take so much trouble in getting it up, and in coming such distances in order to be present; but there was no mock modesty; he accepted their judgment as it was offered. This mixture in his character of profound modesty and humility with perfect consciousness of his own place in the scientific world was remarkable. He thought very little about himself, but when he did think he thought truly and impersonally. Naturally, during his Jubilee he subtracted a great deal from himself and placed it to the count of science. Anyhow, he went through those days most peacefully and emerged from them quite fresh. The following morning he was off early for the annual official Visitation of Greenwich Observatory.

The crowning pleasure in receiving his honours was that my mother, though in weak health, experienced the joy of seeing them bestowed upon him. She died the following December, and afterwards he lived with us.

It has been remarked that he was the first President of the Royal Society, and, far more curious, the first scientific man, who represented either University in Parliament since Sir Isaac Newton; and he was elected after an interval of two centuries, *minus* two months, succeeding Beresford Hope in 1886 when sixty-seven years of age.

It was also subject of remark that, in those times of Irish political disturbance the "House" at last saw a silent Irish member; Sir George Paget, in seconding his nomination, had said that the House would have an opportunity of seeing how good a thing a good Irishman is. It may be remembered how Macaulay records the fact that, when his great predecessor Sir Isaac Newton was Member for his University, he made no speeches: "he sat there in his modest greatness, the unobtrusive but unflinching friend of civil and religious freedom." It has been said that Newton only once opened his lips in the House of Commons, on which occasion he rose and said, "Sir, would you have a window opened ?" Had my father spoken on this subject, he would certainly have asked to have the window shut; for, considering what a hardy man he always was, he had the most curious dislike of open windows all his life.

He had grave doubts as to whether he should accept the invitation to stand for the University. He felt that there were many objections to it, one of them being that the President of the Royal Society should be entirely outside politics; but though he never wavered in his personal political views he felt that the representative of one of the ancient Universities held an exceptional position, and, consenting to stand, he was elected without a contest. He very much enjoyed the debates; certainly he was in the House during a very interesting time in politics, especially for an Irishman. It interested him greatly to hear Gladstone; and, having gone into Parliament with a perfect horror of his politics, it was amusing to note the gradual change that set in, and the enormous admiration which he felt for Gladstone's eloquence and power, and the sense he had of the strong personal magnetic influence which he exercised.

One day especially his silence in the House was remarked. Some scientific question had come up, and still he said nothing. When we afterwards asked him why, he answered that he had been prepared to rise, but that another person had obviously wished to speak and had said enough, although he had treated the subject from a different stand-point from that which he should have himself adopted. Only one member beat him in regularity of attendance, Sir Richard Temple, who, however, lived in London.

To some it may seem a pity that he entered Parliament; for he was so conscientious that what had probably only been intended as a general retainer became a close tie, and his silent services in the House were given at the cost of a great deal of time, which was taken from his own work, while his constant attendance involved considerable bodily fatigue. Altogether, as years ran on, he was left very little time for himself; when his services were requisitioned to assist scientific investigations, the appropriateness was obvious; but Church reform, questions of belief, politics, University legislation, etc., all claimed his time and involved continual committees, and constant streams of letters which he always most dutifully answered, and to very dull people, often at great length.

He was very ambitious about everything he did, and desired to do it thoroughly well and to excel; but because he was not ambitious in the ordinary way and for the usual objects, people were not probably quite aware of it. It was this mixture of intense but exalted ambition joined with great conscientiousness which spurred him on; for though he was eager and energetic by temperament and custom, yet there was a substratum of inertia and procrastination which he always fought against. Often after dinner, reading the newspaper and the subsequent games of backgammon, still basking by the fire, he would say, "How lazy I feel, but I must go and work!" then with a little discontented grunt and lazy stretch he would pull himself together and go. He sometimes spoke of this native slothfulness, but indeed it did not get much chance, and only got its head up in occasional procrastination.

He always had a difficulty in lecturing, and felt considerable anxiety about it. Even his professorial lectures, to which it might have been thought he would have become accustomed by long use, worried him, and he was anxious as to whether he was getting on too quickly, whether his class were following his lectures and his experiments, and was evidently always afraid of sinking his level by the staleness of custom. On our walks, when it was noticeable that he was deeply pondering and must not be disturbed, after he emerged from his long fit of abstraction he would not infrequently say that he had been thinking about his lectures and deciding on his course, and how he should present his subject attractively; and afterwards he would sometimes show that he was depressed about his lectures and thought them a failure. On the other hand, he would be quite gay when he had a specially nice class, and would thoroughly enjoy his course. But any single lecture or course of special or popular lectures was a thorn in his side, and worried and tired him more than anything else. It seemed that he thought that other people had some extremely high standard, would expect so much, and be so likely to be disappointed.

He was always ready to enter into correspondence with anyone on religious subjects, even with uneducated people, and felt great compassion for those who endeavoured to lead upright lives, though unaided by faith in any revealed religion. But he could think of the complete annihilation of evil-doers with what seemed a curious indifference; it appeared that he regarded it as a simple and an almost satisfactory arrangement. When asked how the good who loved bad people were going to manage to be happy in Paradise, he would answer that we only love people for the good we find in them, and that if we were once for all convinced that there was no good in them we should cease to love them. It was in vain to argue the absolute mixture of good and evil in ourselves, and all the difficulties which come from people who are bad because they never really had any chance to be good, and how some spiritual Purgatory might fit the real state of a real world far better, or to use a better word, a state of spiritual evolution after death. But if any came to him with difficulties of conscience or belief he would give much time to writing or talking with them, and would as it were take them by the hand and endeavour to lead their minds into some quiet place.

His views with respect to the condition of the soul after death formerly gave much offence in certain circles; he began such considerations long ago, when thirty-two years of age. On several occasions I have heard him preached against by the clergymen of the churches which we chanced to attend in Vacation. It was curious to note the unruffled calm with which he would listen on such occasions. Once I heard a very humorous sermon preached in the University Church by an Irish Bishop who referred to "the process of ossification which goes on in a Senior Wrangler's heart," and saw Mrs Adams, Mrs Cayley, and my mother peeping at one another under their bonnets.

He was a great student of the Bible, particularly of the New Testament, which he often read in Greek. He would often speak of the journey to Emmaus as a passage which particularly attracted him. Two subjects, already referred to, interested him particularly. One was the question of the Immortality of the Soul, on which he held the belief that it does not inherit immortality as a right, but as a supernatural gift from God, and that Christ died to obtain it, but could only obtain it for those who were good, or being bad had repented before death. The other subject was the State of the Soul between Death and Judgment. This period he believed to be passed in a state of absolute unconsciousness, which he illustrated by having experienced fainting-fits. He held that the soul would awake to the Judgment and gain Eternal Life or absolute annihilation according to its deserts. This seems a rather dreary creed, not likely even to be very consoling for those who were good. But like some other theologians his actions and his character quite belied his creed. Had he become a lawyer, as he was once advised, and had he attained the Bench, I cannot imagine his calm condemnation of even the most hardened sinner to mortal death. It seems certain that he would always have wished him to be imprisoned and to have another chance of being better. Once near the end of his life, on being asked if he did not think that Paradise would consist in a greater power of devoted love and in being always with those we love without any fear of separation, he answered that he thought that perhaps the Soul would be so absolutely centred in God and in worship of His perfection that it would have no perception of anything else.

As a member of the Lay House of Convocation he entered into questions of Church Reform, and was very anxious that some abuses should be corrected, particularly the disparity in the endowment of livings and the difficulty in the removal from their charges of clergy who led evil lives. He desired the necessity for signing the Thirty-nine Articles to be abolished and the Athanasian Creed to be altered. He wrote a short paper on Polygamy in connection with Christian Missions; it seemed to him wrong that women and children should suffer from the introduction of Christian teaching on the subject of Marriage, and he thought that a man who already had several wives ought not to put them away, for he had entered into a bond with them before he was taught Christianity.

In the spring before he died my husband and I were pruning the lower branches of our copper beech, and we noticed that the young leaves exposed to the light were of a copper-brown colour while those sheltered from the light were of a soft bronze-green shade. It seemed to be the action of the light which gave them their deep colour: we thought that he might know why, and that anyhow it would interest him. He had appeared rather slack the day before, and also that morning, saying that he began to feel old and inactive; but when we took some leaves up to his room he at once became quite keen. He said that he did not know the reason, but that he would try and find it out. Soon he came out

on to the lawn, almost running in his eagerness to gather more
leaves ; and he tried experiments with them for several days, but
said that he had not been able to discover the cause. In 1872 he
had worked very hard at elm-leaf experiments.

He talked much with us during the last few months of his
life. He told us that he had never had any experience of what is
called conversion. It seemed that he had grown up from childhood with a steady desire to do what was right. He appeared to
have inherited his tender conscience and high principles, and to
have been carefully trained to subdue himself while still very
young, for the home-life was the reverse of pampering, and both
boys and girls grew up hardy in body and simple in their tastes.

He took to his bed only a few days before his death, but
he had been failing noticeably for some months previously. He
was very patient and struggled on, giving his lectures and doing
his other work as long as he was able. He managed to be present
at the Annual Dinner of the Cambridge Philosophical Society,
held on that occasion in his own College, about a month before
his death, although very ill at the time. He made an admirable
speech, recalling with charming simplicity and courtesy his life-
long connection with the College, to the Mastership of which he
had been called on the last day of his eighty-third year, and with
the Society through which he had published so much of his
scientific work.

We thought that he had received some special illumination
towards the close of his life; he spoke very little about dogma,
but much of the nature and attributes of God, and of how men
may attain to live a higher kind of life while they are still on
earth. He thought less and less of the differences in religious
opinion ; not that he ever minded such differences, for I remember
years before how it grieved him that Mr Huxley should be so
misunderstood, as he regarded him with strong affection and
admiration, considering him one of the truest people whom he
had ever known.

He quite realized his own condition and knew that he had not
long to live; but he evidently wished to live longer and do still
more work, for he was very happy and energetic, and his faculties
were quite unimpaired.

Near the end he was conscious that his life was rapidly drawing
to a close, but his mind remained clear; only during the last few

hours he wandered slightly, and imagined that he was addressing the undergraduates of his College. "Speak," he said, "to the young men," and then, "the way of purity, that is the only way." But he awoke from the state of lethargy into which he had fallen, and during the last half-hour of his life, and when he could no longer speak, he smiled occasional assurance that all was well with him. There was hardly any struggle at the last. It seemed that he had walked with God, and that he was now gently taken away from us.

As he lay dead he looked as though he had seen peace, and yet as if by taking rest in the deepest sleep he was only preparing for some fresh activity. On the night of February the 4th his body lay in the Chapel of Pembroke, where he had so often worshipped; and on the following morning, after a short service, it was borne according to ancient custom round the court of the Chapel, and thence to the University Church.

The University, which had honoured him while living, honoured him dead, and a great company was collected in the church to do him reverence. But I do not remember anything more, except that some of those who most loved him and the members of his College still continued with him and followed him, until after the final prayers his body was laid beside that of his beloved wife, and near two of his children in the Mill Road Cemetery.

So sweetly disposed was he to all with whom he came in contact, that people while they revered were scarcely ever afraid of him, and on entering his society were at once without fear of being misconstrued or misunderstood. He was very decided in his opinions, and having so much business to transact, must often have clashed with the opinions of others; yet I believe that, though anger might be felt towards him at the time, it may be truly said that he made no lasting enemy. The words of Francis Bacon seem to sum up his whole character,—"Certainly it is Heaven on earth to have a man's mind move in Charity, rest in Providence, and turn upon the Poles of Truth."

EARLY LETTERS TO LADY STOKES.

69, ALBERT STREET, REGENT'S PARK,
LONDON, N.W.
Jan. 5/57.

I HAD my first lecture to-day. I got on satisfactorily to myself and I hope interested the men. I shall like my lectures better now for I am coming to electricity which is a more interesting subject to lecture on than mechanics.

At Jermyn St. I got my quarter's salary and found a letter from the Abbé Moigno. It was written to induce me to interest myself in an improvement in printing from a photograph, due to a M. Poitevin (if I have read the name right). The Abbé said that they were endeavouring to place me on the list of candidates for the Corresponding Membership of the French Academy at the coming election. He also called my attention to a paper in Poggendorff by a M. Holtzmann, who had repeated with some variation my experiments on the polarisation of diffracted light, but had arrived at an opposite conclusion as to the direction of vibration. I went to the Athenæum and looked over Holtzmann's paper. I had not time to read it carefully through as my lecture was coming on. I think I must repeat Holtzmann's experiments and my own, before I can come to a mature judgment on the matter.

...It looked funny to see mathematical formulæ in a letter from you. Don't be too ambitious for you will not understand these things. They go much beyond Euclid and Algebra, which will be quite enough for you in the way of pure mathematics; but you may be interested by chemistry and physics.

Jan. 10, 1857.

My cold is gone, reduced to only the dregs of the dregs of a cold, an infinitesimal of the second order. After dinner to-day I finished *The Caxtons.* Tell Mrs. Robinson that I have come to the end; not that it was a task, for I like it very well. I suppose you would laugh at me, being in the habit of devouring a novel at a sitting, or two sittings at the most. I take it more by sips, the way a man takes wine. To-day I had my men to an examination. Alas! I fear they have not been taking in my lectures. It is hard to teach mechanics to men who have had no previous mathematical training. I am now in electricity; a much more interesting subject to lecture on than mechanics.

I doubt if I shall make any further remarks on Prof. Challis's paper. I think I shall send my friend the Abbé Moigno a few lines on M. Holtzmann's experiments to insert in *Cosmos*, a weekly scientific newspaper which he edits. I have got to test the reflecting power of the speculum which Piazzi Smyth used on the Peak of Teneriffe, the reflecting power I mean with regard to the invisible rays. This I hope shortly to do; it won't take long.

I intended to teach you a little Euclid and Algebra just for the sake of strengthening the mind, not for the sake of enabling you to follow my mathematics. I am afraid Euclid and Algebra would not help you much towards understanding what I write on pure mathematics. But I chiefly busy myself about physics, and you will be able to understand something of that after you have seen some experiments.

I am afraid you have puzzled me with your chemistry. The tarnish on the bracelet is, I suppose, a very thin film of sulphuret of silver, and I don't know what will take off that. I tried my hand once on a half-crown belonging to my friend Power of Pembroke. It looked a little suspicious to begin with, and after trying various chemicals I left it looking much more so. It looked about half-way between gold and silver, yet I believe it was perfectly good.

...I send you my smaller colour discs to name your father's by I asked Maxwell to lend your father a copy of his paper if he had one to spare. Please tell me if one arrived.

Jan. 12, 1857.

...I am still on electricity. I have far more and prettier experiments to enliven my lectures by than I had when I was on mechanics. I gave No. 37 to-day and my course consists of 48.

Jan. 17/57.

As no mails go out on Sunday you will not I fear get this for a day and a half after the 55-pager. I am afraid the latter, though it will probably be interesting to you (if I may judge how

I felt about similar letters to me), may have given you some pain.
What gave me occasional pain, being no longer pent up, can now
give me no more, and I feel so happy now that there is nothing to
interfere with my love. The ghosts looked always wrong; but
now how silly and contemptible they appear as well !...

The 55-pager explained to you my motives for the first time,
and it is better you should love me as what I am than as something
else. My letter this morning was for a letter something like
Hofmann's methylethylamylophenylammonium for a word: I guess
you never got a 55-pager before.

Jan. 21/57.

...So then, my dearest Mary, though the ghosts were bad
enough they were not quite so black as you may perhaps have
imagined. It grieves me to think that you should look forward
to but little love. I fancied I should love you as few wives
comparatively get loved, but of course I cannot tell, for I cannot
look into other people's hearts. And as to coldness I cannot tell
you more of my heart than I know myself, and the heart is said to
be deceitful, but it appears to me to be thus. Do you remember
how I told you I broke my arm, and how the muscle between
the elbow and shoulder fell away when the arm was in the sling,
and how rapidly it came to again when the sling was laid aside
and the arm used ? Well, I believe it is even so with my feelings
of affection. I have been living so much alone "amidst thoughts
and making out things" that they are not so strong as those
of one living in the bosom of a family; but they only want an
object on which to exercise themselves. And would it be no
satisfaction to you to think that you were under God the means of
supplying to my character one thing it wants much ? Already
I feel the difference. To me as well as to you it seems quite
natural that people living together, and both trying to do right,
though it may be with many faults, should love each other.

I quite approve of your views, and perhaps they are the truer
on the whole, but I think mine are admissible too. Why should
not love and duty go hand in hand to effect a sacrifice which
love alone might have been insufficient for ? And may not
the love in the end be quite as deep though helped over the
initial difficulty by duty ? I was capable of being moved,
mathematically as it were, by the belief that a particular course

was right; and I do believe that God put these views in my mind, working by means of that which was in me to supply that which was wanting.

...And now the last thing is this; that I do verily hope and expect that I shall be both happier and better for being married to you. It seems to me that I am all the happier from not having made a calculation or estimation as it were of the balance of happiness.

I cannot compare myself with another, but only with myself, and it seems to me that I both do and shall love you better and more surely as it is than I could if I had fallen in love with you in the more ordinary way.

Jan. 24, 1857.

...True, nothing human can be perfect; but any incipient annoyance is prevented by confession from interfering with love.

To reassure you after the great pain I fear I have caused you, I will tell you one thing which will show how I regarded you about two and a half years ago and regard you still. I have never told it to you yet. When the clouds seemed to clear off in that wonderful way, and I saw how I was on the point of sinking into an old bachelor (I mean no disrespect to the *genus*) and felt how much better married life, if carried out as I looked forward to, would be, I felt that perhaps my marriage with you would be even the turning-point of my salvation.

I shall feel anxious to hear that you are again well and happy, but I must have patience.

And now my dearest Mary I hope that henceforth you will be no more afraid of me, but will believe me to be, ever your affectionate, George.

Jan. 27, 1857.

Now that I feel that there is nothing to keep me from opening my heart to you any time in case of any trouble, and find that you are thank God happy again, it seems as if I have got social even in my lonely lodging, and can join your social party at Armagh and form one of the family.

You are quite right in saying that it is well not to go brooding over one's own thoughts and feelings; and in a family that is easy, but *you* don't know what it is to live utterly alone.

Jan. 29, 1857.

...Trials of some kind or other are almost sure to come, and death we know must come at last; but even death itself is only an incident in the bright course if we live as God grant we may. But how many a bright prospect is marred by human fault, from the loss of Paradise downwards. We must bear and forbear, love, comfort, and forgive, until by God's mercy we reach, as I trust we may, that City, into which must enter nothing that defileth, where there shall be no more death, neither sorrow nor crying, neither shall there be any more pain, for the former things are passed away; where, while faith shall have no more exercise, and human knowledge shall have vanished away, love shall abide for ever.

Jan. 30, 1857.

...Last night was a R. S. night. I had a paper to read out on the minute anatomy of the earth-worm and another on a kindred subject. I told General Sabine that in reading such papers I felt as if I were a pair of bellows to be blown for the benefit of the Society. I felt I was a sort of reading machine, the papers were so completely out of my line. Some curious and rather metaphysical discussion arose on one of the papers. What do you think of the doctrine that the queen bee and all the bees of the hive are but one individual?

Jan. 31, 1857. 4 p.m.

I have just finished my Smith's Prize paper, the MS. I mean, and am going to post it to Cambridge to be printed. I have not been out yet to-day, as I wanted to finish it. This paper I find comes in rather awkwardly. It is a paper for the highest Cambridge men, which I have to set to when my thoughts are running in quite a different channel, physical experiments for my lectures at Jermyn St. and so forth. At this time of year at least, Cambridge mathematical books are beginning to look strange to me.

I am going to Cambridge on Monday evening, coming back on Wednesday night, going back to Cambridge on Friday morning, coming back to London on Monday morning.

I have only three more lectures at Jermyn St. which I *must* give, but I shall give some extra ones.

PEMBROKE COLLEGE, CAMBRIDGE,
Feb. 3, 1857.

There are nine men now in my room writing away at my Smith's Prize paper, so having nothing particular to do just at present I take up my pen to write you some news, etc., but my fingers are rather numb with cold and not in first-rate writing order.

On Sunday after Church I walked to the Parks, at least Hyde Park, Kensington Gardens, and Regent's Park. There were hundreds, I suppose thousands, on the ice. I did not see any accidents, but I saw by the papers that numbers got in, but no one was drowned.

On Monday (yesterday) after my lecture I called at the R. S. about a paper, and then came here by the 5 p.m. express which arrived at 7. It seemed very cosy getting back to my rooms again and having friends about me. I spent most of the evening in the rooms of different Fellows talking.

And now I must tell you one thing which may be, but I don't think is likely to be, of importance to us, at least at present. If it comes to anything it may make me wish to put off my marriage, as I mentioned to you a day or two after we were engaged. I should not like what I am going to tell you to go beyond you and your father and Mrs R. at present.

Even before I went to the Observatory last August, I thought that as to the time of our marriage I ought not to put it off merely for the sake of saving more money, when I had what prudent and experienced people considered enough (I thought it might be deemed necessary or rather advisable to save money for two or three years in case it should turn out that you had not any, but it was not so deemed either by your friends or mine, but the contrary), but that I ought to put it off a little if my marriage would throw difficulties in the way of University reform so far as relates to the endowment of my own Professorship. Everybody allows that in the abstract the Professorships ought to be endowed, but where is the money to come from? There seems no source but the appropriation of Fellowships to the endowment of Professorships. This would be a very strong measure, like the alienation of private property; for Fellowships are not endowed by the nation, but given by private individuals who founded them.

Parliament might force such a measure on the University, but there would be strong opposition.

But there is another much milder method which could be put into practice in the case of those who are already Fellows, and could even be extended to those who are not, by the election of them into Fellowships if the Fellows of a College were willing to elect them. It is to dispense with the restriction of celibacy and the possible necessity of taking Holy Orders or else resigning, in the case of Professors who are also Fellows. Thus the Fellowship would in effect become part of the endowment of a Professorship, and yet the intention of the founders is but little interfered with. I believe Heads of Houses had originally no more power of marrying than ordinary Fellows, but they were permitted by a letter of Elizabeth. And if Heads by a letter of Elizabeth, why not Professors by a letter of Victoria, or if our now more limited monarchical authority requires it, by an Act of Parliament?

Now I believe that the Fellows of my College would be willing and glad to endow my Professorship in that way, but I don't know what the Master thinks of it. The subject of the endowment of Professorships is now being discussed by the Council of the University appointed by the recent Act of Parliament. *At present* so far as I know all is vague talk, but a very few weeks may make a difference.

Now, my notion is this. Talk may go on for years and come to nothing, and I don't know whether I ought to put off my marriage for mere talk. But if any scheme were definitely proposed, then I think the case would be altered, and in that case I should be for waiting till it was legalized, provided that were likely to be done before long.

If I were called into residence and my Fellowship were added to the Professorship, my income would be just about what it is independently of my Fellowship which I should have to give up if married at Easter, but our position would be far, far pleasanter. I should be in a fixed and highly respectable position instead of being like a " bookseller's hack " as Airy expressed it to me, " a scientific hack " as one of my intimate friends said to me I was, I should do one thing well (at least I hope so) instead of having so many dissimilar things to attend to that I feel as if I were doing them all badly. I should have (probably) much more leisure for researches, which would then become part of my business, to keep up the reputation of the Chair. I should have pleasant intellectual society. And you, instead of being a nobody buried in big smoky noisy London, would become all at once a full-blown Frau Professorin, with extremely pleasant society at the distance of a short walk from you, living in what I almost consider the country, able to drop in to those lady friends with whom you might become intimate instead of having to take a cab for two or three miles to call on a friend.

I wish you would speak seriously to your father about this: I should like to know what he thinks. At the same time though I have written all this as to what *may* be, I don't think it is at all *likely* to be, i.e., that it is at all likely that there will be anything more than vague talk before Easter.

I cannot tell you what a comfort it is to me now that I have at last done what I so often longed to be permitted to do, and opened my heart to you. It makes me feel that, whether we are married at Easter or wait two or three years, we shall be sure of each other all the same.

There are nine men in for the Smith's Prize this time. I hope to have their papers looked over before I return to London to-morrow night.

By the bye, I intended several times to remind you of the colour discs, but forgot. I should like to have them in case any friend should drop in and like to see the colour top. Never mind if you have left Armagh before this reaches you. Friends' visits at Albert St. are few and far between. Indeed, except Henry and Mary who were staying with me, Hudleston and his brother Henry once, Whitley once to dine with Henry, and Dr Sharkey once on business, I don't recollect a single friend's visit at Albert St. since October when I got there. Yes, I believe I had Fischer once on his way to Scotland.

I have run on and written a pretty good letter after all, and I will now bid you good-bye.

P.S. I must explain that my "castles" meant merely bright anticipations of steady mutual affection. I did not by using that word imply that I looked on them as fancies not to be realized. Your character has been uninterruptedly sunshiny in my eyes. Pembroke did brilliantly this examination. We had three men who went in for mathematical honours and they came out 2nd, 4th and 10th wranglers.

PEMBROKE COLLEGE, CAMBRIDGE,
Feb. 7, 1857.

On Wednesday night I went again to Town by the 7.10 p.m. train for R.S. work next day. It was a Council day, so we breakfasted with Lord Wrottesley* and discussed business of the R.S. till about one; then I called at Jermyn Street, to speak to Sir R. Murchison, went on to the R.S., arranged my papers, and had just a few minutes to write to William before the Council met. We had done earlier than usual, about 20 minutes before 5. Then dinner of the Philosophical Club at 5.30; then evening meeting of the R.S. at 8.30. I had little to do at this but to listen, as it was the Bakerian Lecture, delivered by Faraday *vivâ voce* or I should rather say in person; for on other occasions the papers are read by the Secretary. Next morning I got up at 6 h. 20 m., walked to the station 3 or 4 miles, got a cup of tea and morsel of bread and butter, and got off to Cambridge by the 8 a.m. train, got here (i.e. to the station) at 10 and breakfasted in my own rooms. That day (yesterday) was spent in looking over, i.e. completing the looking over of, the Smith's Prize Papers. William came about 1 p.m. for the meeting of the "Family" Club, so he slept here last night, but he break-fasted with his friends Dr and Mrs Paget, who have recently lost their second (by age) child, having now out of three only the eldest left. We met yesterday at 5.30 p.m. to decide the prizes, at the Vice-Chancellor's, Dr Philpott, Master of St Catharine's Hall. The Master of Trinity (Dr Whewell), Prof. Challis, and myself, examined this time. We gave the two prizes to the first two wranglers, but reversed the order, making Savage, the second wrangler, a Pembroke man, first Smith's Prizeman. The decision over, we dined with the Vice-Chancellor; Professor, Mrs and Miss Willis were of the party. When Professor Willis first shook hands with me before dinner he asked me where Mrs Stokes was. I told him there was no Mrs Stokes yet. At dinner I happened to sit on the Vice-Chancellor's left and Mrs Miller on his right. The subject of houses happened to be mentioned, when Mrs Miller said to the Vice-Chancellor that I should be wanting one shortly. It was the first the Vice-Chancellor had heard of it, so he congratulated me, and Mrs Miller was pleased with having had such a piece of

* President of the Royal Society, 1854–8.

news to tell. Here comes the porter for the letters, so I must conclude, but indeed I have told you most of the news. William went off at 10 a.m. to London. He means to return to Denver on Monday. I am going to return to London on Monday morning. To-day I tried my Ruhmkorff's coil for the first time. I got sparks 0·4 inch long with a Callan's battery of 8 cells in indifferent action. I showed it to several of the Fellows as well as to my old bedmaker. She remarked, "You will have a great many things, sir, to amuse a lady." I suppose she meant you, but nothing had occurred to remind her of you. I could not help thinking to-day how very funny it would be, your overhauling such an old bachelor philosopher's den as my rooms are. I had William last night in my bed in my bedroom, so I slept myself on a stretcher with my head right under a lot of chemicals, my feet nearly opposite a shelf of chemicals or rather set of shelves, and the table of the room so blocked up by bottles, funnels, test-tubes, etc. that there was room for nothing and I put my basin, etc. on a plank lying on two chairs.

As to the time of our marriage, I shall proceed on the supposition that it is to be Easter Tuesday unless something occurs to prevent it. I don't think this at all likely as far as changes are concerned.

Ash Wednesday, Feb. 13th, 1857.

It is high time that I should write you a longer letter than I have done for the last two or three days. One thing which prevented me was not knowing where to address. However I think I will post this to-night, addressing to Armagh, unless I get a letter from you by this afternoon's post leading me to address elsewhere.

If I were to marry *now* I should not think of taking a house in London; for the subject of Professorships is fairly before the Council. From all I have heard that has dropped out about their deliberations, I think it probable that they *will* make recommendations as to endowment and probably even as to the foundation of new Professorships; but whether the Senate will sanction their recommendations remains to be seen. Anyhow I expect that a bold effort will be made towards endowment, whatever may become of it. No talk has been made, so far as I have heard, *in the Council* about endowing them in the way I mentioned; probably it will be proposed to tax the Colleges for the benefit of the whole University. As far as we can guess what is probable, it seems likely enough that the Professorships will be endowed, but it will probably take a few years to effectuate the changes. If that be so I think my plan would be to take a house in Cambridge and take lodgings in London for the winter; to live in fact as I live at present, only substituting a house for a College room, with just *one* great difference, namely having you with me.

There is no harm speculating on the future, so long as we do not so fix our imaginations upon it as to be disappointed should things turn out otherwise. Our present course is to wait till things develop themselves. At present one can only conjecture; by June it seems probable, if it pleases God to prolong our lives to that, that we shall know pretty well how things are likely to go.

I called on the Millers on Monday morning. I wanted to ask Prof. Miller whether anything had been written about the colours of the precious opal, as the Master of Trinity had asked me a question on the subject. I found from him that nothing had been written; it was only conjectured that the colours were due to microscopic crystals of quartz.

Mrs Miller said she had not expected to see me again till I came married. She seemed to have set her heart on being the first to ask you to dinner in Cambridge. I told her of the put off. She spoke of the discussions going on in the Council. Her husband you know is one of those interested in the result, and so you may naturally suppose it is an interesting subject to her as it is to you.

Yesterday I dined with my friend Dr Paget, who as President of the Cambridge Philosophical Society invited the members of the Council. We had a very pleasant party of 15, including Mr Hopkins, Professor Miller, Prof. Challis, and others whom you do not know. To-morrow I am going to dine with Mr Vignoles at Westminster (there is the meeting of the R.S. in the evening) to meet Dr Plarr of Strasburg, who has been writing on the figure of the Earth and was very anxious to see me as I had written on the same subject. I have sent him copies of my papers. I mean to return to Cambridge on Friday morning. I have to examine for the Bell Scholarship next Tuesday in Cambridge, and to give my final examination at Jermyn St. on the following Saturday. So from Thursday afternoon, March 5, till Saturday morning, March 7, inclusive, I expect to be in London, and a letter to reach London then may be addressed Athenaeum Club, Pall Mall, S.W.

It is just 12 and I am going to sit for my photograph* to Power, one of the Fellows who has taken up photography. I shall beg one for you, but I must keep it till I go to Armagh as they are collodion positives on glass. I expect to arrive at Armagh the Saturday before Passion Week.

It is now after dinner. Power took altogether 5 photographs of me. Two of them I think are very good. The afternoon's post brought me your letter of Monday. You don't say how long you are going to stay in Dublin, so I shall send this to Armagh. I return you Mrs Napier's letter as you may like to keep it.

I am putting in a notice of M. Holtzmann's experiments in the next No. of the *Phil. Mag.* Perhaps you will read it, but I don't think you will understand much of it.

I have been very remiss in question-answering. You have asked me 9 marvellous searching questions, which I proceed to answer some of.

1. The virtue you think most of. By think most of I presume is meant think most highly of, rather than reflect oftenest upon. *Answer :*—I think one is more struck with a strong exhibition of virtue, be it of what kind it may, under trying circumstances, than with virtue on account of its being of a particular kind. And with reason; for such a proof indicates the possession of a character which would show itself in other directions if occasion led. Yet comparing virtues in the abstract, I am inclined to think that charity or love (for it is the same word) must rank the highest, for it in a manner includes all others.

2. Favourite hero. Hero I suppose is meant in the usual sense of the word, and I should say in this sense the Duke of Wellington.

3, 4. Prose writer and poet. No decided opinion.

5. Antipathy. *Vide* what was said under virtue. Presuming that you don't mean to go to the assizes but speak of what one meets with in ordinary society, I should be inclined to say a sort of cringing, sneaking, fickle, frivolous, fawning character.

6. Flower. Perhaps a clove-pink.

7. Food. Variety is charming; one would get tired of anything. Though I come from the Emerald Isle I must say the roast beef and plum-pudding of Old England is very good.

---

* Perhaps the frontispiece to this volume.

8. Occupation. Scientific investigations, especially when they lead to discoveries.

9. Colour. I should say a red-purple.

I must now conclude for I want to write to Mr Vignoles, and then I am going to Dr Clark's to tea.

LONDON, *March* 19/57.

When the cat's away the mice may play. You are the cat and I am the poor little mouse. I have been doing what I guess you won't let me do when we are married, sitting up till 3 o'clock in the morning fighting hard against a mathematical difficulty. Some years ago I attacked an integral of Airy's, and after a severe trial reduced it to a readily calculable form. But there was one difficulty about it which, though I tried till I almost made myself ill, I could not get over, and at last I had to give it up and profess myself unable to master it*. I took it up again a few days ago, and after a two or three days' fight, the last of which I sat up till 3, I at last mastered it. I don't say you won't let me work at such things, but you will keep me to more regular hours. A little out of the way now and then does not signify, but there should not be too much of it. It is not the mere sitting up but the hard thinking combined with it.......

PEMBROKE COLLEGE, CAMBRIDGE,
March 28, 1857.

......To-day I have been examining some alkaloids sent me by Dr Herapath of Bristol, and trying some other chemico-optical experiments. I find in horse-chestnut bark a crystallizable substance distinct from æsculine, the solution of which is highly fluorescent. The afternoon post brought me a letter from the Prince of Salm-Horstmar enclosing a specimen of a fluorescent substance he had obtained from the bark of the ash. He sent also a cheque for some money for Darker†.

[* *Math. and Phys. Papers*, Vol. IV. p. 77 (May, 1857). In modern language this memoir relates to the domains of asymptotic solutions of linear differential equations, and their relations to each other and to the regular solutions. Urged on by the then unaccountable discontinuity above referred to, he succeeded unaided in probing the whole matter to the bottom. Eleven years afterwards H. Hankel (*Math. Ann.* i.) arrived at the same results more systematically, by building on Riemann's work.]

† Philosophical instrument maker.

March 31, 1857.

......I feel sure that you will love me more than I deserve; and supposing even that in consequence of the letter you love me a bit less, I set it down as an axiom that it must be that I don't deserve it, and I don't blame you a bit, but hope to deserve it better in future.

I too feel that I have been thinking too much of late, but in a different way, my head running on divergent series, the discontinuity of arbitrary constants, etc., etc.* I was thinking to-day that perhaps I would get you at Armagh to bind up the MS. of the paper I am writing for the Cambridge Philosophical Society. I often thought you would do me good by keeping me from being too engrossed by those things. My only chance of finishing the paper is to work at it in the vacation, for my lectures come on immediately after. Yet I was half inclined to lay all these things aside for the vacation.

......I repay you in kind in one respect; but *my* violets were gathered from the garden where Ridley used to walk up and down learning the Epistles by heart, as he says in his farewell†. But I must stop. The porter has long since been for the letters, but I will take this to the Post-Office myself in hopes, perhaps, of doing something to ease your little troubles. Only just time.

April 1, 1857.

......Secondly, remember you should make allowance for a situation in which you were never placed, that of having a tiny trouble which you were not allowed to mention to any human being, although it was not a thing concerning yourself alone, and being alone, breakfasting alone, dining alone, taking tea alone, week after week, so as to be deprived of the healthy invigorating effects of social intercourse and the mutual interchange of ideas.

April 2, 1857.

......I dare say my letters latterly were somewhat cold; it is likely enough, my head got running so on mathematics of a most transcendental character. I felt at the time how good it would be for me to be prevented from having my mind so engrossed. Yet

* See previous footnote.
† Bp Ridley the martyr.

recollect your "manly" (as I called it, which you thought dubious praise) sentiment about my researches. Again goodbye and God bless you.

38, DAWSON STREET, DUBLIN,
April 6, 1857.

I will take things chronologically, that I may not leave out what I wanted to say. I went with Lizzie to Dr Stokes's* on Saturday night. My sleepiness had gone off: I believe I took a little nap after dinner. Dr Stokes had a small party, not indeed small in all, but about half of it was made up of his own large family. There were there Madame Morosini, an Italian, who sings and plays beautifully, and Mlle C. (I forget the name). Madame Morosini played, but did not sing as she had a cold. Mlle C., who does not speak English, sings most beautifully. Dr Stokes was in great spirits, and once when Mlle C. was singing some Neapolitan ballads, accompanying herself on the piano, he got so charmed that, beginning with moving his hands in time he ended by taking hold of Mr Otway, a barrister, no chicken in years, and skipping about the room, much to Mlle C.'s amusement.......

John, though pretty well recovered, did not venture out yesterday. My mother, Lizzie, and I, went to Church together yesterday morning to St Anne's, nearly opposite. It was, I believe, while he was curate of St Anne's that my father fell in love with and married my mother, then very young. It was sacrament Sunday there as with you.

April 22, 1857.

......The change before you—doubly painful to one so loving and beloved as you—the parting from your friends. But that is a necessity; and though it cannot be but that your father and Mrs R. will feel the loss, still I do believe that it is a comfort to them that you are to be, we all hope happily, married.......

I must soon make up my mind whether to go or not to go to Dr Gladstone's party. To-morrow will be a thorough R.S. day.

PEMBROKE COLLEGE, CAMBRIDGE,
April 24, 1857.

It is really the 25th, as it is now 12.45 at night, but I will write you a line or two before I go to bed. I felt lonely in London on Wednesday evening after the brightness of Armagh, and a trifle

---

* Dr W. Stokes, F.R.S., the eminent Dublin physician.

sleepy into the bargain, and not much disposed to work; so I went to Dr Gladstone's soirée. Crowded rooms as usual; there were several there whom I knew. Yesterday was a complete R.S. day. At breakfast at Lord Wrottesley's I heard some particulars about Sir David Brewster's attachment and marriage. I won't tell you all at present, it is so late. The lady it seems is about 50 years younger than he. Half a century! What a difference!... They appear to have become *bonâ fide* attached to each other, and that pretty quickly. May it be lasting.

......Adams tells me they are still on the Regius Professorships in the Council. I expect they won't come to the Mathematical Professorships for some time. Adams got to-day a letter announcing his election as a Corresponding Member of the French Institute. Tell your father that; it will please him. I have managed somehow or other to get a bit of a cold. I hope my voice will be in trim for Monday's lecture. High time to go to bed. I have done a bit more to my paper. Past one now, good night, I hope you are sound asleep.

ATHENAEUM, LONDON,
May 6, 1857.

......Indeed I feel that you may help me a great deal. If your thoughts wander, where are mine? Indeed they range at large a deal too much. I feel that I am too much engrossed too with my scientific pursuits. They are all very well in their way, nay I even look on them as forming a part of the proper work of my profession; but they occupy too large a share of my mind, and I am often tempted to neglect the work immediately before me in following out something new. I believe that you will be of the greatest use to me, in this respect among others, in keeping me from being so engrossed. But we must take care not to lean too much on each other. It seems to me as if, believing that my marriage to you would be for my great good, I have grown careless, as if that was to do everything for me. We may indeed help each other, but we must not expect too much from that. Nothing can replace individual carefulness, and we must each look for strength where strength is to be found, that we may be able to help each other. Without that it may be but the blind leading the blind.

PEMBROKE COLLEGE,
May 11*th*, 1857.

......If we are married at the time we are at present thinking of, and go to Switzerland as we talked of, I think I will bring a couple of quartz prisms, a quartz lens, and a piece of uranium glass with me, to observe the spectrum on top of the Rigi or Faulhorn.

I read my paper to-night, so don't suppose that you have put me off finishing it. There were no remarks made on it; it was of too transcendental a nature to permit of remarks being made on the spur of the moment.

As I walked out yesterday the nightingales were singing sweetly. I was thinking that would be something new for you. But perhaps you may have heard them already. Probably not, for you have not been much in England, and I don't think they flourish in the Emerald Isle.

The afternoon's post brought me your short note (the morning post brought the letter) announcing that Mrs R. continued improving, and containing its green enclosure. It needed not that now to assure me that you are no longer afraid of me, though it had a symbolical scent. I hope you won't be knocked up by your nursing. I shall be quite content with such short notes while you have so much to fatigue you. But you must not be too fierce about nursing, or instead of being able to nurse others, you will have to get them to nurse you.

May 16, 1857.

......Regius Professorships to Trinity College. They had proposed a plan which Trinity College did not accept, and so they had to go back to old matters. But Adams says he does not think they will get back to the Lucasian Professorship for a good while, as they will probably first discuss the question of trying to get money for the Professorships generally.

On Tuesday I dined at the Frosts; on Wednesday with the Master of Caius; on Thursday at the Philosophical Club in London. Next Tuesday I am engaged to dinner at the Master of Christ's and on Wednesday to the Hopkins's. So you see I am rather gay in that way at present. To-day I declined an invitation to dinner from the Master and Fellows of St Catharine's Hall for the 28th, as I am to be in London on that day.......

I have not forgotten little Jack Horner who sat in a corner, so I comprehend the plum, though nowadays the name Horner is rather associated in my mind with Horner's method of solving numerical equations.

I am afraid I cannot help you a bit in the matter of the governess. You must want a regular blue-stocking, to teach Latin and Greek. Of course Euclid and Algebra must be considered a sort of dust on the balance, a trifle not worth mentioning.

June 3rd, 1857.

......I have got very interested, too interested, in my horse-chestnut bark experiments. One's thoughts run so in one channel, when one is all alone and gets interested in anything particularly. Adams told me to-day they had a meeting of the Council, but were occupied in considering the communication from the Commissioners. I suppose the Council will shortly wind up their deliberations till October; so we must be content to remain uncertain as to our future abode. I have great hopes it may be Cambridge; but there is nothing fresh about that, nor has been this good while. Tell your father that Adams has finished his investigation of the Secular Acceleration of the Moon's Mean Motion, and finds it only half what Laplace made it, so that old eclipses are thrown out considerably.......

[In reply, June 5th, 1857 :—"I am delighted that you have a little time for the Horse Chestnut bark, and I don't see how you can be *too* much interested."]

LONDON, *June* 12, 1857.

To-day has been rather a busy day with me, seeing people on scientific matters, correcting press and so forth. I felt rather fagged, as if I wanted relaxation, and thought if I could only get some talk with you and hear some of your music it would freshen me again and set me up. I ordered a suit for a wedding, as well as 400 cards and envelopes.......

......But that you may not rush to extremes I must tell you that I felt all along that the investigation [about Horse Chestnut bark] in itself was praiseworthy, and it seemed as if there was even in some sense a motive of charity in it, for it might perhaps save chemists who were working at the subject from spending much labour in vain. But on the other hand I had work to do which this was driving into a corner; and I had papers referred to me to look over, and if I had a paper of mine under reference I would like it to be done soon. It did not seem wrong to go on from one experiment to another when perhaps important information might come out. Whatever I might do by daylight, however, by which alone I could work, I felt as if I ought to be working at my papers, etc., in the evening. But when evening came my brain had been well worked and I wanted relaxation, and yet I felt I ought to work but I had no mind for it, could not well settle down to it, perhaps had not, without relaxation, force enough for it. And so when I felt I ought to be working I kept dipping into chemical books à propos to my investigation, and so the thing swallowed up my thoughts. Now I by no means want to pretend that I was not wrong, but only to let you understand better what the nature of the thing was. It began by my softness in thinking I would indulge myself in a few experiments on the Monday after my lectures were over, saying to myself I might treat myself to it in consideration of my lectures, though I felt that that was self-indulgence for I had work to do. And what I ought to have done was perhaps not to refuse to work at all at the experiments, but to break them off with a high hand and stop short in the middle of interesting and perhaps somewhat import-ant investigations, when I found they were interfering with my proper work and swallowing up my thoughts.

And now, my own dear Mary, having told you my weakness in my former letter, and having now explained a little more about it for fear you should be too much troubled, let me tell you this, that the happiness I look forward to in your society was con-ditional on deep mutual love, and on our striving together to live as we ought to live. That was in my mind the one thing needful. Without that, music or accomplishments, or all you could do in that way to make me happy would be but an empty blank. They are all very well in their way but they only come in the second place. When I was engaged in those experiments I felt that it would be very pleasant to have you to take an interest in them and to take down notes for me, but it did not produce the same tender kind of love as when I opened my heart to you with its weaknesses, a feeling deepened still when I found how my faults pained and alarmed you. It seems to me that truth and love must go together. This calls to my mind the passage "speaking the truth in love."

I suppose your father arrived to-night, I must go and see him to-morrow.

And now, my dear Mary, may God bless you and bear you up under the painful and anxious thoughts attendant upon your great change, and the quitting of the home of your childhood and your dear relations. That it is right I for my part, and looking from my side, feel no doubt, although I cannot wonder that my faults should throw doubts on your mind. I should greatly like to get over to Armagh, but I have full as much to do here as will occupy my time. I am not quite certain whether I shall be able to return to Armagh with your father.

And now somehow or other I feel refreshed in my mind. I felt much during the day as if I had overworked myself and wanted relaxation.

LONDON, *June* 13, 1857.

I half expected a letter by the morning's post but none came from you. Perhaps you wrote to Cambridge, if so the letter will be forwarded. Though I wrote last night I write again this morning, fearing that you may be troubled, and hoping I may be able to say something to make you happier. In the first place I may say that I had a good sleep, and felt fresh again and unfagged this morning. You seem to be distressed because you do not feel your heart more full and satisfied and want to know the reason. I can suggest two, first my faults, secondly my absence. You took me for an angel and were disappointed when you found I was not. But even in this I may perhaps give you some comfort. Had I kept my faults more to myself as many men I take it would, remaining more in a state of proud isolation, you might perhaps have gone on thinking me an angel and having large expectations of happiness and even been overflowing in love to me, and yet I should have felt comparatively coldly towards you. But such a state of proud isolation is utterly subversive of my notions of the deep mutual confidence and love of married life. And I find by experience that opening my heart to you with its weakness makes me love you as I never otherwise should have done.

My absence is another obvious cause. The fact is you have seen but little of me personally, and when people are together there are a thousand little things which tend to produce affection which cannot travel by post.

But for my part the belief has never deserted me that God

Himself guided me in the matter, and if you can feel the same then trust quietly in Him. Perhaps these very troubles were given you to put this trust in exercise.......I do not *share* your troubles, for the prospect before me seems brighter and happier than ever since the time when I first seemed to receive the direction as it were fresh from heaven.

LONDON, *June* 18, 1857.

I have just done dinner and will write a few lines to you. I got *two* letters from you to-day. I wanted to post a letter on scientific business at Charing Cross in time for the evening mails, but I passed by my lodging and called in, hoping to find a letter from you. I found one but I did not read it. I felt so *perfectly* at rest that being afraid of being late for the post, I put it in my pocket and went to Charing Cross to post the letter. As it would not be quite decent to read the letter in the street I came here, and read it here. How happy everything now seems. How thankful we should be that these troubles were sent. It is commonly said that when people are engaged they may as well marry at once, for they will never know each other any better till they are married. It has been very different I take it with us. Why? Because we have not shunned to speak or write the plain truth to each other; caring rather to have a clear conscience than fearing the effect which disclosures might produce. It is put down in the *Vicar of Wakefield* (have you ever read it?) that the vicar put off his daughter's marriage, thinking the interval between engagement and marriage the happiest in one's life, and wishing to prolong his daughter's happiness. And so perhaps it often is; people put the best face on it, and then there are bright expectations, and then disappointment. But with us the interval has oftentimes been a period of trouble; yet it seems to me as if these troubles had all now blossomed and put forth their sweetness....... Our troubles have sprung from our faults, yet mostly, almost entirely mine. I feel that, and yet they seem to have been the very way to cure our faults.......away from it [intellectual worry] altogether by music and so forth, refreshing my mind,— keeping my investigations from swallowing up my mind, keeping me awake, impairing my health, and worse still shutting out the view of the things which belong unto our peace. And I felt too that the presence of one I honoured as well as loved would forbid me to indulge in those investigations to the undue postponement of my appointed work...But I do indeed hope and trust that I shall (nay, I must not say *shall*, but *should* if you permitted it) have that higher love which would sweeten uninteresting work by the thought that it is for you....

...I know or fear that I am, or perhaps may almost say used to be, cold. With me, in my life of isolation and abstraction, affection has been in the condition of a virtue requiring cultivation, and of which the culture I fear was much neglected, rather than a feeling springing of itself in the daily interchange of domestic kindnesses. Accordingly I have been in the habit of regarding it, as you said of religious feelings, a thing to be felt rather than expressed. (I feel fearful of expressing too much, lest it should be in fact a lie.)......It is a struggle with both of us between affection on the one side and the things given up on the other. In order that affection may be victorious it must be thorough; a whole love or none. The preparations for our marriage are made; the day is named : but even now refuse me if you wish it.

I verily believe that, setting aside what is positively wrong or selfish, there is nothing so fatal to fulness of love as reserve. It damps the affection of the one who maintains it, and that reacts upon the other.

I have work to do, but writing to you must take precedence of it. That most important day of our lives is now near at hand, unless you say it is not to be. What affects the faithfulness with which we make, and the confidence with which we receive each others' promises, if we are to make and receive them, is most important to our happiness. God be with you. Pray to Him to guide you aright, and if you see it to be right to go on, may He be with us in our journey through life, keeping us united in the ties of the deepest mutual affection and in the ways of His commandments, until it pleases Him to call one of us away ; and at last when this transitory state of our probation is over may we dwell for ever before Him.

I cannot feel as if you would draw back, but you must take your choice.

1 a.m. on Sunday morning...then it is right that you should even now draw back, nor heed though I should go to the grave a thinking machine unenlivened and uncheered and unwarmed by the happiness of domestic affection. But I will not dwell on this

for I do not believe it can be the case: you mentioned it merely
as passing thoughts which troubled you, and which you told
lovingly and conscientiously to me, and I don't love you the less
for having told them out or even for having had them.

Sunday morning now. Your letter set me so thinking about
our feelings that, though it is not far from church time, I will put
my thoughts on paper. You spoke of making me happy. Mary,
I had no prospect of happiness except on condition of your being
happy too, and one part of my own happiness I looked on as
consisting in the endeavour to make you happy. It seems to me
you did not do well to attempt a compromise after the great
letter. It would have been better if you had put it pointedly and
openly to me, could I love you so as to make you happy? And
for want of that you have gone on contenting yourself with the
prospect of a milk and water love, and it is no wonder when your
marriage-day drew nigh you had a desolate feeling of unsatisfac-
tion.......Till your yesterday's letter there was nothing to prevent
the feeling of full satisfaction with which I could make you my
bride, and there is nothing still except your suspicion. The
happiness of deep affection outweighs in my mind the happiness
of the scientific leisure which I give up, but the happiness of
the scientific leisure may outweigh mere milk and water affection.
I feel prepared to make my promises provided you feel prepared
to believe in them; but a great love on the one side requires
a great trust on the other, and you must trust me for love as well
as for everything else, nay, as being about the most important
point of all. There must be no half measures. If you feel that
you cannot trust me, put off our marriage, even for ever if you feel
that you never can. And yet I am your faithful, Gabriel.

[She wrote, Sept. 26, 1857, "When you come you will tell me
if those great pendulum experiments could not be done in April.
You say that there will be a clear month then. I do not think
I could like anything so well as knowing that they were going on;
sitting in the room with my work or books if it were so allowed."]

PEMBROKE COLLEGE, CAMBRIDGE,
Dec. 31*st*, 1857.

MY OWN DEAREST MARY,

Here I sit to write to you my first letter since you
became my wife, in the very place, in the very room, where

I wrote the first letter I ever wrote to you. A very different sort of letter this, just a bit of quiet talk.......I got to College about half-past one. I found Arlett and Ferguson here. In speaking of Professorial matters both declaimed warmly against the absurdity, as seemed to them, of creating a new Professorship before endowing the existing ones. There does not, however, appear the least sign on the part of the University of any movement towards an earlier endowment. It remains to be seen whether the Commissioners will be content. By the Act the power of the University to originate new statutes expires with this year now almost defunct, and then with respect to statutes not yet made they have only the power of accepting or rejecting what the Commissioners offer. The Commission expires Dec. 31, 1858, unless Her Majesty pleases to renew it for one year, which she can do. So we are likely to have an inkling before *very* long of how matters are going.

...I should certainly like very well to be able to settle down in nice quiet Cambridge, but if it may not be we must settle elsewhere and be thankful that we have so much happiness. I think it would be better for me as well as for you to settle in some nice quiet country-like place rather than in a flat or one of Cubitt's houses.

And now I will stop to write to Lizzie before post, and just wish my own Mary a happy new year. Poor Mary! left in the fogs of London while I am enjoying the fresh air and quiet of Cambridge. God keep you. Your loving husband.

PEMBROKE COLLEGE,
Dec. 18, 1857.

MY DEAR STOKES,

I enclose the map of the spectrum which you wish for. I cannot find in the portfolio any other papers relating to the subject. You may well remark on the provision for the Lucasian Professor some eight years hence. Several of us on the Council fought for the common-sense principle of first taking care of what we now have, and adequately endowing the Lucasian before founding the Sadlerian Professorship, but we were beaten. However, I think it very likely that the Commissioners will not be content to look so far forward, and will try to devise some quicker mode of raising the wind.

I am not surprised at Mrs. Stokes falling in love with Highgate, but I wish we could get you away from Hampstead, Highgate and such like places back to old Alma Mater.

Yours very truly,

J. C. ADAMS.

PEMBROKE COLLEGE, OXFORD,
28 *June*, 1860.

DEAREST MARY,

I intended to tell you, but forgot, that Adams told me in the train that there is a comet now visible with a tail 4° or 5° long. It is (or was) about 23° from the Sun, towards the North, so that it does not set all night. I quite forgot it last night, but at any rate there would I suppose have been no chance of seeing it. Indeed I remember now it was raining.

......Section A [British Association] broke up early to-day, though there is ordinarily a crush of matter towards the end. A little after 12 I went off to Section D, where there was a paper by Dr Daubeny, "Remarks on the final causes of the sexuality of plants, with particular reference to Mr Darwin's work On the Origin of Species by Natural Selection," which excited a great deal of interest. The room was filled as full as it would hold, and the paper produced discussion in which Owen and Huxley took part.

Saturday, June 30/60.

I certainly was not *vexed* about the accident; though I would rather it had not happened, still I was more sorry at least at first for the annoyance it was to you. Never mind; I dare say a few shillings will repair the damage, and perhaps even it may do as well as it is and it may be only the appearance that is somewhat spoiled. It is well it was not the quartz.

I heard a day or two ago that our friend Fischer is going to be married in about a month. Better late than never. I see no symptoms of the kind about Adams, unless it be that he got a letter he seemed to read with much interest, and stayed late for dinner in order to write something.

OXFORD, *Sunday, July* 1*st*, 1860.

......After breakfast at [Bartholomew] Price's, Adams and I went to the University Church, where there was the morning prayer as far as the end of the Litany; followed by a sermon, very clever, but rather of the new Oxford School, and having a strong tendency, in my mind, to ignore Providence and the use of prayer, though the subjects were not directly touched upon, the latter at least, and though I dare say the preacher may have had some way in his own mind of reconciling the two. I could not help admiring *my* friend Adams afterwards when we met some other friends who were praising the sermon; for while I remained silent he came clearly and boldly out in defence of what *we* hold to be the orthodox view. I could not help feeling rebuked, and still more when I remembered a sentence in your father's letter; for I felt that I would have let the thing pass in silence when I ought to have spoken, though I should not have put the thing as clearly, had not indeed apprehended it as clearly, as Adams put it. We lunched at Price's and afterwards went to the 2 p.m. University Sermon, a very cold affair, as it seemed to me, and so slowly and mouthily delivered for a sermon addressed to the reason that I could not well follow it. After the sermon we took a walk, and at 4 went to Christ Church Cathedral. After the service we fell in with the Willises (Prof., Mrs and Miss) with whom we went about till 6.30.

Yesterday Adams and I dined with the V.C. There were at dinner Prince Frederic of (I think) Sleswig Holstein, Lord Wrottesley, the Willises, Mr Senior, Mr Kay Shuttleworth, and two or three others. After dinner we went to a great crush *soirée* at Dr Daubeny's.

I should guess two days as the most probable time the Commissioners will sit this week, in which case I should probably be back on Wednesday at 9.15 p.m.......

LENSFIELD COTTAGE, 6th Oct., 1860.

......I own I have felt the want of talking, especially the quiet talk of one's own family. Arlett and Power have come back and I am to dine there to-morrow.

......I paid Miller to-day £3. 4s. 3½d., the remainder of the cost of the "Glasgitter*," making in all £6. 4s. 3½d. It is certainly a splendid one. I feel pretty sure the crystals I mentioned are *not* phylloxanthine, but I have come across quite different in appearance which I believe *are*. I was thinking of writing a bit to your father to tell him about my chlorophyll work, but I dosed after dinner and there would not be time now before post. Your

* Optical diffraction grating.

long letter to Lee reached me to-day. They sent it to the Royal Society and Miller brought it for me here.......Otto Struve mentioned at breakfast to-day a curious case of matrimonial felicity; a couple of whom one was Dutch who could speak no English and the other English who could speak no Dutch, and their servant was a Pole who spoke German. The couple could only communicate with one another in very bad French, and for domestic arrangements the language was I believe an attempt at German....

......Please have a bullock's and a sheep's bladder for me by the time I get back. I can choose whichever seems likely to do best.......

Oct. 12, 1860.

......You may tell your father I am pretty sure there is a second yellow substance in chlorophyll*. The spectrum of Lensfield lawn pretty well satisfied me that there is. I had had suspicions of it before. It was only shortly before I went to London that I observed this spectrum, and therefore I have not yet been able to follow it out.

ROYAL SOCIETY, 12 *Oct.*, 1860.

I am glad of a day for your return being pretty well fixed, but a week seems a good way off. I return to Cambridge to-night, but I own I feel to care very little whether I return or stay till to-morrow. It is not so when a certain person is there, for then I look forward to it with the greatest pleasure. However you or rather I would not gain much if you came back a couple of days earlier, for I am to dine with General Sabine on Wednesday and there is to be a meeting of the Council on Thursday. Don't so fix a day as to cross if it is not settled weather.......

RELATION OF SCIENCE TO THEOLOGY.

LETTERS TO MR ARTHUR H. TABRUM†.

LENSFIELD COTTAGE, CAMBRIDGE,
Jan. 16*th*, 1895.

SIR,

I can reply at once, and with much pleasure, to your enquiries.

(1) As to the statement that "recent scientific research has shown the Bible and religion to be untrue," the answer I should

---

* See *Math. and Phys. Papers*, Vol. IV. p. 236 (1864).
† An official of the London Post Office.

give is simply that the statement is altogether untrue. I know of no sound conclusions of science that are opposed to the Christian religion. There may be wild scientific conjectures put forward by some, chiefly those whose science is only at secondhand, as if they were well-established scientific conclusions, and which may be of such a nature as to involve, on the assumption that they are true, certain religious difficulties; I would not go so far as to speak of opposition, as for the most part religion and science move on such different lines that there is hardly opportunity for opposition.

But if an appearance of opposition may sometimes arise from this cause, it far oftener, I think, arises from the errors of defenders of the faith once delivered to the saints, in putting forward propositions which are mere human accretions to it, and presenting the two as if they had equal claims to acceptance. When I speak of the errors of defenders of the faith I am not thinking of learned theologians of the present day, but rather of those of a bygone age, from whom these human accretions passed into the popular theology, and were supposed to be involved in the Christian faith. This mistaken belief afforded infidels a handle for attacking the faith through the error involved in some of the accretions to it.

To illustrate my meaning I will refer to a proposition dogmatically laid down as part of the Christian faith in a standard book written I believe one or two centuries ago. It is that the Christian doctrine of the future resurrection requires us to believe that all the particles of the present body, however widely separated, even though the body may have been burnt to ashes and the ashes strewn to the winds, will be brought together and will be re-animated to form the future body. I dare say many an infidel lecturer has descanted on the difficulties of believing such a proposition as that. But before the Christian apologist replies by simply falling back on the principle that "with God all things are possible," he would do well to consider whether there is any occasion to defend the proposition at all. My own conviction is that there is no such proposition at all involved in the Scriptural doctrine of the Resurrection. The notion that it is involved in it seems to me intensely silly.

I should doubt if you would find a single theologian at the present day who would regard that proposition as connected in any way with Christianity. But I doubt if infidel lecturers have yet given up harping on it.

I may appear to have been treating a theological fossil as if it were a living animal. But it may serve very well as an illustration of my meaning.

(2) You say "as far as my reading goes I am of opinion that true religion and true science harmonise." I am of the same opinion.

(3) You ask if it has been my experience to find "the greatest scientists irreligious"? That has not been my experience, but the reverse. To confine myself to my own line of mathematical and physical science, and to those who are no longer on earth, though not very many years dead, I could not well select more eminent scientists, of world-wide reputation, than Faraday, Clerk-Maxwell, and Adams, the discoverer of Neptune. I knew all three very well, especially Maxwell and Adams, with whom I was very intimate. I know that they were all deeply religious Christian men. Yours very faithfully,

G. G. STOKES.

P.S. There is nothing private in this letter. You may do anything you like with it.

LENSFIELD COTTAGE, CAMBRIDGE,
29 *April*, 1899.

I now return you James's *Human Immortality*. The materialist makes human nature monistic—body alone, a wonderful material organism, acting merely by the physical forces, and by them alone performing its functions, thinking included. James's book is good as pointing out the insufficiency of materialism, and suggesting a substitute which is a great improvement. But he is I think radically wrong in making human nature bi-partite instead of tri-partite—consisting of a body acted on from without by an individual something in which personal identity is supposed to consist, and which when relieved from the clog of the body, as it is at death, performs its functions all the more actively. Accordingly he supposes that man is by his nature immortal. He shirks the difficulties as to the lost by setting aside that question. He is I think at the bottom of his heart a universalist, though he hardly confesses it even to himself. In fact his speculations draw him strongly in that direction, though he feels that in some way or other that goal cannot be reached.

There is a striking passage, in this connection, in the writings of Justin Martyr, in his dialogue with Trypho the Jew. Justin relates to Trypho the history of his own conversion, which came as the result of a conversation he had had with an old man of meek and venerable manners, whom he fell in with as he was walking in a meadow by the sea-side. From what we know of probable dates, it does not seem impossible that this old man may have talked with St John. In the course of a conversation about the Platonic philosophy, the old man says (I quote from the Clarke translation):

"For to live is not its (the soul's) attribute, as it is God's; but as a man does not live always, and the soul is not for ever conjoined with the body, since, whenever this harmony must be broken up, the soul leaves the body, and the man exists no longer, even so, whenever the soul must cease to exist, the spirit of life is removed from it, and there is no more soul, but it goes back to the place from whence it was taken."

This is in full accord with St Paul's description of man in his entirety as consisting of spirit, soul, and body, and with our Lord's teaching that man can kill the body, and after that has no more that he can do, but God can destroy both body and soul—both body and being—in Gehenna.

LENSFIELD COTTAGE, CAMBRIDGE,
5 *Oct.*, 1899.

DEAR SIR,

The objection you tell me that sceptics raise to the resurrection of Jesus Christ from the dead, on the ground that it would be scientifically impossible, admits of a very short answer, namely, that science has nothing to do with it. Surely no one on the Christian side contends that that resurrection was natural, but supernatural, and as such it lies outside the ken of science altogether. The only logical standpoint for the scientist who would deny the resurrection on the ground you mention is by maintaining that science covers the whole of the complex nature of man, so that he has nothing to do with anything that lies outside the domain of science. Is that assumption reasonable? Biological science concerns itself with the properties of living things, animal or vegetable. But science never has explained, nor does there seem to be the remotest prospect that it ever will be able to

explain, the origin of life.   Science cannot explain the feeling we
have of right and wrong.   Science does not cover the whole of
man's complex nature.

The admission of the resurrection of Jesus Christ, if regarded
as a dry isolated fact, would I think be of little or no value.
It seems to have been God's *design* that it should *not* be so
regarded. We read, " God raised him from the dead and showed
him openly, *not unto all the people*, but unto witnesses chosen
afore of God." Were admission of the fact of the resurrection the
one important thing, the obvious way (if one may so speak without irreverence) to secure it would have been to have shown Him
openly. The evidence for the resurrection of Jesus Christ is
never to be separated from a consideration of the character and
teaching and works of Jesus Christ. The head and the heart
must go together. To demand that the alleged fact of the resurrection shall be accepted or rejected on purely scientific evidence,
is to act like a judge who in a trial in which there were a great
number of witnesses, whose several testimonies would so dovetail
into one another as to produce conviction in the minds of the jury,
should arbitrarily select one of the witnesses (and he perhaps by
no means the most important) and refuse to allow any of the others
to come forward.

25 *July*, 1900.

DEAR SIR,

I had to go by an early train yesterday to London, and
a little beyond, to be present at the funeral of a late friend, and
did not return till latish, so that I could not well have answered
your letter much sooner.

If the sceptics affirm that the present condition of things has
gone on from a past eternity, and is adapted to go on for an
eternity to come, I can only say that they fly in the face of the
best, I might almost say universal, scientific opinion, including
that of some few scientists who appeared to have been sceptics
themselves.

As far as our scientific knowledge goes, matter is indestructible;
it can neither come into existence nor cease to exist. But much
more than the existence of matter is involved in the continuance
of the present state of things. Consider what is required for the
continuance of animal life, say man. Man requires food, and that
food is derived, directly or indirectly, from vegetable life; directly,

as when I eat bread, made from the seed of a grass, indirectly as when I eat mutton, the flesh of an animal that eats grass. With the exception of fungi, which we may regard as vegetables of prey, vegetables absolutely require light. It is only under the influence of light that the leaves of trees are able to perform a function the reverse of combustion: to decompose the carbonic acid of the air, appropriating the carbon, of which by far the greater part of their dry weight consists, and setting free the oxygen. This light they obtain from the radiation from the sun. The sun is continually giving out an enormous amount of energy, which can be measured like the energy given out from a steam-engine.

But where is the supply? In the case of a steam-engine the ultimate source of energy resides in the coals and the oxygen of the air. But what feeds the sun? The solar energy is derived, not from combustion, but from the energy of descending weights: from the energy given out by its contraction, like that of a descending weight. The wound-up weight of a clock is a source of energy, and keeps the clock going; compensates the small friction which would tend to stop it. But when the weight can come to the bottom the supply of energy ceases, and the clock runs down.

Just so, the particles of the sun tend to fall inwards, and to a certain extent do so fall, under the influence of the gravitation arising from the mutual attraction of the particles, which is a force directed nearly towards the centre of the sun. I say "tend" because the fall is in some measure prevented by the lateral velocity, putting the particles in some degree in the condition of planets. But it is only to a certain extent that the inward motion is thus presented; were it otherwise, there could be no permanent giving out of energy. The contraction it is evident cannot go on indefinitely. For the sun, just as for our steam-engines, we are living upon capital; in the latter case we are drawing upon our bank deposit, we are exhausting our coal-fields. At the present day there is a tolerable agreement among scientific men to regard nebulae as suns in process of formation, while among the stars there are a few smaller ones which are blood red. These are generally looked upon as effete stars; stars in process of extinction.

Next as to the origin of life. The doctrine of abiogenesis,

that life can originate from non-life, is pretty well completely knocked on the head. My late friends, Huxley and Tyndall, whom even the sceptics would hardly suspect of being led away from the truth by theological prejudices, are about as strongly against it as any. You only ask about the origination of life. Of course there is a great deal more to be considered than that of its mere origination. There are vast gaps between the life of a grass, that of a fish, that of a bee, that of a man. I do not think that science has succeeded, and I doubt if it will ever succeed, in filling these in. Yet on the other hand what naturalists regard as distinct species of living things, plant or animal, frequently approach each other so nearly that one cannot help thinking that what are commonly called second causes come in.

However, as regards the main question, those who say that the present order of things has gone on from a past eternity, and is calculated to go on for an eternity to come, only, in my opinion, thereby display their own scientific ignorance.

It is high time I should return you your book. I have read the article about my "I," and the Easter Egg. As to the former, it is amusing the way he takes for granted that I knew nothing about Bishop Courtney's nor about Archbishop Whately's book. I have read the former, and have got copies of the passage in the latter referred to in the form of a tract.

As to the "Easter Egg," I have already, I think, discussed with you most of the arguments. In p. 211, about a third of the way down, we see indications of the way in which incautious dogmatism, from which the popular theology is by no means free, puts weapons into the hands of adversaries.

With apologies for having kept the book so long, I remain, yours faithfully.

P.S. I have noted an inaccuracy at the bottom of p. 207. The Paschal lamb need not be a firstling.

3rd August, 1900.

I do not think that what I have written need create in your mind any difficulty. It is merely a question of the meaning that we attach to the word "science." We have first pure science, suppose mathematics, where we deduce certain conclusions from axiomatic premises, and secondly natural science, where we have to deal with external nature, made known to us through our senses. In the latter case invariable sequences give us the idea of cause and effect, and we are led to regard observed phenomena as the effect of such and such causes acting together.

In our laboratories we can resolve water into two gases, oxygen and hydrogen; we can burn carbon in oxygen, forming a heavy gas (carbonic acid) which was not there before, and so on. But the weight of the water which disappeared is just equal to the weight of the oxygen and hydrogen which were produced; the weight of the carbonic acid produced is just equal to the weight of the carbon which disappeared plus that of that portion of the oxygen which is oxygen no longer, but was employed in the formation of carbonic acid. So far as we know scientifically, a certain weight belongs to a certain material, which we can neither increase nor decrease by chemical changes.

We are thus led to the contemplation of, and the quantitative measurement of, what we call matter.

As far then as our *scientific* knowledge can teach us, matter can neither be created nor destroyed.

Science, however, leads us to the contemplation of something more. We have reason to believe that matter does not form a continuous plenum, but consists of ultimate molecules; and further that in matter of a given kind, say the gas hydrogen, the molecules are all just like one another. An analogue of this in the works of man is found only in manufactured articles. Thus the Mint turns out a lot of shillings all like one another. This leads us to the idea that hydrogen, for instance, did not exist just as it is from a past eternity, but was in some way made. But there science comes to the end of her tether; how it was made she cannot inform us.

I quite think that the existence of life is one of the strongest arguments for the existence of a Living Being who is the Author of life. In his Belfast address (*Rep. Brit. Ass.*, 1874) Tyndall in attempting to account scientifically for the origin of life was led to attribute emotion to the ultimate molecules of matter in a fiery mass of gas!

I quite think with you that the great gaps which we find in the series of animated things, both plants and animals, weaken the theory that man came in an unbroken chain from some lowly form of life. Yours very truly.

5 *August*, 1900.

For fear of any misunderstanding, I will just say that the conclusion to which we are led by science—that (say) hydrogen consists of a great number of ultimate molecules which are all alike one another, in that respect resembling manufactured articles, is valid as leading us to say that it must have been caused in some way, instead of having existed as it is now from a past eternity. But *how* it was caused, science cannot say.

Our experiments lead us to affirm that what we call ponderable matter cannot be destroyed or brought into being; to affirm it is a fixed entity. We have evidence of the existence of a *stuff*, as I will call it, which we call the luminiferous ether, which we do not include in the term 'ponderable matter.' We recognise it as possessing inertia, as does ponderable matter. But whether or no ponderable matter is formed from the luminiferous ether, is a matter about which we have absolutely no evidence. We cannot convert hydrogen into ether, or ether into hydrogen, nor, as I observed, do we know whether such conversion is possible, and therefore it is futile to discuss the question whether, if it be, the inertia would remain unchanged.

The 'manufactured article' argument is I think valid as against the notion of ponderable matter, *such as we know it*, having existed from a past eternity. Yours very truly.

7 *August*, 1900.

In any case we must interpret Genesis i. with considerable latitude. And the only reason why this creates a difficulty from a religious point of view is that religious people insisted on a slavish literalism; insisted on theories of verbal inspiration and so forth; indulged in a sort of Bibliolatry; framed a theory that the Bible must be interpreted in a way just like that in which a lawyer would interpret an Act of Parliament; stuck to the letter rather than attempted to catch the spirit; even though it is said that spiritual things are spiritually discerned.

As to the origin of man it is said that God 'formed man of the dust of the ground, and breathed into his nostrils the breath of life, and man became a living soul.' We have here, it seems to me, a recognition of man in the completeness of his tripartite nature. First we have the ponderable matter of which the body is formed; which the physicist can weigh and the chemist can analyse just like the dust of the ground; then we have a divine energy acting on the ponderable matter, 'breathed into his nostrils the breath of life'; and lastly the result of the interaction, 'man became a living soul.' This accords with St Paul's division (1 Thess. v. 23).

This triple division of man's nature was not arrived at by unaided human reason. Plato and very many followers of Plato suppose that man in the completeness of his being consists of body and soul. Man in the pride of his own strength is disposed to think that he has in him a something which he calls soul, which cannot but live for ever; which is half thought of (though expression is not given to such a thought) as almost in a certain sense independent of God. But St Paul says, 'in him we live and move and are.' I think in the N.T. that which a man gives up at death, that which in some way forms the means of our future life, is never called 'soul,' but 'spirit.' Take again the words on the Cross, 'Father, into thy hands I commend my *spirit*.' Dying Stephen said, 'Lord Jesus receive my *spirit*.' Take again, 'Ye are come unto the *spirits* (not souls) of just men made perfect.' And as to the relation of spirit to the future life, see Rom. viii. 10, 11.

Spiritually I think what we are told in Genesis, in the passage I have quoted as to the formation of man, is in full accord with the teaching of the N.T. But *how* God formed man of the dust of the ground we are not told, nor are we, I think, for spiritual purposes concerned to know. We are by no means committed to the supposition that it was something like the way in which a sculptor moulds his clay into a statue. If we say that mental powers and spiritual aspirations such as those of man were conferred on a previously existing animal, the idea may be somewhat grotesque, but I don't think we can say it is irreligious. I don't see that we need trouble ourselves about that. We may simply say, We don't know, and for aught we can see there is no reason why we need know. An M.D. friend of mine long since dead said to me with reference to the similarity of general structure between the body of man and that of lower animals, 'If we saw the keel of a ship laid down, we could not tell what sort of a ship it was going to be.' The conditions of a somewhat similar mode of animal life in man and quadrupeds may require a somewhat similar general plan of structure, which weakens the argument for ancestral derivation. But if we entertain the idea of ancestral derivation as a hypothesis (it is I take it a long way off from being an established theory) we may do so; and we are not to dub a man an atheist for holding that view.

I see nothing against your notion that evolution and special creation may have existed side by side. I have had the same idea in my own mind, and am disposed to look favourably on it.

4 WINDSOR TERRACE, MALAHIDE.
17 *Aug.* 1900.

I meant (but I believe I forgot) to have written about Day 4. Some have supposed that the general progress of creation in Gen. i. is described *as it would appear* to a spectator on earth if such there were. On this supposition there appears to be no difficulty in reconciling the general order which seems on scientific grounds the most probable with that of Gen. i.

Day 1. We have first the creation of ponderable matter; and supposing such matter to exist, and then as now to be subject to gravitation, light would be occasioned by collisions of matter on matter, the greater part of which matter (confining ourselves to the solar system) would be collected in a nebulous mass of vague outline,—our sun in a juvenile stage.

N.B. There would be alternations of day and night, though from a dense mantle of cloud the heavenly bodies could not be *seen.*

There is scientific reason for thinking that the earth was originally in a molten condition—at a temperature so high (above the 'critical temperature' for water, Andrews) that there would be a *continuous* change of condition from water to steam.

On further cooling there would be a separation of water into the two conditions of water and steam. There would now be a separate atmosphere, consisting probably mainly of nitrogen, steam, and carbonic acid gas, separating the water under the firmament, the chemical substance H_2O in the physical condition in which we call it water, from the water above the firmament in the form of an enormous number of minute droplets forming cloud, while all through the atmosphere H_2O would be present in the elastic state which we call steam.

Day 2. Not until the cooling had advanced beyond the

preceding stage could there be water with a free surface, and consequently could we have sea and land.

When the cooling had advanced a good deal the temperature would be low enough to permit of the introduction of vegetation. In the early stages it consisted in great measure of gigantic (as compared with what we have now) cryptogams, the remains of which are preserved in our coal measures. The plants would get diffuse light through the mantle of cloud. Even now many kinds of fern thrive best in shady places.

Day 3. This vegetation, under the influence of light, would decompose the carbonic acid in the atmosphere, appropriating (in some state of combination) the carbon for future use as coal, and introducing into the atmosphere free oxygen without which animal life could not exist.

On further cooling the mantle of cloud would no longer be continuous, but there would be intervals of clear sky, permitting the sun and moon to be seen. They existed ages before, but were invisible on account of cloud, as the sun is, probably, to-day to the inhabitants, if any, of Venus. They are accordingly mentioned as 'set' in the firmament of heaven on the 4th day.

Day 5. The earth having been prepared by cryptogamic (and other) vegetation for the introduction of animal life, we now have animal life, first in connexion with the waters and birds. Among very old remains we have saurians, winged reptiles, the archaeopteryx, which seems to form a sort of link of connexion between a winged reptile and a bird, fishes, etc.

Day 6. Now we have mention of land animals, and last of all Man.

I fail to see any such discrepancy, between what we read in Genesis and what science renders probable, as need create a difficulty. We must remember that on the side of religion we are not concerned with more than a broad outline, and on the side of science, as regards forms of life the records are very imperfect. In consequence of different structures and modes of life one kind of creature may be more likely to leave remains that survive in the rocks than another, so that it may not always be true that the relative antiquity of strata in which remains are found corresponds to the order of introduction of the form of life upon the earth.

I shall probably remain here for three weeks or so.

24 *August*, 1900.

The force (such as it is) of the objection which you represent the sceptics as making turns mainly on the assumption very commonly made, but which I believe to be erroneous and unscriptural, that the soul of man is by its nature immortal. Scripture I think teaches that the *final* end of the wicked is that they are destroyed for ever, not kept alive for ever in a state of misery.

Free agency involves of necessity the possibility of falling away, and the forfeiture, through the agent's own fault, of that eternal life of happiness which he might have had. But there is nothing in this opposed to the statement 'God is love.' Why should God be debarred from creating a free agent, whom He designs for an endless life of happiness, merely because that agent may through his own fault forfeit the happiness for which he was intended? Such is according to the analogy of nature. A tree has many seeds, but not all produce trees themselves. Some come to an untimely end. Is the possibility of this a reason why trees should not be endowed with seed?

It seems to me that the strength (such as it is) of the arguments of the sceptics comes from their introducing into Christianity two things, neither of which belong to it. One is the Turkish fatalism of ultra-Calvinism, the other the Platonic dogma of the immortality of the soul.

LENSFIELD COTTAGE, CAMBRIDGE.
4 *Jan.* 1901.

The question you asked me was I think whether I believed in evolution. It seems to me hardly a correct expression because evolution is not a cause, but the description of a process. Let me illustrate my meaning by a concrete example. It has been known from a very long time that the pole of the heavens, the point, that is, of the heavens about which the stars appear to turn, is not fixed, but slowly changes its position in the starry heavens. It really turns round the pole of the ecliptic, or the point in the starry heavens which is in a direction perpendicular to the plane of the earth's orbit, at the rate of once round in about 25,000 years. The angle between the pole of the heavens, as I called it, that is, the pole of the equator, and the pole of the ecliptic remains nearly invariable—about $23\frac{1}{2}$ degrees. That slow turning of the pole of the equator about the pole of the ecliptic is called (from one of its manifestations) 'the precession of the equinoxes.'

Now suppose a simple-minded religious person had heard that it had been found that the place among the fixed stars occupied by the sun at either (say the vernal) equinox was not quite the same from year to year, but moved very slowly forwards along the path in the heavens of the earth's orbit, so that the equinox was arrived at a little earlier from year to year than it would otherwise have been, he might very likely have simply accepted it as a fact, not thought of attempting to give any reason for it, but contented himself with resting in the idea that God had so ordained it. Nay, dwelling on the idea that 'the heavens declare the glory of God, and the firmament sheweth his handywork,' he might even think it irreverent to seek for any explanation why there should be any such anticipation, from year to year, of the time of the equinox as compared with the position of the sun in his apparent path among the fixed stars.

It was one of the great achievements of Sir Isaac Newton, the discoverer of universal gravitation, to have shown that the precession of the equinoxes was a consequence of universal gravitation. It is a phenomenon *evolved* from gravitation in the sense of *unrolling* in tracing the sequence of cause and effect. The investigation of precession, and of the allied phenomenon of nutation, is now a subject of mathematical calculation. Nobody thinks there is anything at all irreverent in that.

But if precession is referable to gravitation, what, if anything, are we to refer gravitation itself to? Are we bound to say, Gravitation exists because God willed it so, and it would be irreverent to attempt to go further? or are we bound to say, Just as precession is a consequence of gravitation, so gravitation *must* itself be a consequence of something else?

Newton himself was a religious man, and his writings show it. It is shown by the famous Scholium at the end of his immortal work the *Principia*. He wrote largely on prophecies. Yet he thought it axiomatic that since the sun and the earth attract each other there must be something or other between them or they could not do so.

As to the alternative at the foot of p. 2 there is nothing irreverent in seeking to explain a phenomenon as a consequence of something else, which explanation would constitute a passage from one link to another in the chain of causation. Nay a theist might reasonably regard it as irreverent to forbid such an endeavour. For that would come to measuring the mind of the Almighty by our own minds, and presuming to assert that where *we* do not see a prospect of further progress in the chain of (secondary) derivation, no further progress in that direction can be possible.

But evolution, as it is usually treated, is chiefly thought of as something which went on in the past. Further, it is chiefly spoken of with reference to living things, plants, and more especially animals. Can we in any way explain the origin of species? Are we to suppose that each species, or what we regard as a species, originated in the fiat of an almighty power? Or are we to suppose that we are to go indefinitely backwards, and affirm that a chain of secondary causation is to be continued indefinitely backwards, though we can but trace it a little way if at all?

I should say, Neither the one nor the other. The evil (from a theistic point of view) consists in treating evolution as a cause, which it is not, instead of a mode. The former tends towards the denial of a First Cause, the latter may safely be accepted as a working hypothesis; as a guide which *may* lead us from the last link in the chain of causation which hitherto we have been able to reach to a link yet further on, while still leaving what lies beyond hidden in a cloud. The treatment of evolution as a cause, capable of leading us on indefinitely, tends to shut out the idea of a First Cause; its treatment as a possible mode of sequence, leading us a step or two onwards, still leaves the mind directed towards a First Cause, though 'clouds and darkness are round about Him.'

I have endeavoured to put the thing before you as it appears to my own mind. Remember Evolution does not mean a *cause*.

APPRECIATIONS* BY COLLEAGUES.

BY G. D. LIVEING, F.R.S., PROFESSOR OF CHEMISTRY IN THE UNIVERSITY OF CAMBRIDGE, AND FELLOW OF ST JOHN'S COLLEGE.

I MADE acquaintance with Stokes in 1850 at the meetings of the Ray Club. That was a society at Cambridge for the cultivation of Natural Science by friendly intercourse, which had been formed in 1837 in order to fill, so far as that could be done without Henslow's † inspiration, the gap left by the cessation of Henslow's weekly receptions of members of the University interested in Natural History. Natural History was still, when I joined the Club, most frequently the subject of conversation at its gatherings, and it may seem surprising that Stokes, who at that time (1850) was best known as a great mathematician, and had just been elected Lucasian professor, should have been a very regular attendant at the weekly meetings of such a Club. Really, however, his bent was to Natural Philosophy, as his work showed, where his great mathematical ability was employed in handling the problems of Nature. His elder brother, William, a fellow of Caius, had been one of the original promoters of the Club, and was a mineralogist and a chemist with whom I fraternised at once; but I very soon found that George Stokes was equally interested in the same subjects, and quite as ready to discuss, with a beginner, questions connected with them on which probably his own conclusions had been reached by a much shorter induction. He did this, to my great delight, with a deeper insight into nature's mechanism and more suggestive remarks than I had then met with from anyone else except Professor Miller, who also was a member of the Club. Stokes was not content with reaching any empiric law of nature,

* The Obituary Notice of Sir George Stokes written for the Royal Society by Lord Rayleigh in 1903 has been reprinted in *Math. and Phys. Papers*, vol. v. pp. ix—xxv., and is therefore not included here.

[† Rev. J. S. Henslow, of St John's College, Professor of Mineralogy 1822–25, of Botany 1825–61, the friend and mentor of Charles Darwin's early years, resided mainly at his living at Hitcham from 1837 onwards. *Life*, by L. Jenyns, 1862.]

but was always pressing behind the scene to try and see how
the processes were carried out.   It was a pleasure to throw out
a suggestion, and hear what he thought of it, and I feel sure
it was a pleasure to him to help an enquirer.   His thoughts
stimulated mine, and it would be hard to tell how much I learnt
in that way.   In the company of other distinguished men, such as
Sedgwick, Peacock, and Whewell, I have been well content to be
a listener; but the way to learn from Stokes was to get up a
discussion with him.   Never had I any difficulty in getting him
to talk.   His reputation for taciturnity was kept up, if not started,
by people who were afraid to talk to him.   Never was a worse
founded fear, for he was not a specialist who could ride no hobby
but his own, much less was he in any degree egotistic or un-
sympathetic.   He read the newspapers and was generally informed
on the topics of the day, matters and men, and had his opinions
kindly and optimistic about them.   He was willing to exchange
ideas with people who let him see that they had ideas to exchange,
but did not talk for talking's sake, and was not distressed if
people, who had nothing to say, said nothing.   In truth he was
many-sided.   I have been associated with him as co-trustee and
in practical business of other kinds, and never found him a
sleeping partner.   He did his part with caution, sound judgment,
and promptness, at the same time with full consideration for
others, who might be affected by his decision.   Frequently when
I have spoken with too strong reprobation of somebody's action,
he has assented with a softening qualification which implied that
he would think no evil of the actor.   Those who looked for a
brilliant repartee from him, or thought his smile a prelude to
a jocular remark on the follies of society, would, of course, be
disappointed, nevertheless he enjoyed a good story and sometimes
told one.   His smile was no cynical one, but the natural expression
of sympathy with the bright faces about him.   Of late years when
I have had one or other of my nieces living with me and he
has called at my house, they have, on his departure, expressed
astonishment that one whom they had heard spoken of as the
most silent man in the University should keep up a continuous
conversation during the whole time of his visit.   On the ordinary
business and occurrences of the day he would say at once what he
thought and then perhaps drop the subject; while a philosophical,
or any serious question, though he may have said what occurred

to him at the moment about it, did not forthwith pass out of his mind. He would refer to it again and again, sometimes for a whole evening, notwithstanding the interruption of taking part in conversation on various other matters. This persistence in dwelling on a subject on which he had not yet formed a definite conclusion, turning it over repeatedly to get new points of view, was very characteristic, and recalled what Newton said of himself, namely, that he excelled others only in the power of keeping his attention constantly on the problem which he wished to solve. Stokes certainly had, in a remarkable degree, the power of taking pains to do whatever he had in hand as well as he could do it with the means and material at his disposal.

He was generally tolerant of other people's opinions, nevertheless he had a philosophical contempt, shown in his tone rather than in his words, for fanciful suppositions, not proved to have any objective existence, invented to account for facts in nature. Before his explanation of the suspension of clouds appeared, a theory had been current that cloud-drops were bubbles, and in talking with him about his then recently published memoir I mentioned that theory, whereupon he brushed it aside with unwonted brusqueness in the remark, " I can't think how that theory ever came to be accepted—there is absolutely no ground for thinking the drops to be vesicular." His touch indeed had burst that bubble, and no one could put it together again.

The even balance of his mind was that of a strong man. His activity was uninterrupted; he wasted no time in trying to evolve truth out of the speculations of others, or in controversies with fellow workers such as distracted Newton. In truth he never let any occasions of such controversy arise; for he never had any jealousy of other workers in the same field, but was ever ready to communicate to them his views, and let them use them as if they had been their own, repudiating, when it might have been justly claimed, all share in the credit of the outcome. Many brilliant intellects have been lodged in frail bodies, with the result that their wits are sometimes in fine form and at others dull. Stokes' mental and bodily constitutions were closely akin, and had the like patient strength and staying power. His broad chest and firm carriage betokened strength. When he was an undergraduate, and for some time thereafter, the athletic games, now in vogue, were little heard of in the University. There was,

however, more boating and riding and quite as much cricket; but
I never heard of Stokes in connexion with such exercises, nor
did he ever talk to me about them, as he would have done if
he had ever been an active participant in them. One athletic
exercise I know that he enjoyed, namely, swimming. About the
end of the forties an association of graduates and undergraduates
(which later became the Swimming Club) set up a new bathing
place in the Grantchester meadows. Stokes and I were among
the early members of it, and he was noted among us as a bold
and strong swimmer. He has spoken to me with warmth of the
keen enjoyment of a battle with the waves when there was a good
sea on; and he was just as fond of a sharp walk in the face of a
biting "nor'-easter." His bodily organs seemed always to perform
their functions healthily, in spite of hard usage. The fatigues of
long hours and frequent journeys seemed light to him, and he
enjoyed the stimulus of strong tea after dinner without the fear of
lying awake in bed after it.

In spite of his activity his mental attitude was essentially
conservative, though he contrived to keep an open mind on
scientific questions. He did not accept a theory until he had
satisfied himself that there was a sound foundation for it, but
when accepted he did not hesitate to push it to its logical consequences. At the meeting of the British Association at Edinburgh
in 1892, there was a discussion in Section A, on the question
whether the well-known yellow rays of a spirit lamp, with salt
in the spirit, were emitted by the heated sodium chloride or only
by molecules of uncombined sodium. The latter supposition did
not seem to commend itself to many of the chemists present,
but Stokes remarked as we left the room, "Temperature does not
produce any spectrum—a mere motion through space does not
affect the aether—there must be some chemical changes going on
in the flame." This was, no doubt, a logical deduction from the
kinetic theory of gases and the theory of light-waves in an
aethereal medium; and he adhered to it to the end of his life,
for in a conversation I had with him only a few months before
his death he made a similar remark with reference to the spectrum
of carbon, a subject to which he often recurred. Nevertheless
in that same conversation his cautious mental attitude prevented
his giving assent to the argument that, because the thermal
effect of a chemical combination can be shown to be the sum of

several parts derived from the several components and dependent only on their chemical natures, such effect must be derived from changes in the intrinsic energies of the components. He had not tested the theory of intrinsic energies, and so his final remark was, "The heat is more probably due to some form of the potential energy of an attraction."

Stokes' mind certainly affected a great many members of the University, as it affected me. Many have acknowledged how much they owe to him, and his influence has contributed a great deal to raise our school of Natural Philosophy, which produced but few men in the earlier half of last century, to its present fruitful state. But that influence was never, so far as I am aware, exercised consciously with a view to reform, though always with a view to promote true science so far as possible. His instruction was particularly effective because he was, as Newton had been, himself an experimenter. His predecessor as Lucasian professor, Dr King, never lectured ; nor did Babbage the penultimate occupant of the chair. The Plumian professor of Astronomy and experimental Philosophy lectured in the forties on Hydrostatics and Optics, and showed various pieces of apparatus which Airy had used to illustrate his lectures experimentally, but in Challis' hands they were not even working models. We had to get up Natural Philosophy by a painful exercise of the imagination on diagrams and descriptions, and the abstractions formulated by mathematicians to make calculation possible, which presented Nature as a lifeless statue. Imagination could inspire such a statue with life and activity, but it must be the imagination of a mind already acquainted with nature ever moving and energetic. Stokes at once set the study on a new footing, but it was done at the expense of personal trouble undertaken ungrudgingly for very love. The University at that time possessed no laboratories, and provided no appliances for her professors, much less for students, and did not in all cases provide even lecture rooms. If it had not been for Miller, who left at St John's College apparatus for observing the spectrum of Nitrogen Dioxide, and in his lectures showed us conical refraction and other optical properties of crystals, we had no chance of seeing in Cambridge the optical effects of which we were expected to give the theoretical explanation. I never attended Stokes' professorial lectures, for he had not begun them when I took my first degree, but I have often seen his experiments. At first these were made in his own rooms in College, with very simple but effective apparatus, usually put together with his own hands and always so manipulated. He appreciated accurately the circumstances necessary to attain success and spared no pains to ensure it. To the end of his life I do not think that he ever had the help of a trained assistant for his experiments. This independence was characteristic, and the experience gained by it made him such a good critic of other people's experimental work, equally of their methods and of their results.

I have often heard it remarked that Stokes seemed to take all the honours bestowed on him as if they were matters of course. That this was far from being the case I feel sure. He was neither cynical nor thick-skinned, but was naturally sensitive, and showed a delicate consideration for the feelings of others which only those who have fine feelings of their own usually show. He was not really indifferent to the recognition of his merits, though he had the control of his emotions which belongs to a strong character. The attainment of that sort of distinction had never been the object of his life; and when it came it was accepted with pleasure, while the withholding of it would never have disturbed his equanimity. His sensitive nature contributed a great deal to his success in helping the labours of other workers at science, for it enabled him to set errors right without offending the authors. How much more anxious he was to advance knowledge than to win honour for himself is fully proved by the amount of thought and trouble he bestowed on his office of Secretary to the Royal Society. Many whom he has assisted testify to it. Anyone less true to his instinct of duty, less willing to spend his own time and labour without reward, would have let many a mistaken investigator publish his errors, or have quashed discoveries of real value because they were ill-presented. Stokes did neither; he took extraordinary pains with papers on physical subjects presented to the Royal Society, first to ascertain that the experiments described were trustworthy, and then to make sure that he understood the author's theory, consulting other experts when he felt any doubt. Eminently conservative as he was, he nevertheless was in no hurry to reject a theory merely because it was not orthodox, but would suggest methods of testing it, or modifications to meet objections, and so, without throwing good work after bad, he saved a great deal of good work which would else have been wasted because imperfect. That he might have enhanced his reputation as a discoverer by giving the time spent on other people's work to investigations of his own there can be no doubt; but considering how few could have done successfully what he did and how many fewer would have done it, the actual progress of knowledge has, in all probability, been greater through his choice of a path on which he and others pulled willingly together.

By Sir MICHAEL FOSTER, K.C.B., F.R.S.

I saw very little of Stokes (I shall venture in what I have to say, to use the simple name without the titles) until I went to Cambridge in 1870; for in the few relations which up to that I had with the Royal Society I naturally turned to his brother secretary, William Sharpey, who was my great personal friend. Nor in the succeeding ten years did I come into close or frequent touch with him either at Cambridge or at the Royal Society. I joined the latter in 1871; but Sharpey's successor as Secretary was Huxley, and he again was my very close friend; to him I naturally appealed in all matters relating to the Society.

I have however a still vivid recollection of a remarkable feat in the summer of one of the late seventies. It was the last meeting of the Society in June, and at that period the number of papers which, not having yet been read, had to be read at the last meeting was usually very considerable. On that occasion there were I think nearly thirty, and these dealt with very many different branches of science, mathematical, physical, chemical, geological, biological, and other. I remember distinctly how Stokes, standing up, and taking the paper first on the list, gave, in a very few words, a marvellously clear account, such as all of us could understand, of what the paper was, what the author proposed to do in it, and how he did it. Each paper in turn he treated in the same way; of each he gave an exposition, a model of brevity and lucidity; and he seemed to us no less successful with the biological and geographical papers than with those dealing with mathematical and physical themes. As we listened to him, we seemed to grasp quite easily what each author in succession desired to bring before the Society. Only at the end did he seem to fail. With characteristic self-denial he had placed two papers of his own the last on the list. When he came to deal with these his manner somehow changed; and I, at least, went away with the feeling that I had understood every paper except Stokes' own.

In 1881 I was appointed Secretary to the Society, taking the place of Huxley.

The work of the Society may be described as threefold in character. There is what may be called the internal scientific work, all that relates to the communications made to the Society, to the reading of papers, and to their publication in the *Transactions* or *Proceedings*. Then there is the external scientific work, the negociations with the Government or with other bodies, home or foreign, concerning scientific undertakings or scientific questions, the initiation or superintendence of scientific inquiries or expeditions, and the like. Lastly, there is the domestic work, all that has to do with the organisation of the Society itself, its internal economy, the arrangements for the meetings, the library and the publications, and so on.

Though of the two Secretaries one is chosen for his acquaintance with the physical sciences and the other for his acquaintance with biological sciences, I found, on taking office, that the duties devolving on each secretary were not arranged according to this division. One secretary took the whole, or at least the main part of the internal scientific work, both physical and biological, and the other similarly took the external scientific work, both sharing with the Treasurer the domestic work. Huxley, before me, though junior Secretary, had taken the external work, and, on my succeeding him, Stokes, who naturally had the choice as to which duties he should take up, decided to continue to go on with the internal work, which had been under his care for so many years.

I am not sure, but I believe that the division of labour of which I have just spoken was in practice when Stokes became Secretary in 1854. If so, the charge of the publications of the Society was in his hands for the long period from that date until he became President in 1885. And it would be difficult to overstate the amount of the labour which Stokes bestowed on these duties during this long period. Though he naturally consulted his brother-secretary from time to time on the biological questions which were raised by the biological communications, nevertheless

the sole charge of at least the main part of the communications of
all kinds devolved on him, from the receipt of the communication
in manuscript until its publication in the *Transactions* or *Proceedings*. He made it his duty to make himself acquainted, so far as
it was possible for him to do so, not only with the form but also
with the substance of every paper which came in. He spared no
pains in his efforts to secure that the form should be as good as
possible under the circumstances. It is a matter of common
knowledge that a scientific investigator, in making known what
he believes to be a new truth, is more anxious about the substance
of what he has to say than the way in which he says it, so that the
latter at times leaves much to be desired. Stokes, as Editor of the
Proceedings and *Transactions*, did his best to neutralize this tendency, carefully reading through the whole of the proofs, and, as
these passed through his hands, making valuable suggestions to
the author with the view of rendering the meaning of the sentences
more clear. And, recognizing that the obscurity of a passage is
often due to imperfect punctuation, he at times felt it his duty to
offer advice to an author even with regard to his commas and
semicolons. The papers which the Society published during
Stokes' secretaryship are doubtless not in all cases models of lucid
and elegant English; but where there is failure in this respect the
fault cannot be laid at Stokes' door; he, during long years, did his
best to make it otherwise.

Though he thus spent much labour on mere form, and so contributed in an indirect way to the progress of science, which is
dependent on new truths not only being made known, but being
readily understood, he did not neglect substance. And if it seem
to many that his zeal for form brought in reality a great loss to
science, since it took up time which might far more profitably have
been devoted to one or other of the many researches of his own
which he had always on hand, this reproach is not valid, or is not
nearly so valid, as regards his efforts to secure that the substance
of the communications made to the Society should be as good as
possible. Even in the case of biological papers he from time to
time offered to the author valuable criticism, by following which
pitfalls were avoided, and the value of the paper, when it was
eventually published, largely increased. But naturally the aid
thus given was most conspicuous in the case of papers dealing
with subjects with which he was familiar. Again and again, a

young investigator bringing in a crude form the results of his inquiries to the Society received from Stokes, before the account of the research was published, not merely most valuable negative criticism, through which the vexatious promulgation of false or imperfect doctrines was avoided, but inspiring suggestions, through which the inquiry was so modified or extended that the paper when it finally appeared had a worth wholly different from that which it would have possessed had Stokes' hand never have been put to it. It is indeed difficult to say how much science gained through Stokes' secretaryship by his editorial influence on the work of others, stopping that which was not fit to appear, and moulding a crude, imperfect effort into something worthy of being made known.

It must not be supposed, however, that Stokes took no part in the often heavy and responsible duties imposed by the external work of the Society. On the contrary, he was again and again called upon to take charge of matters of this kind, which his special knowledge fitted him to handle in an effective manner. And the Archives of the Society contain copies of many valuable letters and memoranda on astronomical, meteorological, and magnetic questions, addressed to H.M. Government or to other bodies, all of which bear, besides his signature, the mark of the clear, calm statement and accurate reasoning characteristic of everything to which he put his hand.

And in the domestic policy of the Society he took his share along with his brother officers.

When in 1883 the sad, premature death of Spottiswoode suddenly left the Presidential chair vacant, Stokes might very naturally have expected to be called upon to fill the place. The high position which he by that time had for many years held as a leader in investigation, to say nothing of his long services to the Society, pointed to him as Spottiswoode's successor. But it was felt by those who had the affairs of the Society more directly in charge that, were a suitable person available, it would be desirable to follow the custom through which a President whose acquirements are with the physical sciences is succeeded by one whose acquirements are with the biological sciences. Hence their minds turned to Huxley, whose claims to the honour of the chair were, in his own line, no less than those of Stokes.

Spottiswoode's death took place in the month of June, at a

time when the Society was in recess, and the affairs of the Society were wholly in the hands of the permanent officers. The Treasurer, Evans, the Foreign Secretary, A. W. Williamson, and myself took counsel together, consulting so far as at that season of the year we were able to do so, the chief members of Council and the leading Fellows of the Society. The result of our deliberations was to ask Huxley whether he were willing that we should bring forward his name to the Council, with whom lay the duty of nominating some-one for election by the Society as President. Huxley gave us to understand that, highly as he appreciated the honour proposed, he could not accept the nomination unless it were carried out with the cordial support of Stokes. It fell to my lot to make known to Stokes the views of his brother officers. My task, which, had I been dealing with some men, might have been extremely difficult, proved extremely easy; his cordial support was at once unhesitat-ingly given, and Stokes remained the senior Secretary*.

But only for a short time longer. For reasons of health Huxley was led to resign the Presidency in November, 1885 ; and Stokes was of course put in the vacant Chair.

The five years, 1885–90, during which Stokes was President, were not marked by any stirring or disturbing events, so far as the affairs of the Society were concerned. He was never called upon to exert, in any crisis, the autocratic power with which, by tradi-tion, the Society invests its President, or, in any emergency, to take up unexpected responsibilities. The circumstances of the Society during his term of office demanded no more than that he should "pursue the even tenour of his way."

That he did with his habitual calm and wonted dignified quietude. His general mental attitude was that which is usually called conservative, and this he brought to bear on the affairs of the Society, to the Society's gain. His Presidential actions were not of the initiative kind ; he left the unfolding of new projects to other more progressive Fellows of the Society, some of whom were always to be found on each year's Council. He spoke seldom, and never at any great length, from the Chair, with Presidential autho-rity. But his long experience of the affairs of the Society, his great knowledge of what had been done, the exactitude with which he was able to state what could and what could not be done within the laws and customs of the Society, gave the utmost

[* Cf. Prof. Stokes' letter to Dr Romney Robinson, *supra*, p. 39.]

weight to his brief contributions to the discussions of the Council and of the Society. He did not vehemently urge the Society forward, but he used his powers and position to prevent it hurrying in the wrong direction.

One little incident, and so far as I know one only, marred for a little while and to a slight extent the even tenour of his Presidential way. Some Fellows of the Society held very pronounced views about the nature of the office of the President; they regarded it as being what almost might be called sacred in character, not to incur danger of being defiled by touch with common things; they urged that the virtue of the holder of it should be placed beyond all suspicion. Huxley, as may be seen from his *Life and Letters*, held this view very strongly. So soon as he himself became President he at once resigned all positions already held by him, and refused to accept any new ones which he considered inconsistent with the office. He thought that the President of the Royal Society ought not to be a Member of Parliament, since in that position he would be exposed to party temptations and to party public abuse; and ought not to be the President of such a Society as the Victoria Institute, since the fact of the same man holding both offices might be quoted by some as indicating that the Royal Society took up a position concerning the relations of science to religion. And he wrote in *Nature* an unsigned letter urging that Stokes, who was both Member of Parliament and President of the Victoria Institute, ought to be neither. The letter had no effect on Stokes; he continued to hold both offices. Indeed, he hardly seemed to realize the point of view of the writer of the letter; he fancied that it was dictated by other motives. When he learnt later that the letter had been written by Huxley, his dominant feeling was that of astonishment; he laughingly said that he thought that it had been written by a Radical politician.

If as regards the external scientific work of the Society Stokes during his Presidency was of value as a judge and as a critic rather than as an active ruler, his wide knowledge and his clear, critical insight made his position in the Chair one of extreme importance in respect to the Society's internal work. His power of recognising not only the hidden weaknesses but also the often not too obvious worth of communications made to the Society, as well in biological or geological matters involving in any way physical or chemical considerations as in the subjects more clearly his own, strengthened as this was by an absolute single-mindedness of purpose, rarely, one might say never, led astray by any secondary influences, gave a valuable positive weight to the decisions of the Council. It guided them to recognise early and to encourage the worth of the initial efforts of young workers who had not presented their work in the best possible way, and who might have been hardly judged by those who could not distinguish so clearly and patiently as did Stokes, the essential from the passing and the irrelevant.

By Sir W. HUGGINS, O.M., K.C.B., Past Pres. R.S.

I WILLINGLY adopt your suggestion to put down in a few sentences, from my own personal reminiscences, an expression of my appreciation of the personal character and of the very high intellectual qualities of Sir George Stokes, whose acquaintance and friendship for more than forty years I regarded indeed as a great privilege and possession. I became acquainted with him in the early sixties, a few years after his appointment as Secretary of the Royal Society, in his official character in connection with papers which I communicated to the Society. This acquaintance grew more intimate when I became a member of the Council in 1866. Soon I found that I had gained a true friend, whose ready sympathy and advice during frequent correspondence almost to the time of his death, I regarded as precious beyond words.

One of the most distinguishing characteristic qualities of Sir George, was the generous way in which he was always ready to lay aside at once, for the moment, his own scientific work, and give his whole attention and full sympathy to any point of scientific theory or experiment about which his correspondent had sought his counsel. Notwithstanding the many heavy duties resting upon him, his reply came nearly always without delay by an early post.

Not less remarkable than his ready intellectual sympathy, was the clear mental grasp with which he discussed the point submitted to him, and the new light with which, in a few simple sentences he surrounded it. His scientific instinct was of a very high order, and all but unerring. In a quite unusual degree he combined with unsurpassed mastery and originality in the application of mathematical methods to physical problems, a very high endowment of that scientific imagination, or to use Bacon's words, that "learned sagacity in the procedure from one experiment to another" upon which rests success and progress in experimental science.

His own experiments, like those of Faraday, were characterized by the extreme simplicity which is possible only to a master-mind.

I consider myself fortunate, indeed, to have had such a man as correspondent and friend for so many years; I cannot even estimate my indebtedness to him.

In later years, Sir George had the consciousness, not too common, that with increasing age some failure of his powers was not improbable. In one of his letters he spoke of the decrease of the sensitiveness of his eyes to violet light, and asked if I had the same experience. In other letters he hoped that he had not forgotten any points, and asked me "to jog his memory" if this were the case. Due, no doubt, in part to his methodical habits, I never noticed the smallest sign of forgetfulness.

I hesitate to add any lighter reminiscences ; but in the first type-written letter I received from him I was amused by the comment "that now I shall be able to read my own letters."

He was much interested, many years ago, when after a quiet dinner here, I asked him to examine a mastiff which I then possessed, in elementary mathematics. I claimed for my dog, whom I had named Kepler, that he would answer with accuracy any algebraical questions, of which the answers could be given by a number of barks not exceeding five or six. The dog answered successfully every question which Sir George put. He thoughtfully remarked that there was clearly mind of some kind at work. My own explanation of the dog's success is that he was able to read in my face when he had barked the right number of times, though I endeavoured to avoid giving him any sign.

To sum up;—in Sir George Stokes a great man indeed has passed from us, whose greatness is most appreciated by those who knew him best.

By the Right Reverend GEORGE FORREST BROWNE,
D.D., Lord Bishop of Bristol, Hon. Fellow of St
Catharine's College.

At a time when it seemed not improbable that a proposal
would be made to deprive the Universities of their representation
in the House of Commons, some of the University Conservatives
raised the question—Was it right that Cambridge should send
as its representatives men whose first business it was to be
politicians? They agreed among themselves that one of the
two representatives ought to be directly concerned in the teaching or in the management of the University; better still, if
practicable, both. Professor Stokes was approached privately.
Agreeing heartily in the principle laid down, and encouraged
by the thought that his predecessor Sir Isaac Newton had held
the position now offered to him, he set himself with characteristic
simplicity and thoroughness to the one primary question, Could
he fully perform all the duties of his Professorship if he undertook
this new duty? It was with him much more than a question of
trains, and beds, and hours. That side of the question was of
grave importance to a man of his age and habits, and, it is
unnecessary to say, it was worked out and proved with a minuteness of detail which was a lesson for life to the younger mind
of the intermediary. But would the physical labour and hurry
unfit him for guiding the intellects of the very flower of the
mathematical students of the University? His physical endurance
was remarkable. The idea of hurry no one ever associated with
his calm demeanour. And he could always sleep, placidly and
soundly, whatever was going on. He said yes.

His course as Representative was one of careful sacrifice of
personal comfort, and unceasing watchfulness to see whether any
part of any question before the House touched the interests of the
University itself or of the graduates as a whole. There is no
doubt that he keenly enjoyed the position, laborious as he made
it by his journeys to and fro by night and day. He did not add
to the strain by making speeches. "No man," it was well said of
him, "has been silent on so many platforms." He eventually
retired, on grounds which every one could understand and appreciate, namely, the labour of the double position for a man of his

age, and the feeling that someone else should have the opportunity
of enjoying the honour. But there is no harm in saying, now,
that there was another reason. On a question which was sure to
come to the front, a question of deep interest to the majority
of the graduates, though not touching the University itself or the
Colleges, he had made up his mind, after much thought, that if
he voted he must vote against the view of the majority, and that
he could not conscientiously abstain from voting. That was the
culminating reason for his retirement from a position which has
become greatly stronger because he has held it. The other
Universities of the kingdom have not been slow to follow the
example of Cambridge. On his retirement, it may be added,
Cambridge set another excellent example. The Conservatives
offered the position to one who was known as a University
Liberal, Dr R. C. Jebb, the Regius Professor of Greek, who had
assured the same intermediary that on three points of policy,
which alone were mentioned to him, his views were in general
accord with theirs. It is greatly to be regretted that the death
of Sir Richard Jebb has prevented the fulfilment of his promise to
write an appreciation of his distinguished predecessor in the
representation of the University.

As a member of the University of Cambridge Commission,
1877—1881, Professor Stokes took on all questions the side of
moderate reform, with emphasis on the "moderate" rather than
on the "reform." That remark does not mean more than this,
that while he required strong evidence in favour of reform as
contrasted with modification, moderation was with him a fundamental demand. In the arrangement of Professorial stipends,
he constantly urged that while a "modest provision" should be
made for the maintenance of a "modest household," there should
be no stipend large enough in itself to form the attraction for
candidature. The honour of holding such a position should be
the primary attraction. In common with other Commissioners
he looked with a longing and hopeless eye to the system of the
Scottish Universities, where large pensions for retired Professors
are a charge upon the national funds. That the Colleges should
make some direct contribution to the income of the University he
readily allowed. But he maintained that a grave depression of
agricultural prospects had set in, and that those of the Colleges
which depend chiefly upon land and tithe, and not on house

property, all, that is, but two or three, would become less able to make contribution without hampering their own work. This was one of the chief controversies which divided a generally harmonious Commission. The contention waxed rather than waned. It was decided that all Colleges must pay an annual contribution, increasing each five years, assessed upon their income, but that the agricultural depression might go further than it had gone, and power should therefore be given to the Chancellor of the University to stop for a period the quinquennial increment (£5000) of the College contributions. The final fight was on the question, should the ultimate maximum of the contribution be £25,000 or £30,000 ? Professor Stokes was determined that £25,000 was— to say the least—as much as could be borne by the Colleges, and that some of the Colleges would be seriously injured by the necessity of paying their quota towards that amount. The chairman of the Commission, Sir Alexander Cockburn, had ceased for some time to attend the meetings of the Commission, feeling out of harmony with its proposals. Professor Stokes took part in a representation to him, through one of the Secretaries, that at the next meeting it would be decided whether £25,000 or £30,000 should be the limit, and that his vote would settle it one way or the other. This was put to him by the Secretary as a piece of business information, not as a request for his attendance or for his support of the smaller sum. Sir Alexander said that he would certainly attend, and made a very frank utterance as to his vote. That was on a Friday or Saturday. The meeting was on the following Tuesday. Sir Alexander died suddenly on the intervening Sunday, and the ultimate contribution of the Colleges stands at £30,000 a year.

As regards the Statutes of the Colleges, Professor Stokes, who took very little part as a rule in the detailed discussion of various points, always became vocal when questions affecting the tenure of the offices of Tutor and Dean and Chaplain were under consideration. There was a regular iteration of such epithets as parental, kindly, genial, firm, and—oftener than all—tactful. It was clear that if it could have been put into a Statute he would have had an enactment of this kind—" when once you have got a tactful Tutor, don't think of superannuation "; and he would have given a very short shrift to an untactful person in that position. He did all he could to add to the status and importance of the office of Chaplain and the disciplinary office of Dean. He thought that the offices should be combined, and that a special appointment of a specially suitable man should be made, not of necessity a man fully up to the Fellowship standard. If such a man filled the office well, he held that he had done great service to the best interests of the College, a service which deserved recognition by election to a Fellowship. The religious interests of the young men stood very high in Professor Stokes's estimate of the duties of a place of higher and highest education.

A large experience of Professor Stokes as a member of the Council of the Senate, as also of many Syndicates and other administrative bodies, has left a general impression of the presence of lucid simplicity and the absence of subtlety. As it used to be a matter of weekly interest to note the different ways in which Dr Westcott and Dr Lightfoot, sitting next each other, would deal with a subject, so it was with Professor Stokes and Professor Cayley. Both alike were accustomed to let the ordinary business of the Council or other body take its course without their inter-vention. They were very unsuspicious of concealed purposes in the management of affairs. When Professor Stokes had to take part in something which called for his intervention, he would make a remark which—as new-comers naturally said—any one might have made. The listener respectfully supposed that it came from profound depths and was moulded into the very simplest form by an intellect of unusual power. Still, thoughts and words alike were free—it may be more true to say were freed—from all appearance of difficulty. He seldom joined in the general discussion of a question, unless it was one which belonged specially to his own sphere of work. He never played the useful part of one who helps others to decide against a view by the manner in which he maintains it. When he expressed an opinion, he spoke with a gravity of manner which shewed his sense of the responsibility of expressing an opinion. But he usually listened and was silent till the time came to vote. Then it was seen how highly sensitive the balance of his mind was. A chairman had often to wait for his vote, and more frequently than with other men to record him as not voting. One seemed to see a testing instrument of the finest delicacy under careful experiment. A grain more, or a grain less, he could vote yes, or no; failing the grain, neither.

Professor Stokes's abstraction of mind and manner gave the impression of a mind formed into compartments, varying greatly in size, with no intercommunication. But no matter which compartment he might have in use at the time, all was orderly, kindly, reverent, charming, loveable. Much there was to enjoy in the present; much, of general impression, there is to enjoy in the retrospect; comparatively little that has stamped itself in minute detail on the memory. For among the many compartments of his mind, there was not a compartment which supplied incisive speech or epigram. There was a compartment of genial humour; and when that compartment was open and in use, there was a smile on the face, and a playfulness of gesture with the hand, and a laugh in the voice, which no one will forget.

His demeanour towards religious services, religious questions, religious things, was itself religious. He took special interest in the difficulties of sceptics. In the later years of his life he noted two facts of observation. The one was, that the doctrine of an endless eternity of punishment by torture was a moving cause of sceptical objections against Christianity as a whole; the other, that neither he nor such of his acquaintance as had entered upon the subject with him accepted the doctrine as part of the faith delivered to the saints. He took pains to press the second of these observed facts upon the class of persons concerned in the first. The result was a series of letters, artlessly written at odd times as occasion served; written conversationally, without copies of former letters, and thus with much repetition. They were published in 1897 (James Nisbet and Co.) under the title of *Conditional Immortality*.

Sir George Stokes, as he had then become, set out with the principle that the "natural immortality" of the soul was a dogma taken from Platonism, and that the combination with it of the Scriptural revelation of the "finality of the perdition of the lost" had created the doctrine of "endless torments." By "natural immortality" is meant, of course, that every soul must go on for ever, that no soul can come to an end. Equally of course, the question raises no doubt of the existence of the soul after the death of the body; it is concerned only with the eternity of such existence. While he freely used illustrations drawn from physical science, he was very clear against drawing argument from the physical to the spiritual. To one who put an argument for the

immortality of the soul from the physical law of the conservation
of energy, he replied in words of a directness that Dr Temple
might have used,—" I do not think there is anything at all in
it "; and he proceeded to give interesting reasons for his opinion.
Indeed he went further than this, and would accept neither
physical nor metaphysical arguments for the survival of the soul,
in place of the revelation made by the Son of God.

Dealing with the soul as separate from the body, his scheme
was as follows. The Scriptural account of the creation and fall
of man is a guide to the truth. Man's soul was intended to
be immortal, the immortality depending upon the condition of
obedience. Man's will violated the condition of obedience, and
so man's soul lost the offer of immortality. Not that the souls
of men from Adam to Christ did not live after the death of the
body, but they would not continue to exist eternally. The death
of Christ—he was clear against curious examination into the
mystery of the Atonement—restored to the soul of man not
immortality but the offer of immortality. One of his favourite
phrases was "Life in Christ"—no other revealed eternity of
existence. "The wages of sin is death, the gift of God is eternal
life through Jesus Christ our Lord." We must understand that
he attached vast importance to the preaching to the spirits in
prison.

He confessed that of the large number of those who had told
him that they did not hold the doctrine of "endless torments,"
the majority believed that all souls would eventually be saved.
For himself, he felt it to be much more difficult to make that
view accord with the statements of Scripture than the view which
he himself held, that at the end—the words are his own—they
that have done evil will come forth unto a resurrection of judgment; they will be consumed like tares burnt up in the fire;
consigned to a second death from which there is no resurrection.
He told with evident pleasure that a Cambridge friend, to whom
he had stated his scheme, remarked to him afterwards that
"reading the Bible with that idea in the head is like turning
a key in an oiled lock."

It is clear that in this scheme there is room for endless speculation as to the condition and treatment of the soul between the
death of the body and the final judgment. With such speculation
he would seem to have had no sympathy. Purgatorial flames

were as far from his conceptions as were endless torments, if we may judge by his silence. As relating to this point, evidently one of very great importance, a passage may be quoted in conclusion from one of his letters, without criticism or comment.

"What does St John say of conformity to Christ? 'We know "that when He shall appear we shall be like Him, FOR we shall "see Him as He is.' I have long thought that this revelation "would be the means of causing the character of each man to "take its final set for good or for evil. Just as in this life the "preaching of the Gospel may be a savour of life unto life or "a savour of death unto death, so the revelation of the last day "may produce these opposite sets in the character, and the direc- "tion that the set will take may be determined by the character "that has been formed during the state of probation. On those "in whom the set will be for good the gift of eternal life in its "final inalienable possession will be bestowed. As to those in "whom it will be for evil, the general teaching of Scripture "is, I think, that they will be destroyed like tares burnt up "in the fire."

BIOGRAPHICAL AND SCIENTIFIC EVENTS.

1841. Senior Wrangler; first Smith's Prizeman.
    Fellowship at Pembroke; vacated on marriage, 1857; re-elected under new statute, 1862.
1849. Lucasian Professor of Mathematics.
1851. Fellow of the Royal Society.
    Manchester Lit. and Phil. Soc., Hon. Member: Wilde Medal, 1897.
1852. Rumford Medal of the Royal Society.
1854. Elected Secretary to the Royal Society.
1856. Haidinger Medal of Imperial Vienna Academy.
    Rotterdam, Bataafsch Genootschap, Correspondent.
1864. Hon. F.R.S., Edinburgh.
    Royal Academy of Göttingen, Correspondent.
1867. Royal Society of Sciences of Upsala, Member.
1869. President of the British Association at Exeter.
1871. Hon. LL.D., Edinburgh.
1873. Royal Irish Academy, Hon. Member.
1877. Gauss Medal of the Academy of Göttingen.

1878. Société Française de Physique, Member.
1879. Royal Prussian Order *Pour le Mérite*.
       Académie des Sciences de l'Institut de France, Correspondent succeeding Ångström.
1880. Ophthalmological Society, Hon. Member.
1882. Imperial Academy of Vienna, Foreign Associate.
       Royal Academy of Göttingen, Associate.
       Madras Harbour Commission.
1883. National Academy of Sciences, Washington, Foreign Associate.
1883 to 1885. Burnett Lecturer, University of Aberdeen.
1885 to 1890. President of the Royal Society.
1886. Royal Institution, Actonian Prize. LL.D., Aberdeen.
1887 to 1891. M.P. for Cambridge University.
1887. Philosophical Society of Glasgow, Hon. Member.
1888. Munich Academy of Sciences, Foreign Member.
       R. Accademia dei Lincei, Rome, Foreign Member.
1889. Baronet of the United Kingdom.
       American Philosophical Society, Member.
1891 to 1893. Gifford Lecturer, University of Edinburgh.
1892. Royal Berlin Academy of Sciences, Corresponding Member.
       Institution of Civil Engineers, Hon. Member.
1893. Copley Medal of the Royal Society.
       University College, London, Life Governor.
1896. Imperial Academy of Sciences of Vienna, Foreign Hon. Member.
1897. Imperial Society of Naturalists of Moscow, Foreign Member.
1899. On the occasion of the celebration of his Jubilee by the University of Cambridge he received
       The Arago Medal of the French Academy, struck only on special occasions,
       The Portrait Commemoration Medal struck in Gold by the University of Cambridge,
       Royal Academy of Belgium, Diploma of Associate Member,
       and congratulatory addresses from scientific bodies in all parts of the world.
1900. Elected one of the (8) foreign members of the French Academy of Sciences, in succession to Weierstrass.

1900. Royal Academy of Turin, Foreign Member.
Royal Academy of Belgium, Associate.
1901. The Helmholtz Medal, Royal Academy of Sciences, Berlin.
1902. Elected Master of Pembroke College, Cambridge.
Società Italiana delle Scienze, Foreign Member (one of 12).
Doctor of Mathematics of the University of Christiania, on
the occasion of the centenary of Abel's birth.
1904. Bronze memorial medallion placed in Westminster Abbey.

In addition the following may be noted:
Member of the Governing Body and *ex officio* Fellow of
Eton College.
Member of the Board of Visitors of Greenwich Observatory
for 44 years.
Historic Society of Lancashire and Cheshire, Hon. Member.
President of the Victoria Institute, London.
Received Honorary Degrees from the Universities of
Oxford (D.C.L.), Edinburgh (LL.D.), Glasgow (LL.D.),
Aberdeen (LL.D.), Dublin (LL.D.), Victoria (Sc.D.),
Cambridge (Sc.D., LL.D.), Christiania (Math.D.).

GENEALOGY.

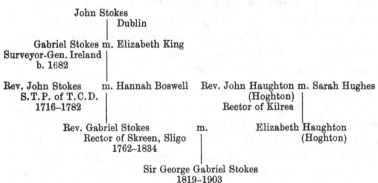

John Stokes
| Dublin

Gabriel Stokes m. Elizabeth King
Surveyor-Gen. Ireland |
b. 1682

Rev. John Stokes   m. Hannah Boswell   Rev. John Haughton m. Sarah Hughes
S.T.P. of T.C.D. | (Hoghton) |
1716–1782   Rector of Kilrea |

Rev. Gabriel Stokes   m.   Elizabeth Haughton
Rector of Skreen, Sligo | (Hoghton)
1762–1834

Sir George Gabriel Stokes
1819–1903

SECTION II.

GENERAL SCIENTIFIC CAREER.

It has been thought that but slight advantage would arise from attempting any connected description of Sir George Stokes' scientific work. The published collection of his *Mathematical and Physical Papers* in five volumes, taken in connexion with the historical analyses by Lord Rayleigh, Lord Kelvin, and others, forms ample material for the mathematician or physicist; while an account, to be interesting to the general reader, would have to be too long. Moreover in his *Burnett Lectures on Light*, and the other scientific addresses mentioned in the Preface to Vol. v. of his *Papers*, he has himself gone over much of the ground of his researches, in a manner which forms a model for popular and at the same time accurate exposition.

His scientific relation to the times in which he lived comes out most clearly in his correspondence. It has been felt to be inadvisable to break up the longer series of scientific letters; but some of the shorter groups, and those illustrating the events of his life, are here set forth in order of time, without any attempt at making up a continuous narrative.

A main impression derived from a general survey of his public activity, as revealed by the many thousands of letters that exist, is the state of dependence upon him which was the normal condition, as regards scientific investigations, of most of the numerous Committees of which he was a member. Other persons find it hard to make progress with their own problems : and this liability to become the chronic receptacle for all the unsolved difficulties which had been encountered by public committees, while they proceeded on their way, may well have induced the habits of silent meditation, as well as the reluctance to tackle new extensive undertakings, which have been noticed in the preceding section. It has been remarked that Prof. Stokes' normal course when a scientific difficulty arose was silence, while others talked round the subject, to which he had been contributing nothing: but at the next meeting, when his colleagues may have forgotten all about it, he would be ready with a considered plan for advancing the problem by definite modes of attack.

To the period soon after taking his degree belong a series of letters relating to physical science, of a didactic or tutorial character, written in fluent Italian in 1841, to the Marchese di Spineto, who is mentioned in Todhunter's edition of Whewell's *Correspondence* as an active resident in Cambridge about ten years before.

It has been remarked that Prof. Stokes to some extent resembled Gauss, in recasting and polishing his researches before publication, so as to present a logically-connected whole, leaving perhaps little trace of the process of discovery. In at least one case, documents exist revealing the thorough way in which he could become absorbed in a problem which attracted while it baffled him. Early in 1848 Prof. Baden Powell, of Oxford, communicated to him a new phenomenon of interference which had arrested his attention, and which proved to be a case of Fox Talbot's bands, formed, however, by a fluid prism in which the retarding plate was immersed. This was the origin of a memoir[*], in which Stokes took up the theory of these bands, condensing and elucidating the explanation of the phenomenon that had been already partially given by Airy. This memoir, as published, was divided into Sections I and II. The letters which follow are taken from correspondence relating to a second part of it, containing Sections III, IV, and V, which had been offered to the Royal Society through Prof. Baden Powell[†].

The correspondence with Prof. Powell on the whole subject had been arranged and fastened together, apparently at the time, and it reveals the manner in which the theory expanded in Prof. Stokes' hands. When ultimately Prof. Powell rather complained of the '*embarras de richesses*,' the idea of writing a separate paper seems to have originated.

The explanation regarding the inadequacy of the then famous

---

[*] *Phil. Trans.*, May 25, 1848; *Math. and Phys. Papers*, II. pp. 14—35.

[†] This paper remains in the Archives of the Royal Society; but a brief abstract (reprinted here) appeared at the time in the *Phil. Mag.* and in *Royal Society Abstracts*, Vol. VI.

prediction of conical refraction, to serve by itself as a guarantee of the validity of Fresnel's wave-surface, which is also printed *infra* as written out for the benefit of Prof. Stokes' critics, did not see the light until 1862, when it was incorporated in somewhat different terms in the "Report on Double Refraction"*. The argument has gained in interest, as since that time the existence of a fine dark line (Poggendorff's) in the middle of the section of the refracted cone of light has been detected by careful focussing. It is this fine line that marks the path of the ray which is spread out into a cone and so enfeebled in intensity; it therefore represents the real conical refraction, while the bright cone around it arises from ordinary refraction of rays adjacent to this critical one†. This subject occurred only incidentally in Prof. Stokes' manuscript, as a reason supporting the urgency of further exact tests of the form of the wave-surface, for which he then proposed the use of Talbot's bands formed by thin plates of the crystal. The more searching method of prismatic refraction had been proposed by him already in 1846, and reprinted in the 'Report' in 1862; as is well known, it has been carried out very thoroughly by himself and other experimenters‡. Remarks similar to the above apply to the other proposals made in this memoir for the utilisation of Talbot's bands§, namely that they would have been convenient at that time for rapid handling, but would hardly be thought of for exact measurement now that experimental resources are so much improved.

The problem of this memoir which, according to the letters *infra*, gave Prof. Stokes most trouble, was the explanation of a group of monochromatic bands, seen near the edge of the field of an approximately achromatised prism-pair, and changing colour with movement of the eye. He concluded that they must arise from diffraction: and he readily saw that the narrowness of one of

---

* See *Math. and Phys. Papers*, IV. p. 184.

† For detailed discussion of the dynamics of the subject and the influence on it of magnetic excitement cf. Larmor, *Proc. Lond. Math. Soc.*, 1893, pp. 273—90, and especially Voigt, *Ann. der Phys.* 1905—6.

‡ Cf. *Math. and Phys. Papers*, Vol. IV. (1872), p. 336. The details of the experimental investigation by Prof. Stokes, of which this brief abstract alone was ever published, exist substantially complete in manuscript, and are being prepared for press by Mr Glazebrook.

§ Cf. Introduction to Part I. of the Memoir, *Math. and Phys. Papers*, Vol. II. p. 15.

the prisms used, a mere facet on a plate, served as a slit restricting the beam of light. The explanation, which he calls obvious in the letter, came when he recognised that, for each direction of vision, light in the neighbourhood of a definite wave-length is completely achromatised over a considerable range by the prism-pair; and in fact the phenomenon is an example, perhaps the earliest explicit one, of what are now well known as achromatised diffraction-bands. This explanation introduced the consideration of the curve of dispersion of the prism-pair, in which dispersion is plotted against wave-length. Possibly he was thence led directly to the brilliantly simple explanation of Fraunhofer's empirical rule for the best achromatisation of an astronomical double object-glass*, in which the true principles for improving chromatic correction, since followed up with well-known success at Jena and elsewhere, were laid down.

The unilateral character of Talbot's bands depends on the circumstance that it is the intensities, not the amplitudes, of the bands of adjacent colours that must be superposed. The matter has recently become much simpler, in general terms almost intuitive, through the happy introduction by Prof. Schuster† of Stokes' principle of recurring groups of disturbance, arising from concomitant and nearly homogeneous undulations, as a graphical substitute for the somewhat complex abstract calculations of the memoir.

[From *Royal Society Abstracts*, Vol. v. pp. 795—6, Jan. 25, 1849.]

Supplement to a paper "On the Theory of certain Bands seen in the Spectrum." By G. G. Stokes, Esq., M.A., Fellow of Pembroke College, Cambridge. Communicated by the Rev. Baden Powell, M.A., F.R.S.

The principal object of the author in this communication is to point out some practical applications of the interference bands recently discovered by Professor Powell, the theory of which was considered by the author in the paper to which the present is a supplement. The bands seem specially adapted to the determination of the dispersion in media which cannot be procured in sufficient purity to exhibit the fixed lines of the spectrum. The ordinary

---

* "On the Achromatism of a Double Object-Glass," *Brit. Assoc.* 1855, *Math. and Phys. Papers*, Vol. iv. p. 64; "On the Principles of the Chemical Correction of Object-Glasses," *Photographic Society*, 1873, *loc. cit.* pp. 344—54.

† *Phil. Mag.* vii. 1904; "Theory of Optics," § 186.

experiments of interference allow of the determination of refractive indices with great precision; but in attempting to determine in this way the dispersion of the retarding plate employed, there is the want of a definite object to observe in connection with the different parts of the spectrum. In Professor Powell's experiment, the wire of the telescope, placed in coincidence with one of the fixed lines of the spectrum previously to the insertion of the retarding plate into the fluid, marks the place of the fixed line, and so affords a definite object to observe, when the retarding plate is inserted into the fluid and the spectrum is consequently traversed by bands of interference.

The practical applications considered by the author are principally four. In the first, the variation of the refractive index of the plate in passing from one fixed line to another is determined, the absolute refractive index for some one fixed line being supposed accurately known. The observation consists in counting the number of bands seen between two fixed lines of the spectrum, the fractions of a band-interval at the two extremities being measured or estimated.

In the second application, the absolute refractive index of the plate is determined for some one fixed line of the spectrum. The observation consists in counting the number of bands which move across the wire of the telescope, previously placed in coincidence with the fixed line in question, when the plate is inclined to the incident light.

The third application is to the determination of the change in the refractive index of the fluid, for any fixed line of the spectrum, produced by a change in the temperature. The observation consists in counting the number of bands which move across the wire of the telescope while the temperature sinks from one observed value to another, the temperature being noted by means of a delicate thermometer which remains in the fluid. For this observation a knowledge of the refractive index of the retarding plate is not required.

The fourth application is to the determination of the change of velocity of the light corresponding to any fixed line of the spectrum, when the direction of the refracted wave changes with reference to certain fixed lines in the plate, which is here supposed to belong to a doubly refracting crystal. The observation consists in counting the bands as they pass the wire when the plate is inclined. It requires that the plate should be mounted on a graduated instrument. It would be possible in this way to determine, by observation alone, the wave surface belonging to each fixed line of the spectrum.

While considering the theory of Professor Powell's bands, the author was led to perceive the explanation of certain bands, described by Professor Powell, which are seen in the secondary spectrum formed by two prisms which produce a partial achromatism. Although the account of these bands has been published many years, they do not seem hitherto to have attracted attention. It is easily shown by common optics that when two colours are united by means of two prisms, the deviation, regarded as a function of the refractive index, the angle of incidence being given, is a maximum or minimum for some intermediate colour. For the latter colour, two portions of light of consecutive degrees of refrangibility come out parallel; and therefore the diffraction bands belonging to different kinds of light, of very nearly the same refrangibility with the one in question, are superposed in such a manner that the dark and bright bands respectively coincide. Thus distinct bands are visible in the secondary spectrum; although none would be seen in the spectrum formed by a single prism, in consequence of the mixture of the bright and dark bands belonging to different kinds of light of nearly the same degree of refrangibility. The diffraction bands here spoken of are of very sensible breadth, in consequence of the small width of the aperture employed in the actual experiment.

When a spectrum is viewed through a narrow slit half covered by a plate of mica, the edge of which bisects the slit longitudinally and is held parallel to the fixed lines of the spectrum, the bands described by Sir David Brewster are seen, provided the mica plate lie at the side at which the blue end of the spectrum is seen, and provided the thickness of the plate and the breadth of the slit lie within certain limits. When these bands are invisible in consequence of the slit being too narrow, or the spectrum too broad, it follows from theory that the bands ought to appear when the slit and plate are turned round the axis of the eye, so that the edge of the plate is no longer parallel to the fixed lines of the spectrum. The author has verified this conclusion by experiment, employing plates adapted to observations with the naked eye, which are best suited to the purpose.

To Prof. S. H. CHRISTIE, Sec. R.S.

PEMBROKE COLLEGE, CAMBRIDGE,
March 15*th*, 1849.

MY DEAR SIR,

Some days ago I heard from Prof. Powell in a general way the wishes of the Council of the R.S. with respect to my paper. I immediately wrote to Prof. Powell explaining my views on the subject, which you have perhaps learned before this.

I could not but feel in writing my paper that some of the experimental methods proposed might be practically impossible, in consequence of some unforeseen difficulty attending the actual working out of the experiment, and one or two were obviously at least very laborious. I sent, however, my paper as it stands to Prof. Powell, requesting him at the same time to strike out without hesitation whatever he thought not likely to be really useful. As he was used to experiments I thought that he would be a better judge in the matter than I could be. Prof. Powell, however, let everything stand, most probably objecting, from a feeling of delicacy towards me, to strike out anything he could not absolutely pronounce to be impracticable.

I fully enter into the feeling of the Council that it is objectionable in general to publish an account of proposed experimental methods which have not been actually tried, and on this very ground I had great misgivings about sending my paper at all. Judging, however, from the apparently fortuitous manner in which the several applications occurred to me, while my attention was strongly directed to the subject in consequence of Prof. Powell's letters, from the apparent ease of some of the experiments, and from the accuracy of the results to which they seemed likely to lead, so far as I was competent to judge, I ventured, with the encouragement of Prof. Powell, to submit my paper to the R.S. At the same time I must confess that I had exceeded all due bounds, both in details and in proposing one or two experiments which from their difficulty are not likely ever to be carried out.

After what I have said it is almost needless to observe that I am quite ready to remodel, and materially curtail my paper. I should propose to strike out all experiments in which an angular motion of the plate is required, as well as much of the detail of the others. The only part which I would retain at full is my explanation of the bands seen with a pair of achromatising prisms. The explanation, to be sure, when once suggested, is almost ridiculously simple, but it cost me a good deal of thought before I struck into the right road. It is strange that the phenomenon should have been before the public for about ten years without having hitherto attracted attention. I would also retain at full the account of an experiment which I have tried, and described near the end of my paper *.

As to conical refraction, I think the referees must have mistaken my meaning. I should be obliged by your forwarding to them the accompanying explanation. I make this request, not with any view to publication, for I think the passage in question may well be struck out, as not being of much importance, but merely for the sake of my own character; for I should be sorry to be thought guilty of so gross a blunder as to suppose the conical refraction either was seen in fact, or ought to be seen according to theory, in any direction indifferently within a biaxal crystal.

Should you conceive my proposal to meet the views of the Council, I would be glad to receive back my paper in order to remodel it.

As you will probably have to communicate to the Council my willingness to remodel my paper, may I request you at the same time to make my apology for the nature of the paper I sent ?

I am, dear Sir, yours very truly,

G. G. STOKES.

March 16*th*, 1849. Explanation of a passage relating to Conical Refraction in the supplement to a paper " On the theory of certain bands seen in the spectrum."

Everybody who has studied the undulatory theory of light allows, and must allow, that the singular phenomenon of conical refraction is a most striking verification of the *general* correctness of the form of the surface which Fresnel conceived to be the wavesurface. For my own part nothing gave me so much confidence in the general correctness of the wave-surface as actually witnessing the phenomenon. I regard the phenomenon as an absolute

---

* The effect of placing the plate and slit with their edges oblique to the direction of the dispersion.

proof that the true wave-surface, whatever it may be, does actually turn itself, so to speak, inside out in the neighbourhood of four points. If we enquire in what way such a turning inside out is geometrically possible, we shall see that it may take place either in a sort of double curve, or in a sort of double funnel, or in a sort of double funnel with a cusp edge. The second supposition must be rejected because it would then be possible (at least in directions lying within certain limits) to draw four tangent planes, parallel to one another, or rather four such planes in the positive direction, and as many more in the negative, which would require a quadruple refraction. The third supposition must also be rejected because then there would not even be an *approach towards* such a phenomenon as external conical refraction (and observation shows that there must be at *least* such an approach), not to speak of the complexity and *bizarrerie* of the law which must connect the wave-velocity with the direction, in order to render such a form of wave-surface possible. We are thus, without making any supposition as to the accuracy of Fresnel's wave-surface, driven to the first supposition, which necessitates the existence of external conical refraction. I do not of course mean to say that Fresnel's wave-surface *may not be exact*; all that I say is that it seems to me to be unphilosophical to infer its *exactness*, or even near approach to exactness, from the observed existence of external conical refraction.

Let us turn now to internal conical refraction. Suppose the true wave-surface not to be touched by a tangent plane along a circle, or even a plane curve of any kind, but only to be very nearly so touched. Then a curve of double curvature may be traced on the wave-surface, which shall be very nearly a circle, and such that the tangent planes at its different points shall be all but coincident.

Conceive now the plane waves within the crystal, and then the plane incident waves corresponding to them. Since the refraction of a wave from one medium into another depends on nothing but the wave-velocities, it is evident that the directions of the incident waves must be all but coincident. Passing now from waves to rays, we see that the incident rays necessary to form the refracting cone, corresponding to the curve considered, form a cone of such an exceedingly small vertical angle as to be experimentally inseparable from a ray ; that is to say, the smallest pencil we can actually employ will contain all the rays necessary to form the entire refracted cone. Consequently if the wave surface be touched nearly, though not exactly, in a plane curve, we shall still have internal conical refraction, so far as experiment can decide. Nevertheless if we consider the way in which it is geometrically possible for the wave-surface to *fall over*, so to speak, we may see without much difficulty that the non-existence of quadruple refraction renders it necessary for the wave-surface to be touched along its tangent plane along a plane curve; though whether that curve be or be not a circle cannot of course be made out by such general reasoning.

It follows, I think, from the preceding considerations, that the mere *existence* of the very remarkable phenomena of conical and cylindrical refraction is no great evidence of the exactness of Fresnel's wave-surface. The evidence of exactness which is furnished by these phenomena depends, as seems to me, on the near coincidence of the diameter of the ring, and the direction in which it is seen, with the results calculated from theory, by means of numerical values of the three optical constants of the medium obtained quite independently, and on the agreement of the form of the ring and the mode of its polarization with the results predicted by theory. But surely, from the very nature of the observations, the test is not sufficiently searching to warrant us in concluding that Fresnel's wave-surface is exact for each colour in particular.

There is yet another conceivable mode of passage, which is by something like a pair of opposite cones in which the vertex has been replaced by a short ridge, so that instead of a point of union between the opposite sheets there is a line of union. But by considering the geometry of such a figure it might be shown that the supposition would require a certain wave-velocity to be equal to zero, which is absurd.

Feb. 20th, 1850.

DEAR SIR,

I fully appreciate the force of the objection to the printing of my paper, arising from its being in a great measure devoted to the mere suggestion of experiments. As far as I can conjecture from the tone of your letter, the Committee have no wish that the paper should appear, only they think it might possibly appear if remodelled. For my own part I should, I think, rather prefer that my name should not appear in connection with a paper having so little to recommend it. It might do very well as a piece of a long investigation, but is not of sufficient dignity to be presented alone, especially when in so doing I go out of my usual course, which is to present my investigations to the Camb. Phil. Soc. I think under the circumstances the wishes of all parties would be best consulted by allowing the matter to drop. I sincerely regret the trouble I have already caused, both to you and to the members of the Committee.

Of course I am ready to abide by my former offer if the Committee really wish it; but that, if I conjecture aright, is not the case. Should my present suggestion meet the views of the Committee, I shall consider myself at liberty to send my explanation of the bands seen with a pair of achromatising prisms to the editors of the *Phil. Mag.*, or to communicate it to the British Association.

With regard to the chromatic variation of $(\mu' - \mu)/\lambda$, which it is suggested ought to be μ'/λ, I may be permitted to remark that my method is to determine the chromatic variation of μ'/λ by combining the plate to be examined with a *known* fluid, or at least one of which the refractive index can be determined at the time of the experiment. Thus μ will be known from observations of the usual kind, λ is known from Fraunhofer's very accurate determinations, and therefore the chromatic variation of μ/λ will be known; and as the chromatic variation of $(\mu' - \mu)/\lambda$ is determined from the bands, that of μ'/λ follows.

The following letter from Prof. R. Willis has been taken from documents relating to Prof. Stokes' well-known memoir, "Discussion of a Differential Equation relating to the Breaking of Railway Bridges under travelling loads *."

PARKER'S PIECE, *July* 21, 1849.

DEAR STOKES,

I find that in practice the values of β are much greater than those which you have supposed. I have just returned from inspecting a bridge of 70 feet span, the statical deflection of which

---

* *Camb. Phil. Trans.* 1849: *Math. and Phys. Papers*, Vol. II. pp. 178—220.

with the greatest load was only ·215 inch, and taking the velocity at 50 feet per second, β if I mistake not would be nearly = 200. In our experiments $\beta = 8$, and thereabouts. It would therefore greatly improve the practical value of your paper if the curve were calculated for a few values of β greater than unity, and half the number of points would be quite sufficient. You have taken 50 in the length ; 25 at most would enable the curve to be shown.

If you are too tired of the subject to undertake this, which I fear is the case, I will request you to show me the short way to do it ; as it is clear that all the practical cases belong to values of β greater than the fractional ones which you have chosen, and pro- bably, as you have stated, with such large values of β your solutions would agree very nearly with the results, notwithstanding the neglect of the inertia of the bars.

I am just starting for Salisbury, where I shall spend the next week. I should have told you that in the case of the bridge, 70 feet span, the engine and tender produced a deflection of 0·215 inch when placed at rest over the middle of the bridge, but when a velocity of 55 feet per second was given to the engine we obtained a deflection of 0·245 inch, that is, an increase of ¼ upon the statical deflection. The rails were slightly curved upwards, which probably diminished the increase of deflection.

My attention has just been directed to a paper in the *Engineers' Journal* of last September upon this subject. I send it to you, but if I mistake not you will find it full of erroneous conclusions.

I am much interested with the papers you showed me in the *Mathematical Journal*.

I remain, Yours most truly,

R. WILLIS.

The following letters show that the project of a sea-level canal across the Isthmus of Panama was already to the fore in 1852, and that the scientific problems connected with it had been carefully considered. The results here given with regard to tidal currents in such a canal have become again of special interest, in connection with the findings of the International Commission of Engineers which recently reported to the American Government.

To LIONEL GISBORNE, Esq.

PEMBROKE COLLEGE, CAMBRIDGE,
March 30th, 1852.

SIR,—My friend Mr Griffith wrote to me two or three days ago requesting my consideration of a question which you had proposed to him; namely, what would be the greatest velocity in a ship canal connecting the Atlantic and Pacific Oceans, the section of the canal being as in the figure, the length 30 miles, and the height of the tides 25 feet at the Pacific side, and only 3 feet at the Atlantic side?

I have considered the problem to a certain extent, but find it would be very difficult to obtain a solution pretending to much exactness. The thing would not I hope be too difficult to master, but it would require more time; and as Mr Griffith informs me that you sail on the 2nd of next month I think it best to inform you at once of the results which I have arrived at.

In the first place I regarded the motion as a purely tidal phenomenon, neglecting altogether the resistance of the bed of the current. On this supposition it came out, taking a very rough solution, that the greatest velocity would be almost $10\frac{1}{4}$ miles an hour. With a velocity which, were it not for friction, might attain such a magnitude as this, it is plain that friction and the resistance of the bed, &c., must be of the utmost importance. The phase of high water would take almost $1^h\ 15^m$ to travel the length of the canal, and this time is but small compared with the tidal period, $12^h\ 20^m$ nearly. Hence it is far more nearly exact to regard the canal as a river, in which the velocity is determined by a certain slope, namely, the difference of elevations at the two extremities, distributed over the whole length, than to regard it as a frictionless channel connecting two oceans. In the present case, supposing the times of high water the same at the two ends,—which would not, it is true, be exact, but which is not of much consequence on account of the smallness of the tide at the Atlantic side,—the slope may amount to $\frac{1}{2}$ (25—3) or 11 feet in 30 miles. Making a calculation accordingly, and introducing a certain correction which I cannot well explain here, I found for the greatest velocity about $4\frac{1}{5}$ miles an hour. This result was obtained by taking into account the resistance of the bed, and not considering the inertia of the water. It would no doubt be sensibly modified by inertia, though I am still strongly inclined to think that it is not very far from the mark. I think it probable that the effect of inertia would be still further to reduce the velocity found above, perhaps to 3 miles an hour.

My conclusions therefore are,

(1) It is not safe at present to take the greatest velocity at less than about 4½ miles an hour.

(2) It is however probable that if the problem were solved complete the greatest velocity would be found to be less, perhaps 3 miles an hour.

(3) If it should be seriously proposed to carry the plan into effect, leaving the ends of the canal open, the velocity of the water ought to be made the subject of a very careful mathematical investigation.

FROM MR GISBORNE.

41, CRAVEN STREET, STRAND,
13 *Sept.* 1852.

The enclosed note will explain to you why your letter to Mr Griffith (who is in Germany) was opened and not answered to the address you give. I enclose you the letter you were good enough to write to me before I left for South America, which contains your results as to the velocity in a suppositious case of a ship canal. I trust to be able to enclose you a copy of my report with this, if not I shall do so to-morrow. The cross-section I recommend is as over—the length 30 miles, with the tide in the Pacific rising and falling 11½ feet above and below the Atlantic, where the total difference between high and low water is only about 16 inches. In spring tides the difference of level may sometimes be 15 feet, and if you look into the question perhaps it would be interesting to find the velocity in that case also. As my report is, for *the present*, only for private circulation, please consider it as confidential.

FROM SIR J. F. W. HERSCHEL.

32, HARLEY STREET,
Nov. 7, 1852.

DEAR SIR,

I am much obliged by your fine specimen of "Canary" glass, and shall try the spirit-lamp experiment forthwith.

I am sorry my report on your paper was not shown to you, as there were some things in it I wished you to see*. I must congratulate you on the singularly fertile field of photological research you have opened out and in which you are advancing with such rapid paces. The extension of the spectrum upwards is very extraordinary as a physical fact, though not an unexpected result *à priori*.

Yours very truly,

J. F. W. HERSCHEL.

It will be of interest to reprint the condensed abstracts of Sir John Herschel's papers on phenomena of fluorescence, as given. in *Royal Society Abstracts*, Vol. v. pp. 547 and 549.

Feb. 13, 1845.

The MARQUIS OF NORTHAMPTON, President, in the Chair.

Ἀμόρφωτα, No. 1. "On a case of Superficial Colour presented by a Homogeneous Liquid internally colourless." By Sir John Frederick William Herschel, Bart., F.R.S., &c.

The author observed that a solution of sulphate of quinine in tartaric acid, largely diluted, although perfectly transparent and colourless when held between the eye and the light, or a white object, yet exhibits in certain aspects, and under certain incidences of the light, an extremely vivid and beautiful celestial blue colour,—apparently resulting from the action of the strata which the light first penetrates on entering the liquid, and which, if not strictly superficial, at least exert their peculiar power of analysing the incident rays, and dispersing those producing the observed tint, only through a very small depth within the medium. The thinnest film of the liquid seems quite as effective in producing this superficial colour as a considerable thickness.

* In his report, which contains a detailed discussion of the subject, Sir J. Herschel "considers this paper to be one of the most remarkable and important contributions to physical optics which have appeared for a long time." There is a postscript to the report as follows: "Perhaps I may be allowed to mention that in my experiments on sulphate of quinine which formed the basis of my papers on epipolic dispersion, I was led up to the very threshold of Mr Stokes' discovery by the following facts, which I find recorded under the date Feb. 25, 1845, with two special *Nota Bene* attached:—viz. that on transmitting through various liquids the light incident on the quiniferous solution, the epipolic blue gleam was quite extinguished when the incident light had passed through spirit of turpentine having a very slight yellowish tinge, though if passed through water purposely rendered a little *more yellow* the gleam was produced."

April 3, 1845.

Ἀμόρφωτα, No. 2. " On the Epipolic Dispersion of Light; being a Supplement to a paper entitled ' On a case of Superficial Colour presented by a Homogeneous Liquid internally colourless.' " By Sir John Frederick William Herschel, Bart., F.R.S., &c.

The author inquires whether the peculiar coloured dispersion of white light, intromitted into a solution of sulphate of quinine, is the result of an analysis of the incident light into two distinct species, or merely of a simple subdivision analogous to that which takes place in partial reflexion as exemplified in the colours of thin plates. He endeavours to ascertain the laws which regulate this singular mode of dispersion, which for brevity he terms *epipolic* on account of the proximity of the seat of dispersion to the intromitting surface of the fluid. It might have been expected that by passing the same incident beam successively through many such dispersive surfaces, the whole of the blue rays would at length be separated from it, and an orange or red residual beam be left: but the author establishes, by numerous experiments, the general fact, that an *epipolical beam of light*, meaning thereby a beam which has been once transmitted through a quiniferous solution and undergone its dispersing action, is incapable of farther undergoing epipolic dispersion.

There were only two liquids, out of all those examined by the author, namely oil of turpentine and pyroxylic spirit, which, when interposed in the incident beam, act like the solutions of quinine in preventing the formation of the blue film: and the only solid, in which the author discovered a similar power of epipolic dispersion, is the green fluor of Alston Moor, and which by this action exhibits at its surface a fine deep blue colour.

It appears from the Abstract of Prof. Stokes' memoir, reprinted in *Math. and Phys. Papers*, Vol. III, that Sir D. Brewster had observed the phenomenon some years previous to Sir J. Herschel*.

The general account of fluorescence given by its discoverer at one of the public lectures at the Meeting of the British Association at Belfast in 1852 has already been referred to (*supra*, p. 10). In further illustration of the great impression made at the time by this discovery, an extract from a speech reviewing the work of the Association, made at the concluding meeting by Dr Romney Robinson, the President of the Royal Irish Academy, may here be added.

" Of the communications brought before Section A, most certainly that of Professor Stokes, which the public has heard

---

* For later correspondence with Prof. E. Wiedemann on fluorescence, see *infra*.

here in a more popular form, holds the highest place, from the singular light which it has thrown on that class of phenomena first brought into notice by Herschel and Brewster. I consider that this may be termed the third epoch in the history of light—the first, being that in which Newton discovered the decomposition of the solar ray by means of the prism; the second, the discovery of the polarisation of light by Malus—and I regard this as not inferior in importance to the other two. Already does it open to the eye that views its development a course of which we can scarcely contemplate the end, and which is not more promising in its theoretical results than it is in the host of practical applications of which it is susceptible."

The following letter from the Abbé Moigno probably relates to some proposal for a similar lecture to be delivered in Paris.

FROM M. L'ABBÉ MOIGNO.

3 juillet, lundi, 1854.

MON CHER MR STOKES,

J' accepterais sans répugnance, s'il fallait s'y résigner, de faire toute notre belle série d'expériences dans une séance particulière de la Section de Physique; mais il me semble qu'il vaudrait mille fois mieux en faire l'objet d'une séance générale du soir; et je vous conjure instamment de nous faire atteindre ce but. L'objection, qu'il faut que dans les séances publiques tout soit expliqué en français, serait levée par un moyen bien simple. Vous, Monsieur, qui occupez dans l'optique un rang si éminent, chargez-vous d'être notre interprète, et faites vous-même cette grande et magnifique leçon. Je préparerai tout, avec M. Duboscq, je rédigerai les programmes, je vous donnerai tous les détails nécessaires, et le soir venu je m'effacerai complètement en me cachant derrière vous. Nous avons tant amélioré nos expériences depuis votre passage à Paris, que vous pouvez compter sur un succès grandiose et complet.

FROM SIR W. ROWAN HAMILTON.

OBSERVATORY, NEAR DUBLIN.
December 17th, 1857.

MY DEAR STOKES,

By an evening postman I received a short time ago your interesting and welcome letter, and have treated myself (as you see) to a large ruled sheet, for the purpose of comfortably replying to it: but shall not expect you to reply on a scale by any means so large. In fact, I had been wishing to write to you, for some time past, but chose to wait till I should hear of your present address.

I trust that you received, through Dr Robinson, a private copy of my last printed paper : in formula (26) of which, I forget whether I marked with a pen the correction of a slight erratum, by inserting the factor 2 before $\sqrt{20\pi}$, under the numerator $\sin 86°\,49'\,52''$. A glance at the second line of that formula will prove that this mistranscript, for I have too great confidence in the Red Lion Court Office to call it a misprint, has not at all affected the results.

A second copy of your own important Paper, "On the Numerical Calculation of a Class of Definite Integrals and Infinite Series" (Cambridge, 1850), reached me by post nearly a month ago, and I was much obliged thereby. A former copy had been kindly sent by you, but at the time I was so much occupied with other things, that although I did run my eye over it, I could only *remember* the *existence* of that *Note*, (in page 20 of the private copies,) wherein you were pleased to acknowledge that I had previously published, in my Essay on Fluctuating Functions, an expression equivalent to that which you have numbered (52)*, and have obtained by a quite different method. The first copy went astray in very good company, I can assure you; but I trust that it will yet turn up; and now that I *have* the second, I will try to keep firm hold of it. I regretted that I could not venture to *print any reference* to it, from *memory*; but may hope that I shall have some opportunity soon. Meanwhile, as regards its relations to my *recent* paper, it seems that there has been only an *affinity*, and that I have not any *anticipation* to acknowledge ; for my own late investigations had (I think) a different aim. Just to prove that, since the duplicate arrived, I have read some parts of it with care, let me remark that in lately applying a method of my own to verify your series (62) for the function (57)

[* The asymptotic expansion for the Bessel functions of order zero. *Math. and Phys. Papers*, Vol. II. p. 352. The differential equations to which his method applies were generalised so as to include all of this type in the Supplement to the Memoir on Discontinuity, *Trans. Camb. Phil. Soc.*, 1868 ; *Math. and Phys. Papers*, Vol. IV. p. 285, see footnotes pp. 289, 298, 80.]

of your Third Example, or rather (for my own satisfaction) to deduce that series anew, I was led to the expression,

$$v = bx^{\frac{1}{2}} \sum_{m=0}^{m=\infty} [m + \tfrac{1}{2}]^{2m} [0]^{-m} (-2x)^{-m} \cos\left(x + c - \frac{m\pi}{2}\right), \quad (a)$$

where b and c are constants, as giving the law of the descending series for the integral v of the differential equation (58) of your Paper, which may be written thus,

$$(D^2 - x^{-1}D + 1)\,v = 0. \tag{b}$$

Comparing (a) with (62), I next write,

$$v = bx^{\frac{1}{2}} \{R \cos(x+c) + S \sin(x+c)\}, \tag{c}$$

and have

$$R = \sum_{m=0}^{m=\infty} [2m + \tfrac{1}{2}]^{4m} [0]^{-2m} (-4x^2)^{-m},$$

$$S = (-2x)^{-1} \sum_{m=0}^{m=\infty} [2m + \tfrac{3}{2}]^{4m+2} [0]^{-(2m+1)} (-4x^2)^{-m}. \tag{d}$$

If then we write

$$R = 1 - \frac{r_2}{1\,.\,2\,(8x)^2} + \frac{r_4}{1\,.\,2\,.\,3\,.\,4\,(8x)^4} - \cdots,$$

$$S = \frac{r_1}{1\,.\,8x} - \frac{r_3}{1\,.\,2\,.\,3\,(8x)^3} + \cdots, \tag{e}$$

the expression (a) gives, generally,

$$r_m = (-4)^m [m + \tfrac{1}{2}]^{2m} = (-1)^m (2m+1)(2m-1) \ldots (3-2m), \tag{f}$$

and therefore,

$$\left. \begin{aligned} + r_{2m} &= (4m+1)(4m-1)(4m-3) \ldots \\ &\qquad 5\,.\,3\,.\,1\,.\,-1\,.\,-3 \ldots (5-4m)(3-4m), \\ \text{and} \quad -r_{2m+1} &= (4m+3)(4m+1)(4m-1) \ldots \\ &\qquad 5\,.\,3\,.\,1\,.\,-1 \ldots (3-4m)(1-4m), \end{aligned} \right\} \quad (g)$$

each being thus given as the product of an even number of regularly decreasing odd numbers.

For instance,

$$\left. \begin{aligned} r_2 &= 5\,.\,3\,.\,1\,.\,-1; \quad r_4 = 9\,.\,7\,.\,5\,.\,3\,.\,1\,.\,-1\,.\,-3-5, \&\text{c.} \\ -r_1 &= 3\,.\,1; \qquad\quad -r_3 = 7\,.\,5\,.\,3\,.\,1\,.\,-1\,.\,-3, \&\text{c.} \end{aligned} \right\} \quad (h)$$

these *numbers* (as I had the satisfaction to observe) agreeing with those of your formula (60), but the *law* having been independently deduced, through the expression (*a*), which is included in the following, for the integral* of the differential equation

$$(D^2 - ax^{-1} D + 1) v = 0, \qquad (i)$$

$$v = bx^{\frac{a}{2}} \sum_{m=0}^{m=\infty} \left[m + \frac{a}{2} \right]^{2m} [0]^{-m} (-2x)^{-m} \cos\left(x + c - \frac{m\pi}{2}\right). \quad (j)$$

Now, as regards some recent results, which appear to be thought my own, by you and De Morgan, and by some of my mathematical friends in Dublin. About the integral

$$\int_0^\infty f(t)\, dt = \tfrac{1}{2},$$

I am glad that you do not take me as having laid any stress upon it, for I think it *very* likely that it has been somewhere published; and, as you remark, the investigation of its value is very easy. What I wanted it for, in my paper, was chiefly to make sure of the *rigorous transformation*,

$$(I_t ft =) \int_t^\infty ft\, dt = (\tfrac{1}{2} - I_t ft =) = \tfrac{1}{2} - \int_0^t ft\, dt;$$

and thereby to confirm, by arithmetical experiment, the propriety of the partly periodical and semiconvergent development (belonging to a great class treated of in your paper, and elsewhere), of which I have as yet published only the first terms, but am in possession of the law, for the *approximate calculation* of the *numerical value* of the lately mentioned integral $\left(\int_t^\infty ft\, dt\right)$, when *t* is real, positive, and large. Perhaps I may be disposed to write down something more about the series last alluded to; but I shall pass on now to the 4th page of this sheet, leaving the few lines below, for a sort of Note to be filled in.

* I do not know whether this form (*j*) for the integral of (*i*) has been published, but probably it has occurred to you. The series terminates, when *a* is an even whole number. As a verification, since

$$\left[m + \frac{a}{2} \right]^{2m} = \left[m - \frac{a}{2} - 1 \right]^{2m},$$

(*j*) shows that if *v* satisfy (*i*), it satisfies also

$$(D^2 + (a+2) x^{-1} D + 1) x^{-a-1} v = 0, \qquad (k)$$

whatever constant *a* may be.

NOTE.

Dec. 19/57.

Not to delay posting this letter too long, I shall merely mention here, that I formed a table of constant coefficients, connected by a sufficiently simple law, for the developments of the *successive integrals*,

$$\left(\int_t^\infty dt\right)^1 ft, \ \left(\int_t^\infty dt\right)^2 ft, \dots$$

as far as

$$\left(\int_t^\infty dt\right)^7 ft,$$

under forms analogous to that which Poisson suggested, and you and I worked out, for the integral ft itself. A fundamental equation, which assisted me in this and in other similar researches, may be thus stated: "If

$$v_n = \sum_{m=-\infty}^{m=\infty} [a]^m \, b_{n,m} \, x^{a-m} \cos\left(x + c - \frac{(n-m)\,\pi}{2}\right); \qquad (l)$$

and if

$$b_{n,m} = b_{n,\,m-1} + b_{n-1,\,m}, \qquad (m)$$

while a and c are constant; then

$$Dv_n = v_{n-1}." \qquad (n)$$

Although, in the *arithmetical* aspect of my little paper, the definite integral from zero to infinity, $\int_0^\infty$ &c. $= \frac{1}{2}$, served chiefly (as has been owned) to *connect* the two *parts* of that integral, $\int_0^t$ &c., and $\int_t^\infty$ &c.; yet it long ago struck me (while pursuing some *physical researches* which I never brought to the point of publication, but which it is *not quite impossible* that I may, at some time or other, resume, sufficiently at least to give an intelligible account of their general character and *aim*), as a remarkable thing, that the *constant* $\frac{1}{2}$ should here be precisely the *coefficient of the first negative power of the symbol I* (or I_t), *in the descending development of that finite* (but irrational) *algebraical function* $(1 + 4I^2)^{-\frac{1}{2}}$; or rather here, $(4I^2 + 1)^{-\frac{1}{2}}$, to which I had been led to reduce the *summation of the known ascending series* for the definite integral ft. I thought that I saw reasons for this being *not an accidental thing*; and was encouraged by that thought, to make long ago some very laborious numerical calculations, whereof some have been reproduced, with extensions of accuracy,—see, for instance, the formula (26)‴, for

$$\pi^{-1} \int_0^{\frac{\pi}{2}} d\omega \sec \omega \sin (40 \cos \omega),$$

$$= + 3\,772\,428\,770\,670\,800 \cdot 537\,7058$$

$$- 3\,772\,428\,770\,679\,799 \cdot 974\,8177$$

$$= + 0 \cdot 562\,888\,1,$$

which I believe to be correct in its last figure. And I long ago verified a suspicion which soon arose, (in part suggested by a *physical anticipation*,)* that the *succeeding symbolical terms*, of the descending development of $(4I^2 + 1)^{-\frac{1}{2}}$, had all (so far as I examined, and I went a good way in doing so) *useful relations to the numerical values* of $I^n ft$, or of $\left(\int_0^t dt \right)^n ft$, when t was a *large* number. But it is only *lately* (since the last Dublin Meeting of the British Association), that the thought of calculating the values of $\left(\int_t^\infty dt \right)^n ft$ occurred to me. For *two* very general and very recent *Theorems* respecting *such* multiple or repeated integrals, I refer you to another sheet which has been begun; but close the present sheet by saying that, with regards to Mrs Stokes,

I remain, very faithfully yours,

WM R. HAMILTON.

Professor George G. Stokes.

---

* It was partly connected with a Theory of Darkness (Scotology), which at one time interested me. Using finite differences, instead of partial differentials, I wanted to understand *how an extensive but finite vibration could be propagated sensibly forward, and not at the same time sensibly backward*. And I was led (I think as early as 1835) to perceive that this must depend on the approximate equality of two definite integrals, as regarded their numerical values, say *A* and *B*, which produced *sensible light* by their *sum*, *A* + *B*, but *sensible darkness* by their *difference*, *A* – *B*. This *expectation* (it was more than a *guess*) led gradually to the discovery of *abstract relations* between certain formulæ of definite integration.

LETTERS FROM THE PRINCE OF SALM-HORSTMAR.

COESFELD, NEAR MÜNSTER, IN PRUSSIA.
1856.

Pardon me, if I take the liberty to trouble you with a request.

Your experiments on the fluorescence have so highly excited my interest, that I cannot omit applying to you for help, inasmuch, that I beg you to send me a glass dyed yellow by means of *silver-oxyd*, as I found it impossible to get such a glass in Germany. At the same time you would greatly oblige me by letting me know the address, where this glass is to be got in England.

Respectfully yours,

PRINCE OF SALM-HORSTMAR.

October 27, 1856.

I beg your pardon, Sir, for my tardy answer, however I too was absent from home several times.

I am most obliged to you for the specimen of the fluorescent substance from Mr Schunck [?] and all your very interesting communications, and especially for the trouble you took in occasioning the yellow glasses to be sent to me by Mr Darker, which I got of late to my great satisfaction.

Let me say in answer to the doubt you express with reference to the explanation of one of the results in my papers on fluorescence: that the ethereal solution of the green colouring matter of the true infusion (Euclena viridis) has *not* been kept in the perfect dark, but that the light to which it was exposed was but very little. As the volatilization was already completed in a few days, it would be very striking that the light should have been the cause of disappearance, the more so as the green colouring does not discolour so quickly besides. However, a delusion might be possible, if the substance should have expanded very thinly in the test-tube, therefore I will repeat the experiment.

I am very glad that your observations on the direction of the emission of the fluorescent light from solutions do correspond with my own later observations.

With many thanks for all your trouble and kindness.

24 *March*, 1857.

I send you hereby a bill of exchange for £2. 1s. 3d. for London, which sum Mr Darker did put erroneously in your account—for the yellow glasses which I have received in the last year.

Further I send you a crystallised substance, to which I have given the name of *Fraxin*, having discovered it in the bark of *fraxinus excelsior.* A *diluted* solution of it in warm water or spirit, containing any trace of *lime* or another *alkali*, will produce a *blue fluorescence* in the day-light, and a *yellow* fluorescence in a box of cobalt-glass. A solution in spirit of wine will prove good for a longer space of time.

An extract of the same bark in cold alcohol gives a *red* fluorescence, particularly striking if the light be concentrated by a burning-glass. The substance producing this red light is a *yellow fluid* substance, which is precipitated by adding cold water to the extract (concentrated).

I have found out that the red fluorescence from the solution of Aesculatin in sulphite of ammonia, saturated with baryta, is originated or better said produced by the *green* part of the spectrum,—which is remarkable.

12 *Juni*, 1857.

Ich danke Ihnen sehr für Ihren brief vom 5ten Mai den ich nach rückkehr von einer reise erhielt.—Sie haben mich auf meinen irrthum aufmerksam gemacht in hinsicht der rothen fluorescenz der geistigen auflösung aus der rinde von Fraxinus, wofür ich Ihnen sehr dankbar bin, denn die rothe fluorescenz ist *nur* eine folge des Chlorophyls. Ich habe diesen irrthum jetzt in Poggendorff's Annalen berichtigt durch eine nachschrift.

Ich habe versucht mir Ihr Paviin aus der rinde von Aesc. Pavia darzustellen, aber es gelang mir nicht genügend, weil der baum schon in der blüthe war wo die rinde nur sehr wenig Paviin enthält, so dass ich *zu wenig* erhielt um zu versuchen ob Paviin mit verdünnter Schwefelsäure Aesculetin giebt*.

Ich muss nun warten bis im frühjahr um den versuch zu wiederholen.

Ich danke Ihnen auch sehr für die Warnung mich nicht durch Kobald-glass irre machen zu lassen bei beurtheilung der fluorescenz, und für Ihre bemerkung dass daher die von mir beobachtete

---

* See *Math. and Phys. Papers*, Vol. IV. p. 112 (1859), and p. 119 (1860), in which fraxin is identified with paviin, in two communications to the Chemical Society of London. In a letter of date April 21, 1857, Prof. Stokes' correspondent had given his reasons for identifying fraxin with aesculin instead of paviin. This letter ends with the question "May I ask you whether on further occasion I may write to you in German, with Latin letters?"

gelbe fluorescenz der fraxin-auflösung eine optische täuschung ist wegen der neigung des auges die complementare farbe zu sehen—oder die wirkung des contrastes.—Ich habe ein mittel gefunden dieses direct zu beweisen, dadurch dass man in ein gelbes glas ein kleines *loch* macht um das auge gegen das blaue licht des Kobaltglas Kasten's zu schützen. Ich wünsche dass Sie meinen irrthum wegen der gelben fluorescenz des fraxins im blauen licht in Poggendorff's Annalen berichtigen.

Ich bitte Sie mir zu sagen was Sie sagen zu der beobachtung von Holtzmann in Poggendorff Annal. von 1856, p. 446, wonach das polarisirte licht *in* der polarisationsebene schwingen soll?— Professor Beer in Bonn schreibt mir darüber, dass die grossen *Abweichungen von den theoretischen werthen* und der vollständige widerspruch mit den resultaten Ihrer versuche—auf einer oder der anderen seite eine fehlerquelle voraussetze welche erst aufgefunden werden muss, um sagen zu können ob die Versuche Holtzmann's etwas über die grundvorstellung von der lage der schwingungen hier entscheiden.—Ich hoffe Sie werden die versuche von Holtzmann mit einem russgitter selbst anstellen und Ihre resultate dann bekannt machen.

Darker hat mir den empfang der zahlung angezeigt.

Mit vieler achtung Ihr ergebener

FÜRST ZU SALM.

15 *Februar* 1860.

GEEHRTESTER HERR PROFESSOR!

Ich danke Ihnen vielmal für Ihre antwort vom 8ten februar deren inhalt mich sehr erfreut hat. Ich habe die mir gütigst geschickte probe (die mit der *gelben* farbe) gleich auf der stelle chemisch geprüft und gefunden,—dass ihr aus Aesculus-hypocastanum-rinde dargestelltes *Paviin* in der that *Fraxin* ist, denn es gibt *gleiche* reactionen mit *eisenchlorid*, mit barium oxydhydrat und strontium-carbonat und, was die hauptsache ist, es wird durch verdünntes *acidum sulphuricum* in der wärme von 80° R^{re} in Fraxetin verwandelt in 6 minuten—*welches* ich Ihnen hierbei schicke. Zum vergleichen lege ich auch eine probe von meinem *aus Fraxin* dargestellten *Fraxetin* hier bei. Ihre entdeckung auf optischem weg ist sehr interessant weil Sie die identität beider substanzen gleich *richtig voraus* geschlossen haben und dieser schluss Sie zu der chemischen darstellung aus der rinde von Aesc. hypocastanum geführt—wo es die chemiker *als solches* nicht vor Ihnen erkannt haben. Die wissenschaft ist Ihnen also vielen dank schuldig und die chemie wird jezt die fluorescenz immer mehr gebrauchen *lernen* müssen.—Sie haben aber der chemie eine methode gelehrt diesen stoff vom Aesculin zu scheiden, und ich bin Ihnen sehr dankbar für die mir gemachte Mittheilung der methode und die lezte spur von *blei* lässt sich sehr leicht durch Schwefel-Wasserstoff entfernen aus dem Paviin, oder Fraxin, weil Kohlensaures blei etwas löslich ist.

Sie werden finden, dass *die beiden* hier überschickten proben des Fraxetin, in ihrer auflösung in Wasser was eine spur von Kalk enthält, ganz *gleiche* optische eigenschaften haben, so wie ihre chemischen reactionen gegen eisenchlorid, baryt und strontium ganz übereinstimmen mit dem was ich in Poggendorffs Annalen über fraxetin mitgetheilt habe. Beide proben sind in wasser einmal umkrystallisirt nach dem Waschen. Ich hatte keinen grösseren vorrath mehr :—desgleichen von einer probe fraxetin welches vor einem Jahr in gesättigtem Alkohol krystallisirt ist und aus *Eschen-Rinde*fraxin durch verdünntes Acidum sulphuricum dargestellt.

Ich bin Ihnen sehr vielen dank schuldig für die mir geschickten beiden druckschriften über die darstellung des Paviins und über die methoden mit hülfe der fluorescenz dergleichen Körper auch in gefärbten flüssigkeiten auch dann zu finden wenn zwei verschieden-fluorescirende Körper gemengt darin enthalten sind.

9 *Maerz* 1860.

GEEHRTESTER HERR PROFESSOR,

Ich danke Ihnen verbindlichst für Ihren brief und die darin enthaltenen sehr interessanten mittheilungen.

Ich erlaube mir eine frage über einen gegenstand betreffend die fluorescenz-erzeugenden strahlen der Sonne. Ich wünsche zu wissen ob Sie bei *keinem* fluorescenz-Spectrum *das verschwinden* einer Fraunhoferschen linie gefunden haben?—Würde auch nur *eine* solche linie verschwinden;—so würde daraus folgen dass *licht* in der dunkeln linie des Sonnenspectrum's enthalten ist. Man findet zwar im photographirten Spectrum vom Müller, dass nach einer bestrahlung von 15 Sec. dauer, viele Fraunhofersche linien *verschwinden,* aber ich glaube dass die ursache *nicht innerhalb* der

Fraunhoferschen linien zu suchen ist, sondern *ausserhalb*, indem die in der empfindlichen chemischen Substanz durch die chemischen strahlen inducirte chemische *bewegung* der *zersetzung*, sich nach verlauf von 15 Sec. auch ihre chemische bewegung mittheilen kann den atomen *gleicher* substanz welche sich im dunkeln raum der Fraunhoferschen Linien befinden. Mich interessirt diese frage um so mehr, weil Kirchhoff (in dem Monatsberichte der Berliner Akademie der Wissenschaften vom October und September) aus den beiden Linien D von Fraunhofer, verglichen mit den entsprechenden *hellen* Linien im Spectrum einer Kochsalz-flamme, durch seine *versuche* den schluss zieht, dass *Soda* in der Atmosphäre *der Sonne* ist. Es fragt sich aber ob wir eine fluorescente Substanz haben, welche von den Strahlen des Sonnenspectrums NEBEN D afficirt werden und ob D dunkel bleibt, und ob ferner die hellen Linien D im Spectrum der Kochsalz-flamme eine fluorescenz erregen oder nicht, denn wenn sie eine *dunkle* linie geben auf dem passenden fluorescirenden Stoff und die linien D des spectrum's der Sonne gleiche negative reaction zeigen mittelst der fluorescenz, so würde dieses günstig sein für Kirchhoff. Mir scheint es wenigstens nothwendig zu seyn, erst direct nachzuweisen, dass im dunkeln raum der linien D ein auf das Auge oder auf eine fluorescirende substanz wirksames *Agens* sich befindet. Kirchhoff hat zwar bewiesen dass die hellen linien D im Kochsalzspectrum als *dunkle* linien erscheinen, *wenn* das Sonnenspectrum *durch* eine Kochsalzflamme geht, und dass sie alsdann noch schwärzer erscheinen als die im Sonnenspectrum *ohne* die flamme. Dieser erklärt Kirchhoff dadurch, dass die *flamme* des Kochsalzes die von der Soda in der Sonne ausgehenden wellenlängen in D, *absorbirt* oder vernichtet, *weil* sie *von gleicher* wellenlänge sind (in entgegengesezter richtung interferiren).

14 *Maerz* 1860.

GEEHRTESTER HERR PROFESSOR!

Ich theile Ihnen etwas mit was mir Rochleder geantwortet hat als ich ihm schrieb Ihre mittheilungen über das Paviin in der rinde der rosskastanie. Sie werden darin eine vollständige bestätigung der analytischen identität von Paviin und Fraxin finden. Rochleder antwortet mir, dass er vor mehreren jahren bei der reinigung des Aesculins einen krystallinischen Körper gefunden habe welchen Herr Kawalier im laboratorium von

Rochleder analysirt hat durch verbrennung; er habe aber bisher "diese Untersuchung und Analyse noch nicht bekannt gemacht, weil er bis zur vollendung seiner *ganzen arbeit* über die substanzen alle die sich *im rosskartanienbaum,* von der frucht, rinde, blätter und blüthen, finden, warten wollte. Er werde aber jetzt in den nächsten tagen die ganze arbeit der Akademie in Wien einschicken und mir einen separaten abdruck davon zuschicken." Sobald ich denselben erhalte, werde ich Ihnen denselben zuschicken, und kann Ihnen nur einstweilen sagen dass die zahlen der Analyse ganz mit denen vom fraxin überein zu stimmen scheinen, denn Rochleder nennt diese substanz geradezu *Fraxin* in seinem brief.

Verzeihen Sie wenn ich Sie mit obiger angelegenheit belästige indem ich von Ihrer *grossen güte* gebrauch mache.

Mit hoher Achtung, Ihr dankbar ergebener

8 *August* 1860.

GEEHRTESTER HERR PROFESSOR!

Ich muss Ihnen das verfahren mittheilen um eine flüssigkeit mit *schön grüner* fluorescenz zu bereiten. Sie nehmen Schwefelsäure die durch erhitzen so frei von Wasser gemacht ist, dass bei dem erhitzen eine glasscheibe nicht mehr mit Wasserdaempfen beschlägt, und alsdann lässt man die Säure in einer verkorkten flasche kalt werden. Von dieser Schwefelsäure giessen Sie etwas in ein probierglas und setzen einige tropfen *Ochsen-galle* hinzu und rühren um, so erhalten Sie gleich die schöne grüne fluorescenz. Es ist nicht nöthig dass die galle frisch ist, sie darf auch schon riechend sein. Diese fluorescenz-flüssigkeit können Sie nun *offen* stehen lassen wochen lang *ohne verminderung der fluorescenz.*

Ich glaube dass diese erscheinung Ihnen neu ist, deshalb theile ich es Ihnen mit.

2 *März* 1865.

Ihre höchst wichtige Entdeckung der für den electrischen Funken specifischen ultra-violetten ·Wellenlängen, welche die Länge des Spectrums der Sonne so ungemein weit uebergreifen, hat mich in die exaltirteste Freude versetzt.—Welche Freude müssen Sie empfunden haben als Sie das *neue Wunder* vor sich sahen mit Hülfe Ihrer früher entdeckten absorbirenden Medien.—

Es kann also in der Sonne kein electrisches Licht seyn. Oder die Atmosphäre der Sonne hat *nicht* die Gase der unsrigen. Oder die Temperatûr der Sonne ist *niedriger.*—Schade dass man die ultra-rothen Wellen nicht auch mit einem solchen Absorbens näher untersuchen kann.—Haben Sie das Spectrum des Blitzes noch nicht im ultra-violetten Sinn geprüft?—Ob es nicht durch Luftschichten grosser Dimensionen absorbirt wird?

In October, 1854, Prof. Stokes accepted the invitation of the Royal Society, conveyed through their President, Col. Sabine, to take up the post of Secretary on the Mathematical and Physical side, in succession to Samuel Hunter Christie, who had held it since 1837. The Biological Secretary, Dr W. Sharpey, had taken office at the previous anniversary, Nov. 30, 1853, and remained his colleague until 1872, when he was succeeded by Prof. Huxley; on the latter becoming President in 1881 his successor was Prof. Michael Foster, who remained Prof. Stokes' collaborator during the remainder of his service until he was transferred to the office of President in 1885. During this long period Stokes worked with the greatest smoothness and good-will with his various colleagues. Instances are numerous, throughout a voluminous correspondence, of the cordial understanding that existed, and of the unhesitating satisfaction of the Royal Society with its good fortune, in retaining the services of an officer distinguished by so much scientific renown and by so much zeal and success in forwarding the work of his fellows.

The work of the Royal Society had been rendered more convenient by Prof. Stokes' engagement as Lecturer at the School of Mines. But when the University of Cambridge was able to assign some endowment to the Lucasian Chair in 1859, he felt that it had now a prior call upon his time, and accordingly he resigned that lectureship. In a letter to his father-in-law, Dr Robinson, at a later period, he expresses regret that the responsibility for a family makes it necessary for him to continue to devote his time to the work of the Royal Society, to the neglect of his own private researches.

In 1858 he served, at the instance of the Chairman, Lord Wrottesley, Pres. R.S., as Secretary of the Royal Commission on the Ordnance Survey.

October 16*th,* 1854.

MY DEAR COL. SABINE,

Unless the duties of the Secretary of the Royal Society are much more onerous than I anticipate, I know of no reason why I should not undertake them, in case the Society please to elect me at the next anniversary. Indeed, I should like very well to hold the office. I shall now be resident in London a good part of that portion of the year during which the Society meets; and as to my lectures here in May, I now give my men but five days in the week, and by omitting Thursday in place of Saturday I should be able to go to London for the meetings.

Believe me, Yours very truly,

G. G. STOKES.

Aug. 10, 1859.

MY DEAR SIR RODERICK [MURCHISON],

I have for some time been rather in a state of suspense as to whether or not to undertake next winter the duties of Lecturer at Jermyn Street, Secretary of the Cambridge University Commission, and Secretary of the Royal Society. It would be sharp work certainly, but might be managed by putting on more pressure; but there is such a thing as boilers bursting. However, it is not a failure of health I fear, for I have hitherto been strong; but I feel that I have more work on me than I could execute to my satisfaction. I thought it might be executed if I managed to get an early hour at Jermyn Street, and made some trials to get one. To a certain extent I succeeded; but pending the final decision I felt so strongly that I might be taking more duties on me than I could properly fulfil, that I came to a sort of resolution with myself that if there were *any* hitch in the arrangement about early hours I would resign or at least tender my resignation. Yesterday's post brought me a letter which showed that there was a hitch, and accordingly I have executed my intention by writing to the Lord President of the Council. I enclose you a copy of the letter.

I hope you do not think that I dislike our School at Jermyn Street, and hold it cheap, because I am ready to resign my Lectureship there for a Secretaryship which expires on Dec. 31, 1860. On the contrary it is not without many feelings of regret that I think of leaving you. I have made many friendships which would make me feel as if I belonged to the institution, even if I had ceased to be Lecturer. I have improved myself by having my attention called to the practical manipulation of instruments, belonging to branches of Physics to which hitherto I had not paid much attention. But I own that I should like, if I could, to have more time for original research, from which at present from the multiplicity of my offices I am almost wholly shut out. I own too that I feel a sort of divided allegiance to the University of Cambridge and the School of Mines, being Professor of Mathematics at one place and of Physics at the other, and would rather serve one master. I told you that legislation was in train at Cambridge which would have the effect of calling me into residence (supposing I were still alive and well) in the year '66 at the very latest; and the University would have the power (and it seems likely they would exercise it) of instituting the new state of things earlier. Hence it is only a question whether I should resign now or a few years hence. On the other hand I could not properly, even if I were disposed, resign the Secretaryship of the Commission.

The upshot of the whole is this. In the prospect of what I should otherwise have to do next winter I am anxious to be relieved of my duties at Jermyn Street. But I should be sorry by unqualified resignation to damage the institution; as the effect might be to oblige the appointment to be made in a hurry, and so to saddle the institution with a man not so good as could have been got had there been more time to look out. If however I cannot be excused by being relieved of my duties next winter, which was my object in tendering my resignation at present, I think it but fair that I should be looked on as not having tendered it, and left to resign at such future time as I see occasion.

I told you I was not afraid of my health failing, but of my having more work than I could execute to my satisfaction; but my wife has got a woman's tenderness for her husband's health, and for that and other reasons is very anxious I should resign. So I have given her a *carte blanche* to plead her own cause, not that it depends on you except as one whom Lord Granville will probably consult as to a successor.

The following letter from A. W. Hofmann, then Professor at the London College of Chemistry, deals with a matter which is sometimes a source of difficulty, namely the delay involved in submitting scientific memoirs to the judgment of referees, who are often busy men unable to take the matter up at once.

9, FITZROY SQUARE.

July 6, 1860.

MY DEAR PROFESSOR STOKES,

I am afraid you will consider me a great bore, but I hope at the same time you will excuse my troubling you once more in this matter, which is really of great importance for me. If there are definite rules which prevent you from referring my paper before the time when the Society meets again, of course there is an end to the transaction.

If on the other hand there are no strict rules laid down—and from the form in which you express yourself in your note I incline towards this, in my case, pardonable belief—you will I hope help me in this difficulty.

You know how rapid is the circulation in Chemistry at the present moment, and how much a paper loses by coming *post factum*. The paper contains the result of the last three years' work, and especially the theoretical conclusions*. Now, say you, the paper stands over the vacation, it may be that I lose six months of most valuable time. It is true some of the principal facts are published in the *Proceedings*, but this just enables others to build upon those facts the very conclusions at which I arrive. All this inconvenience would disappear if the paper was just demanded back from the referees, and ordered in case of recommendation to be put in type. A line from you would accomplish this miracle.

If you will be so good as to let me know when you will be in town, I would call on you in order to give you additional explanations.

I am very sorry to hear what you write about Prof. Kopp's paper†. If it be finally rejected, the *Transactions* will lose what at all times will be considered an ornament.

Hoping you will excuse this long note,

I remain, my dear Sir, Yours very sincerely,

A. W. HOFMANN.

---

* In *Phil. Trans.* 1861, pp. 409–534 are Hofmann's three memoirs, ' Contributions to the History of the Phosphorus Bases,' all communicated on June 21, 1860.

† ' On the relation between boiling-point and composition in organic compounds,' *Phil. Trans.* 1861, pp. 257–276, communicated by Hofmann.

The following letters refer to an engraving taken from a drawing in the Pepysian Collection at Cambridge, which forms a frontispiece to the Newton-Cotes *Correspondence*, edited by J. Edleston, demy 8vo., London, 1850. A portrait by Kneller, painted in 1689, in which year Sir Isaac Newton was in London attending the Convention Parliament, the *Principia* having been published two years previously, was identified by Dr S. Crompton in 1866 at the Art Treasures' Exhibition. It belonged to Lord Portsmouth, and proved to have come into the family by the marriage of his ancestor Lord Lymington with Newton's niece Miss Conduitt. It has been engraved by T. O. Barlow. Dr Crompton considers that the India ink drawing in the Pepysian Collection is a copy from this picture by an inferior hand, doubtless made for Pepys. See *Proc. Manchester Lit. and Phil. Soc.*, Oct. 1866.

FROM PROF. E. DU BOIS-REYMOND.

BERLIN, 19 PUTTKAMMERSTR.

Nov. 13*th*, 1858.

DEAR SIR,—In remembrance of the kindness you shewed me during my stay in England three years ago, I now venture to intrude upon you with a request which, however, I hope you will consider as not addressed to you if its fulfilling should cause to you the least trouble.

I possess an exceedingly fine engraving of Sir Isaak Newton, shewing him half-length, in his dressing-gown, at an earlier age than that at which he usually is represented. The engraved area is elliptical, $5\frac{1}{2}$ inches long and $4\frac{1}{3}$ inches wide; under it is to be seen Newton's autograph,

IS. NEWTON

"HYPOTHESES NON FINGO"

and the following inscription in small capitals:

Engraved by T. Outrim from the original drawing in the Pepysian Collection at Cambridge.

This engraving was presented to me by the MM. Schlagintweit after their first journey to England, I should think about '51; and they told me that it was intended as a frontispiece for some hitherto unpublished papers or letters of Sir Isaak Newton.

The exquisite beauty of this likeness has so much struck some of my scientific friends that they ardently wish to have an opportunity of purchasing copies of it; and so I have been induced to try to obtain information about the two following points:

1. Whether it would be possible, for love and money, to procure some more copies of the above-mentioned engraving; how this could be managed, and what would be the charge.

2. Whether, in case copies could not be obtained any more from the original plate, there would be any legal objection to a Prussian engraver copying the copy in my possession for the purpose of selling his imitation.

As I understand you are still a Professor in Cambridge, I have thought there could be no person likely to be better informed with respect to everything connected with the memory of Newton, and possessed of greater advantages for procuring any desirable knowledge, than you, and therefore I have resolved upon applying to your kindness.

I need not tell you that I should be exceedingly happy if there should be some opportunity for me to be of any service to you here.

Believe me, my dear Sir, Ever yours most truly,

E. DU BOIS-REYMOND.

BERLIN, 19 PUTTKAMMERSTR.

October 14th, 1859.

MY DEAR SIR,—I will not endeavour to excuse myself from having deferred so long answering your letter, and thanking you for all the trouble you have taken in finding out the engravers of Newton's portrait. I received your letter when I was just beginning my course of lectures and found myself unable to attend to anything else, and from that time till now I have frequently, and with increasing remorse, thought of writing to you, but have never mustered leisure enough to do so. Yet what I had to tell you was very short and simple, viz., that if you can, without losing too much of your invaluable time, have a few impressions (say a dozen, if the copy does not cost more than two shillings, and half-

a-dozen, if more) taken from the plate, you will greatly oblige several admirers of your illustrious predecessor in the Lucasian professorship. The plan of having a new plate engraved here after the copy in my possession has been altogether abandoned.
Believe me, my dear Sir, Yours very thankfully,

E. DU BOIS-REYMOND.

The first of the following letters from Prof. Magnus announces the election of Prof. Stokes as Foreign Associate of the Berlin Academy.

BERLIN.

7 *April* 1859.

GEEHRTER HERR,

Die Art in welcher Sie die Physik durch Ihre vortrefflichen Arbeiten im Gebiete der Optik gefördert haben, hat in meinen Freunden, wie in mir, schon seit längerer Zeit den Wunsch rege gemacht Sie in engerer Verbindung mit unserer Academie zu wissen. Es freut mich Ihnen anzeigen zu können, dass in der soeben stattgehabten Sitzung Ihre Wahl als correspondirendes Mitglied für das Fach der Physik erfolgt ist. Sie werden in einiger Zeit das Diplom erhalten, ich wollte indess nicht unterlassen Sie von dem Ergebniss dieser Wahl so bald als möglich in Kenntniss zu setzen.

Mit der grössten Hochachtung, Ihr ergebenster,

G. MAGNUS.

Prof. Dove und Prof. Poggendorff tragen mir auf Sie bestens zu grüssen.

FRIEDRICHSRODE IN THÜRINGEN.

8 *Sept.* 1859.

VEREHRTER HERR,

Hierbei erlaube ich mir Ihnen eine Einladung zu einer Stiftung zu übersenden, die bestimmt ist den Namen Alexander von Humboldt's zu ehren, die aber, wie ich hoffe, zugleich von dem grössten Nutzen für die Naturwissenschaften überhaupt, und besonders auch für die Physik, werden kann. Ich bitte Sie sich der Sache anzunehmen, und womöglich dahin zu wirken dass in England ein Comite sich bilde das Beiträge sammelt. Die Stiftung wird auf keine Weise beschränkt werden, sie kommt den Talenten aller Nationen zu gut! Sie würden mich verbinden wenn Sie die Güte hätten mich gelegentlich zu benachrichtigen über den Fortgang der Sache in Ihrem Vaterlande.— Sollten Sie eine grössere Anzahl der gedruckten Einladungen zu erhalten wünschen, so bitte ich mir gefälligst zu schreiben. In etwa 8 Tagen bin ich in Berlin zurück.

Hochachtungsvoll, ergebener,

G. MAGNUS.

CORRESPONDENCE WITH MICHAEL FARADAY.

Stokes and Faraday would naturally be congenial spirits. Their close acquaintance doubtless dates from the years when Stokes lectured at the School of Mines. It was in preparing for his lecture at the Royal Institution, "On the Change of Refrangibility of Light and the exhibition thereby of the Chemical Rays," in Feb. 1853, that Stokes discovered that the spectrum of the electric spark was six or eight times as long as the visible spectrum, whereas in sunlight the higher rays have disappeared by absorption by the atmosphere. The earlier of the following letters relate to Faraday's memoir on the colours of finely divided gold*. The others relate to the well-known incident of the withdrawal by Faraday of one of his papers submitted to the Royal Society, and place it in a light creditable to both persons concerned. See Dr Bence Jones' *Life of Faraday*, Vol. II. pp. 247-8, from which the following extract is taken:

"The experiments which he made on this subject are recorded in the twenty-fourth series of *Experimental Researches* received by the Royal Society, August 1, 1850, on the possible relation of gravity to electricity. He finishes the paper thus : ' Here end my trials for the present. The results are negative : they do not shake my strong feeling of the existence of a relation between gravity and electricity, though they give no proof that such a relation exists.' Ten years afterwards he says the same thing, in the very last paper that he wrote, so constant was he even in science when he had made up his mind."

* This subject of the colorations produced by fine metallic particles or thin films, whether by transmission or selective reflexion, occupied much of Prof. Stokes' attention. For its present state of development cf. two papers in *Phil. Trans.*, 1904-5, by Mr J. C. M. Garnett.

ROYAL INSTITUTION,
17 *June*, 1856.

MY DEAR SIR,

I am much obliged for your letter. Though I have
a strong impression on the side of the question which admits that
finely divided particles of gold may transmit ruby light, yet as I
said I am by no means certain. I mean to work out the point.
Electric explosions of gold wire seem to present an easy way of
settling the question, as they may be made to occur in hydrogen,
carbonic acid, and other gases*; but I want to idle for a time, so I
shall put your letter with my experimental notes and resume both
together.

Ever truly yours,

M. FARADAY.

6 *Dec.,* 1856.

MY DEAR STOKES,

Though my paper is very long I must add a little
more. The accompanying comes in at the account of the effect of
heat on gold. If you are inclined to look for the connexion, you
will easily find it by the pencil references at beginning and end.
I am going out of town again immediately, but shall return
permanently at the end of next week and then will try to
see you.

14, BELLEVUE TERRACE, SOUTHSEA, PORTSMOUTH,
8 *June*, 1860.

MY DEAR FARADAY,

I found your paper at the R. S. and took it here
to read†. I am nearly sure you asked me to read it and give you
my opinion about it. I will answer on that supposition.

I own my own opinion is against sending it in for the *Transactions*. It might have done as coming in incidentally, in the body
of a paper containing positive results, but it seems to me it
would hardly do for an independent communication to the *Transactions*, a communication I mean made at one time, though
forming part of a train of experimental enquiry. If such negative

---

* Compare the recent preparation of emulsions of metals by Bredig, by electric
sparking between these metals as terminals under water.

† See Dr Bence Jones' *Life of Faraday*, Vol. II. p. 412.

results had the effect of correcting a commonly entertained expectation, or if the author's previous labours had led those who had followed them to regard a positive result as probable, or even not unlikely, the case might be different. But to my mind the antecedent probability of a positive result was too slender to justify the publication, in such a solemn manner as in the *Transactions*, of a negative result.

I should not myself expect a change in the temperature or electric state of a body, even if we could transfer it to a place where gravity was only half what it is at the surface of the earth ; but even if a change were to be expected under these circumstances we could hardly expect to render it sensible in merely passing from the bottom to the top of a tower*. To my mind the antecedent probability of a positive result is the product of two (to my mind small) fractions expressing the separate probabilities.

I write on the supposition that the change to be expected was a change in the gravitating relations of the experimental mass— [attributable] to a change for example from a place of strong to a place of weak gravity—and not merely to a motion with or in opposition to the force of gravity. Such I take to be your view.

A sentence at the top of p. 2 will require modification. " The so-called variation of gravitating force by change of distance, can only be taken into account in either astronomical or cosmical phenomena ; neither of which can be made the subject of experiment." This statement is too absolute, because the change *is* taken into account in Cavendish's experiment.

I don't think there would be any objection to the paper's appearing in the *Proceedings*. I should be glad if you would take the opinion of some one else.

I remain here till Tuesday, when I go to town.

Yours very truly,
G. G. STOKES.

ROYAL INSTITUTION,
11 *June*, 1860.

MY DEAR STOKES,

I am very grateful for your kindness, though I had not ventured to presume on troubling you except by the general

---

* Lord Kelvin has stated somewhere that the experiments were made in the shot tower on the Surrey side of the Thames.

question whether the account was worth appearance in the *Proceedings* or anywhere else at the R. S. I quite go with you in all you say, and think that the paper had better be withdrawn altogether if it can be. I want no other opinion than yours and my own.

I hope my acknowledgments will catch you at Southsea or at least follow you to London safely.

Ever your very obliged,
M. FARADAY.

84, BROOK ST., W.,
May 5,....

MY DEAR SIR,

Many thanks for your note about the Faraday MS. I shall be very glad if, after you have read it and re-read your note to Faraday, you can give me some note as to what Faraday probably wrote to you when he asked for the paper to be returned to him.

I am sorry to give you so much trouble, but there are many things very characteristic of Faraday in the paper.

Believe me, yours most truly,
J. BENCE JONES.

ROYAL INSTITUTION,
6 *Oct.*, 1862.

MY DEAR STOKES,

I think so highly of the honor done me by your University* that I am not willing to omit mention of it in the Royal Society list. Might I ask you to instruct Mr White what to say and where, in the list attached to my name.

I have recently been named by the King of Italy Knight of the Order of St Maurice and Lazarus, which as a mark of respect to him and his reasons I think ought to go in too. If you think so, will you speak of it at the same time.

Ever yours,
M. FARADAY.

---

* Faraday was nominated by the Duke of Devonshire for the degree of LL.D. at Cambridge, on the occasion of his installation as Chancellor; new Statutes, abolishing the previously existing religious test, had made it possible for him to accept the honour. See Bence Jones' *Life of Faraday*, Vol. II. p. 449.

Prof. Stokes in various connexions has referred to his friendly relations with Sir David Brewster, which were not affected by the circumstance that Sir David was apparently never able to accept the undulatory theory of light. The papers referred to in the first of the following letters relate to polarization and scattering of light transmitted through thin plates of agate cut at right angles to the striation, leading up to the surface colours exhibited by pieces of ancient glass which have become laminated by weathering; the imperfection of the polarization by reflexion from diamond is also announced. The second letter refers to Brewster's observations on crystalline reflexion, the results of which had formed a main basis for MacCullagh's fundamental theoretical work on that subject.

ALLERLY, MELROSE,
Dec. 15*th*, 1860.

MY DEAR MR STOKES,

You are welcome to do whatever you think proper about my paper; and I think if you look into my paper of 1813 in the *Phil. Trans.* 1814, pp. 225, 227, and into Arago's paper on Polarisation in *Encyc. Brit.*, and Sir John Herschel's observations on both, you will see the connexion between my paper of 1813 and the present.

I consider the discovery contained in my present paper as the most important I have made, and one which had never been anticipated. The labour it cost me has been very great, as any person will find who may repeat my experiments—if they can find the means of repeating them.

That a bundle of films should produce coloured polarisation like crystalline plates, and that one of the interfering pencils should consist of a number of pencils of different origin, is certainly a very remarkable result.

I believe that Dr Lloyd has found that this is deducible from theory.

I am, ever most truly yours,
D. BREWSTER.

DUNCLIFFE GARDENS,
March 8*th*, 1861.

My experiments on Crystalline Reflexion are not published. I have a small 4to volume filled with them, which I will give, along with the specimens of Calc Spar with which they are made, to you or any one who will take up the subject.

I abandoned the subject to Professor MacCullagh, who had a

fine instrument constructed for him in Dublin on purpose to investigate the subject.

Some of my results are very curious, but the want of time and of a good instrument induced me to leave the subject in Prof. MacCullagh's hands*.

My volume of experiments is at my residence in the country, otherwise I would have mentioned to you some of the more remarkable results.

ALLERLY, MELROSE,
Dec. 8th, 1864.

I would esteem it a particular favour if you could lend me any specimens of grooved glass which give good diffracted spectra. I possess the very finest specimens upon steel, executed by the late Sir John Barton; but I have been stopped in prosecuting a very curious discovery by the imperfection of my specimens of grooved glass.

ENGLISH CYCLOPÆDIA.

In 1859-61 the *English Cyclopædia of Arts and Sciences* was brought out by Charles Knight, on the basis of the previous *Penny Cyclopædia*. The articles on physical subjects were revised and brought up to date by Prof. Stokes; and a number of new ones, mainly in the domain of Physical Optics, were written by him, a list of the latter being prefixed to the last volume (VIII.). It seems from the correspondence relating to it that he took this task seriously, and that it gave him considerable trouble, various articles being expected from him that were not written. A list follows of the articles thus contributed, revealing the master hand by the keen luminous reflexions and historical appreciations which they contain, though detailed mathematical analysis is avoided; the pages refer to the compact quarto single columns.

Absorption of Light	Vol. I.	1859,	pp. 22—3
Elliptic Polarization	„ III.	1860,	„ 842—3
Interference	„ IV.	1860,	p. 927
Optics (in part)	„ VI.	1861,	pp. 65—6
Polarization of Light	„ VI.	1861,	„ 858—63
Rainbow	„ VI.	1861,	„ 931—5
Sight	„ VII.	1861,	„ 555—62
Spectrum	„ VII.	1861,	p. 701
Undulatory Theory	„ VIII.	1861,	pp. 466—83

* See MacCullagh's 'Collected Papers,' *passim*.

ELASTICITY OF SOLIDS.

One of the classical performances of Prof. Stokes was the placing of the theory of Elasticity on an exact basis, independently of special laws of action between molecules of ultra-ideal simplicity, such as those on which the equations had been founded by Poisson, Cauchy, and their successors. This investigation came in his early productive period (*Camb. Phil. Trans.*, 1845), and is characteristically appended to the memoir in which he developed similar reasoning to establish * the equations of the motion of fluids as affected by viscosity. In this memoir he asserted the existence of two independent elastic constants for isotropic matter, including substances like india-rubber, in opposition to the school of Poisson who, proceeding from the special hypothesis of intermolecular attraction, arrived at a definite relation connecting these two constants. The question is perhaps still to be considered as unsettled, or rather it has been diverted into the question of what types of bodies have a right to be called simply isotropic.

The rari-constant hypothesis found its most distinguished advocate in the illustrious Saint-Venant, whose work in freeing elastic theory from unnecessary restrictions runs parallel with Stokes', and who has left so great a mark on the subject in many other directions.

The following letter, possibly preparatory to the very complete Historical Introduction and Appendices to his edition of Navier's *Leçons* (1864, pp. 90—311 †), shows how, in advancing subjects, a point of view which is familiar and obvious in one school may differ essentially from the natural course of thought in another. The practical British method of development in mathematical physics, by fusing analysis with direct physical perception or intuition, still occasionally presents similar difficulties to minds trained in a more formal mathematical discipline.

---

* For previous history see his *Brit. Assoc. Report on Hydrodynamics*, 1846, Part vi, reprinted in *Math. and Phys. Papers*, Vol. i.

† In this work the consideration of Prof. Stokes' memoir, as also the controversy mentioned in the text, is reserved wholly for Appendix V. The editor finds (pp. 720, 755) no sufficient *proof* in that memoir that the isochronism of vibrations demands linear relations between stress and strain. The detailed criticism of Prof. Stokes' principles of elastic theory (and Maxwell's) is in pp. 732-742.

From M. BARRÉ DE SAINT VENANT.

Vendôme (France),
22 janvier 1862.

Monsieur,

J'ai lu à plusieurs reprises quelques-uns de vos beaux et savants mémoires, et j'y ai toujours trouvé instruction et grand plaisir, bien que je sois peu familier avec votre langue. Mais il y en a un sur lequel je viens solliciter de votre obligeance quelques explications ; elles me sont nécessaires pour le comprendre, ce qui tient peut-être à ce que je l'ai eu trop peu de temps entre les mains lors de mon dernier voyage à Paris.

C'est le mémoire *On the theories of internal friction of fluids in motion, and of the equilibrium and motion of elastic bodies*, lu le 14 avril 1845 et inséré au tome 8ᵉ de Cambridge 1849.

D'abord, de la page 287 jusqu'au Supplément sur quelques cas du mouvement des fluides, qui commence à la page 409, *je n'ai rien trouvé qui soit relatif aux corps solides élastiques.* Et cependant c'est bien ce mémoire là que cite M. Maxwell au tome XX. des *Trans. Royal Society*, Edinburgh, 1853, p. 89, où il dit que vous vous appuyez sur le fait général de l'isochronisme des petites oscillations pour égaler les pressions à des fonctions du premier dégré des déplacements *, et que vous résolvez finalement les équations dans trois cas†, 1°, un corps pressé également sur toute sa surface, 2°, une tige étendue, 3°, un cylindre tordu.

Je n'ai rien trouvé de tout cela au tome 8ᵉ des Transactions de Cambridge, pages 287 à 409.

Est-ce que j'aurais dû le chercher ailleurs ?

Et puis, Monsieur, je vois que vous arrivez d'une manière simple et ingénieuse, sans considérer comme Cauchy une surface ellipsoïdale et ses rayons vecteurs normaux, à l'équation du 3ᵉ degré

$$\left(e - \frac{du}{dx}\right)\left(e - \frac{dv}{dy}\right)\left(e - \frac{dw}{dz}\right) - \left(e - \frac{du}{dx}\right)\left(\frac{\frac{dv}{dz} + \frac{dw}{dy}}{2}\right)^2 - \&c. = 0,$$

qui donne, dans les solides, les trois *dilatations principales*, et, dans les fluides, les trois vitesses d'extension principales.

Mais j'avoue ne comprendre pas bien sur quel principe vous

---

* Stated in § 15 of the memoir.   † See § 20 of the memoir.

vous basez pour y arriver ainsi. Je vois bien que si u, v, w sont les déplacements du point P dans les sens x, y, z (je dis *déplacements* comme s'il s'agissait d'un solide, parce que le raisonnement doit être le même que pour les *vitesses* dans un fluide), on a pour les déplacements relatifs des points P et P'

$$\frac{du}{dx}x' + \frac{du}{dy}y' + \frac{du}{dz}z', \quad \frac{dv}{dx}x' + \ldots, \quad \frac{dw}{dx}x' + \ldots$$

Je vois aussi qu' à ces déplacements relatifs provenant des déplacements absolus u, v, w, vous en ajoutez d'autres

$$\omega'''y' - \omega''z', \quad \omega'z' - \omega'''x', \quad \omega''x' - \omega'y'$$

provenant de trois petites rotations arbitraires ω', ω'', ω''' autour de Px, Py, Pz, ce qui vous donne pour les déplacements totaux

$$U = \frac{du}{dx}x' + \left(\frac{du}{dy} + \omega'''\right)y' + \left(\frac{du}{dz} - \omega''\right)z', \quad V = \ldots, \quad W = \ldots,$$

et que vous déterminez ces rotations arbitraires de manière qu'on ait

$$\frac{dV}{dz'} = \frac{dW}{dy'}, \quad \frac{dW}{dx'} = \frac{dU}{dz'}, \quad \frac{dU}{dy'} = \frac{dV}{dx'},$$

ce qui vous donne

$$\omega' = \tfrac{1}{2}\left(\frac{dw}{dy} - \frac{dv}{dz}\right), \quad \omega'' = \tfrac{1}{2}\left(\frac{du}{dz} - \frac{dw}{dx}\right), \quad \omega''' = \tfrac{1}{2}\left(\frac{dv}{dx} - \frac{du}{dy}\right),$$

en sorte que ω', ω'', ω''' pris en signe contraire sont, comme l'a démontré Cauchy au tome $2^{\text{ème}}$, page 321, des *Exercices d'analyse et de physique mathématique, les rotations moyennes* [*] du système autour de Px, Py, Pz en vertu des déplacements u, v, w, et comme on le démontre facilement d'une manière directe.

D'où

$$U = \frac{du}{dx}x' + \tfrac{1}{2}\left(\frac{du}{dy} + \frac{dv}{dx}\right)y' + \tfrac{1}{2}\left(\frac{dw}{dx} + \frac{du}{dz}\right)z', \quad V = \ldots, \quad W = \ldots;$$

expressions qui sont de même forme que les numérateurs des trois fractions algébriques, dont les dénominateurs sont x', y', z', et dont

---

[*] This reference to Cauchy, of date soon after 1840, does not according to St Venant carry priority in the application of the notion of the differential rotation in fluid motion, as he describes Prof. Stokes' analysis of 1849 (*Leçons de Navier*, p. 733) as 'nouvelle et remarquable.'

The definition of (ω', ω'', ω''') given by Prof. Stokes in § 14 in terms of the angular momentum in a small portion of the fluid that is spherical at the moment under consideration, and his indication of a proof that this angular momentum remains momentarily constant, carry us in fact to the very threshold of the fundamental theory of vortex motion discovered by Helmholtz in 1858.

la considération de l'ellipsoïde démontre l'égalité à chacune des trois dilatations principales e (*Exercices de Mathématiques* de Cauchy, 2ᵉ année, 1827, p. 63 et 68), en sorte qu'on a bien

$$\frac{U}{x'} = \frac{V}{y'} = \frac{W}{z'} = e,$$

équations entre lesquelles l'élimination de x', y', z' est facile et donne précisément l'équation du 3ᵉ degré en e; et les cosinus des angles des directions des trois dilatations principales avec x, y, z sont

$$\frac{x'}{\sqrt{x'^2 + y'^2 + z'^2}}, \quad \frac{y'}{\sqrt{}}, \quad \frac{z'}{\sqrt{}}.$$

Mais je ne vois pas, Monsieur, dans votre raisonnement le motif de tirer cette conclusion (sans doute faute d'avoir pu l'étudier suffisamment). La justesse du résultat prouve la justesse du raisonnement, mais j'avoue ne pas le comprendre. Je ne vois pas comment, lors que les conditions

$$\frac{dV}{dz} = \frac{dW}{dy}, \quad \frac{dW}{dx} = \frac{dU}{dz}, \quad \frac{dU}{dy} = \frac{dV}{dx},$$

sont remplies, c'est à dire lorsque les *rotations moyennes* autour de Px, Py, Pz sont nulles, les quotients

$$\frac{U}{x'}, \quad \frac{V}{y'}, \quad \frac{W}{z'}$$

doivent donner * la valeur d'une des dilatations principales, et x', y', z' doivent être proportionnels aux cosinus des angles de sa direction avec les x, y, z.

Veuillez excuser, Monsieur, l'importunité de ma demande. Mais l'intérêt que ces matières là ont pour moi et l'habilité avec laquelle vous les traitez me fait beaucoup désirer de bien comprendre les considérations que vous présentez.

Veuillez agréer, Monsieur, l'expression de mes sentiments de haute considération.

de St VENANT,

auteur de divers mémoires (entre autres, sur la torsion, *savants étrangers*, tome XIV, et Sur la flexion, *Journal Liouville*, 1856).

M. Stokes, de la Société Royale, professeur à Cambridge.

---

* This arises from a misinterpretation of the rather difficultly expressed argument in Prof. Stokes' § 2, in which the principal directions are determined from the fact that for points along each of them the relative displacement is radial.

P.S. Je ne sais si vous avez lu une *Note sur la dynamique des fluides*, insérée aux Comptes rendus des séances de l'Académie des Sciences, le 27 novembre 1843, tome XVII, page 1240. J'y arrive par une autre hypothèse aux mêmes équations que vous *.

P.S. Est-il possible, Monsieur, de se procurer chez un libraire votre mémoire de 1845 sur les fluides et les solides élastiques, sans acheter tout le volume de 1849 de Cambridge?

While Prof. Stokes was occupied in 1862 in mapping out ultra-violet spectra by fluorescence† he discovered that Prof. W. Allen Miller had been going over the same ground by photography. As these were very early examples of the determination of spectra, and are described in Prof. Kayser's *Handbuch* in connexion with the history of the subject, the following letters seem deserving of record.

FROM PROFESSOR W. ALLEN MILLER.

KING'S COLLEGE, LONDON.
March 2/62.

I am rather sorry to find we have been so closely treading on each other. I have, since the meeting at Manchester, been following up the photographic examination of the spectrum under a variety of circumstances, and have obtained photographs of the spectra produced by most of the common metals, as well as of a large number of salts, altogether nearly 150 of one sort or other; but I had cut out work in this direction which I thought would occupy my leisure for three or four months longer.

If you propose to read your paper on Thursday I should like to add a short one, and would bring down my photographs, though I should certainly have wished to have had my experiments more complete. Do not however let me *in the least* influence you, as you are ready and I am not; only if you do read your notice I will add a short statement of what I have been about, and will show some of my results. I am very much obliged to you for so kindly mentioning your intentions to me.

---

* Referred to in Prof. Stokes' *Report on Hydrodynamics*, 1846.

† 'On the long spectrum of the Electric Light,' *Phil. Trans.* 1862, read June 19; *Math. and Phys. Papers*, Vol. IV. pp. 203–233, see p. 208 in above connexion.

April 25, 1862.

I am looking forward to the completion of my paper in time for the present session. Pray do not allow me in the least to interfere with your own wishes. I am afraid I have already occasioned you inconvenience by your kind consideration for me. I am working on steadily, multiplying my experiments, and improving the definition of some of my spectra.

Whichever way you decide, will you give me a few days' notice when you propose to bring the matter forward, as I should like even if I am not finished to give a short note for the proceedings?

I have not succeeded in getting a good impression of the highly refrangible lines of aluminium, but have not made any trials with that special object. With zinc and cadmium I obtain all you have sketched, and several others which you have, I suppose, seen, but purposely omitted.

I have obtained also clear actinitic spectra of tellurium and of arsenic, and indeed of all the metals which can be easily procured. I find the subject expand as I work upon it, and think it will be better to subdivide my results, and perhaps give *two* papers instead of one. If, as I hope, the spectra will print pretty sharply, a good many illustrations will be needed, indeed they constitute the interest of the whole.

I find hydrogen, nitrogen, oxygen, carbonic acid, and carbonic oxide are all *diactinic* in moderate thicknesses. A bad term, but one wants something to express transparency to chemical action and to rays which produce fluorescence. Can you suggest anything better?

April 29, 1862.

I shall be happy to wait till the end of the session.

I have already obtained a great number of results on the transparency of different media, such as water, alcohol, ether, bisulphide of carbon, oil of turpentine, chloroform, benzol, &c., besides solutions of various salts in water, including sulphates, nitrates, sulphites, acetates, oxalates, chlorides, tartrates, cyanides, and solutions of all the ordinary bases and acids. Most of these I had done before I spoke to you on the subject; I have since repeated a few where I had doubt. I quite agree with you as to the superiority of the fluorescent observations in these cases, and had I been aware at first how easily they are made, should not have wasted time in photographing them. I have got about 200 photographs of different media, having commenced with a view of getting a good dispersive and non-absorbent material for a photographic prism. I think that by multiplying my experiments I have, although by perhaps an unnecessarily laborious method, got over the difficulty of impurities—thus I have compared carbonates, sulphates, chlorides, iodides, bromides and fluorides of the same bases, and have obtained, where the acids were transparent, concurrent results in most cases. My experiments have generally been made on saturated solutions. Besides these experiments I have made a series on reflections from metallic and other surfaces, on transmitted light through various gases, on sparks produced in various gases, and on spectra from lights of different kinds; the latter are at present very incomplete. I have still a large number of experiments on hand.

Perhaps I shall see you at the R. S. on Thursday. I hope we shall not be obliged to send in our papers on the last day, "the murder of the innocents," as one of our friends facetiously terms it.

Aug. 23, 1862.

I am much obliged to you for the trouble you have taken respecting my paper. Do not let yours be delayed for me, as it was owing to my fault in not completing the MS. that the report was not ready earlier.

I have been arranging the different spectra for the engraver. Two plates will, I think, be required, including between 50 and 60 spectra; 28 or 30 will go on a quarto plate, in double rows made the long way of the plate.

I was anxious to have seen proofs of the photographs from the engraver. I do not think he will have much difficulty in copying them accurately, but I think I should see proofs from time to time.

I have no experience of Bunsen's Battery; I always use Grove's, and am quite satisfied with it. The cost is not much greater, and it is more durable and very easy to keep in order. I am told the connections soon get eaten through in Bunsen's, and the spongy charcoal retains the nitric acid, and gives off fumes slowly when it is dismounted.

Dr Andrews of Belfast however uses Bunsen's.

Of interest about this time, as indicating the extent of his scientific acquaintance, is a list of persons who were to receive private copies of the British Association Report, 1862, on Double Refraction.

Adams	Ferguson	Latham	Sedgwick
Ainslie	Ferrers	Lloyd	Sharpey
Airy	Fischer	Lubbock	Smith
Andrews	Forbes	Lunn	Spottiswoode
Arlett	France		Stoney
Akin	Frost	Mathison	Sykes
	Fuller	Maxwell	Sylvester
Babbage		Miller	Stevelly
Beatson	Gaskin	Miller, W. A.	Stokes, W. H.
Blomfield	Gray	Moon	Stokes, H. G.
Boole	Goodwin		Stokes, Mrs
Brayley	Griffin	Newman	Stokes, Archd.
Brewster	Grove		Stokes, Hudleston
Brougham	Guest	Phear	
Budd		Phelps	Tait
	Hamilton	Phillips (Pres. Qu.)	Thomson, J.
Campion	Haughton	Power, Ja.	Thomson, W.
Cartmell	Hays	Power, Jn	Thurtell
Cayley	Hennessy	Potter	Todhunter
Challis	Herschel	Potts	Tozer
Cookson	Hind	Preston	Tyndall
Cory	Holdich	Price	
Cowie	Hopkins		Walker
		Rankine	Walton
De Morgan	Jellett	Robinson	Wheatstone
Donkin	Joule	Romer	Whewell
Drozier		Rosse	Willis
	Kelland	Routh	Wolstenholme
Earnshaw	Kingsley	Russell	Worcester
Ellis (Sidney)	Kirkman		Wrottesley
		Sabine	
Faraday	Lamb [Corpus]	Savage	

Prof. Stokes was a member of the committee appointed by the War Office in Feb. 1864, at the instance of the British Association, "to inquire into the properties of gun-cotton as a substitute for gunpowder for military and mining operations." In their Report (Dec. 1867) they say that "The general result of the series of experiments thus instituted by the committee cannot be better stated than is done in the concluding portion of Prof. Stokes' report" on pp. 15—17, which in five propositions summarises the results obtained by gun-experiments.

In connexion with Prof. Stokes' discoveries on the mode of oxidation of blood, the following letters from Dr Hoppe-Seyler seem of permanent interest. In reply to a request for advice, Dr F. G. Hopkins, F.R.S., sends the following remarks on them.

" Points to note with regard to the letters and their references to Stokes' work seem to be these. In the first letter Hoppe appears to claim knowledge of the effects of reduction upon oxyhaemoglobin, though he had not published observations. It is remarkable however that two years had elapsed from the time when H. described the spectrum of oxyhaemoglobin before Stokes' paper appeared; and the proof of the peculiar relationships of the pigment to oxygen was of such fundamental importance that it is hard to believe that Hoppe, if fully in possession of the facts, would have withheld publication. One cannot but agree with Gamgee (Schäfer's *Text-Book*, Vol. I. p. 208) that Stokes' 'combination of chemical and optical methods in this research... shed a flood of light on phenomena which had till then been shrouded in darkness....' Indeed, Hoppe says in the letter that his unpublished researches were not identical with those of Stokes, and I suspect they had not led to any such definite results.

"I am not sure whether the *Phil. Mag.* paper (Nov. 1864) contained the whole matter of the *Proceedings Roy. Soc.* paper*. H. appears to have received the former from S. If it does, it is odd that Hoppe should make no reference in the letter to Stokes' discovery of 'red haematin' (haemochromogen). This is a derivation which took a prominent place in his subsequent work and writings.

"As to Hoppe's second letter, it should have removed all doubt from the English worker's mind concerning the identity of 'blood crystals' and the pigment. It advances evidence which has never been shaken."

Tübingen, Württemberg,
1 *December* 1864.

Hochgeehrter Herr,

Veranlasst durch Ihre interessanten Mittheilungen im *Philosophical Magazine*, November 1864, p. 391, erlaube ich mir Ihnen beiliegende kurze Publicationen von mir zu übersenden, da ich aus Ihren Arbeiten ersehe, dass meine neueren Mittheilungen

---

* Republished in full in *Phil. Mag.*, Nov. 1864.

Ihnen unbekannt gewesen sind. So wenig ich auf den Namen Haemoglobin bestehe, so halte ich ihn doch, nachdem ich den Körper chemisch ausreichend untersucht habe, wegen seiner Spaltungsproducte und seines Vorkommens für den geeignetsten. Die von Ihnen geschilderten Veränderungen des Haemoglobins durch Reduction sind mir wohlbekannt, aber noch nicht veröffentlicht. Meine Versuche, die mit den Ihrigen nicht identisch sind, habe ich jetzt zur Veröffentlichung abgesendet. Die Gegenwart des reducirten Blutfarbstoffs im venösen Blute lässt sich durch Untersuchung der vom Blute am wenigsten absorbirten Lichtstrahlen leicht nachweisen, während zahlreiche deutsche Arbeiten über den Sauerstoffgehalt des venösen Blutes bei verschiedenen Zuständen Gehalt dieses Blutes an sauerstoffhaltigem Haemoglobin, von mir *Oxyhaemoglobin* genannt, bereits mit derselben Sicherheit nachgewiesen haben, als es Ihre und meine optischen Untersuchungen des Blutes gethan haben. Ich bin mit der Untersuchung des Blutfarbstoffs noch unausgesetzt beschäftigt und hoffe im nächsten Jahre meine Arbeiten *in extenso* herausgeben zu können; ich werde mir dann erlauben Ihnen dieselben zuzusenden. Ich bedaure sehr, dass ich keinen Separatabdruck meiner zweiten Mittheilung über den Blutfarbstoff mehr besitze, dieselbe befindet sich in Virchow Archiv für pathologische Anatomie etc., Band 29, pag. 233; die übrigen bis jetzt von mir veröffentlichten Notizen lege ich bei. Schliesslich bitte ich zu entschuldigen dass ich Ihnen deutsch schreibe; ich wäre im Englischen vielleicht weniger verständlich gewesen.

Mit ausgezeichneter Hochachtung,

HOPPE-SEYLER.

HOCHGEEHRTER HERR,   31 *März* 1865.

Indem ich Ihnen für Ihr geschätztes Schreiben und übersendeten Schriften meinen besten Dank sage, kann ich nicht umhin Ihnen einerseits mein Bedauern darüber auszudrücken, dass Sie die Krystallsubstanz der Blutkrystalle für etwas vom Farbstoffe des Blutes verschiedenes halten, und andererseits meine wesentlichsten Gründe aufzuzählen, warum ich diese Identität behaupte. Ich bin völlig einverstanden damit, dass der Name Haemoglobin, den ich aus der Bezeichnung von Berzelius Haematoglobulin abgekürzt habe (Berzelius hatte bereits erkannt, dass

Haematin mit einem Eiweissstoffe, Globulin, im Blute in
Verbindung sei) nur Gültigkeit haben kann für den Begriff, den
ich damit verbunden habe; ich habe damit die Krystalle des
Blutes bezeichnet, die auch abgesehn von den optischen Eigenschaften sich von allen andern Körpern chemisch völlig unterscheiden.

Dass nun speciell der Farbstoff den Krystallen zugehört
ergiebt sich aus folgenden Verhältnissen :

1. Die Krystalle können noch so oft 6—8 mal umkrystallisirt
werden, sie behalten gleichen Gehalt an Eisen und gleiche
Intensität der Färbung ihrer Lösung, so lange sie unzersetzt
sind.

2. Eine warm gesättigte Lösung der Krystalle in Wasser
giebt beim Erkalten Krystallisation von demselben Verhältniss
der Farbenintensität zum festen Rückstand wie die Mutterlauge.
Beim Verdunsten liefert die Mutterlauge reine Krystalle, nichts
ausserdem.

3. Jede chemische Einwirkung, welche die optischen Eigenschaften des Farbstoffs ändert, hebt auch die Krystallisirbarkeit
auf. Ausgenommen sind hier Kohlenoxyd und reducirende Stoffe
wie Schwefelammonium, welche die Krystallisirbarkeit nicht auf-
heben aber auch den Farbstoff nicht zerstören. Das Haemoglobin
krystallisirt nach Behandlung mit Kohlenoxyd leichter als ohne
dies, mit Sauerstoff leichter als ohne diesen.

4. Wenn man durch verdünnte Säuren oder durch Stehen-
lassen der feuchten Krystalle über 0° den Farbstoff sich ändern
lässt, so entstehn zugleich mehrere Eiweissstoffe.

5. Alle bekannten thierischen Farbstoffe und ebenso alle
Albuminstoffe sind fällbar durch Acetum Plumbi und Ammoniak,
nur nicht der Farbstoff des Blutes, i.e. seine Krystallsubstanz, so
lange sie unzersetzt ist.

6. Bei der spontanen Zerlegung, Erhitzen, Behandlung mit
Schwefelwasserstoff ändert sich stets der Farbstoff während die
Krystallisirbarkeit und die chemischen Eigenthümlichkeiten dieser
Substanz verloren gehn. Unter 0° getrocknet wird beim Erhitzen
über 100° weder die Krystallisationsfähigkeit noch das Verhalten
im Spectrum geändert.

Woher kommen nun die Angaben von Lehmann, Funke und
Anderen, dass man die Blutkrystalle farblos erhalten könne ?
Alle diejenigen, welche dies behaupten, haben diesen Körper nicht

genügend gekannt. Lässt man Blutkrystalle feucht einige Zeit stehn und spült sie dann mit Wasser ab, so bleibt ein fibrinartiger Körper mit Spuren von Haematin in der Form der Krystalle zürück, während Haematin, Albumin und unzersetztes Haemoglobin gelöst werden. Dies ist die einzige Erklärung die ich für derartige Angaben kenne und doch glaube ich nicht dass Jemand in dem Umfange die Bildung und Eigenschaften der Blutkrystalle untersucht hat als ich seit 15 Jahren, seit ich mit meinen Freunden Kunde und Funke bei Lehmann sie untersuchte. Ueber die Bildung der Krystalle sind gleichfalls ganz irrige Vorstellungen allgemein. Das Blut einer grossen Anzahl Thiere krystallisirt ohne Weiteres, wenn man die Blutkörperchen in Wasser löst, mit Aether schüttelt, filtrirt, das Filtrat mit ¼ Volumen Alkohol versetzt, und 24 Stunden bei −1° bis −10° C. stehn lässt.

Die Beobachtungen von Senarmont über künstlichen Pleochroïsmus waren mir wohl bekannt; ich habe deswegen diese Eigenschaft der Blutkrystalle nie als Beweis davon angesehn, dass der Farbstoff selbst krystallisirt sei.

Ich habe noch hinzuzufügen, dass die Blutkrystalle, besonders wenn sie klein sind, hell aussehn, weil sie mehr Krystallwasser enthalten als Wasser in den Blutkörperchen enthalten ist; sie werden aus den Blutkörperchen durch Lösung in Wasser oder im wasserreicheren Serum erhalten und enthalten etwa ebensoviel Krystallwasser als Haemoglobin. So lange man mir nun nicht farblose Krystalle aus dem Blute vorlegt (und zwar nicht mikroscopisch) die man ohne Zerlegung umkrystallisiren und auf ihre Eigenschaften untersuchen kann, so wie ich es mit dem mehrmals umkrystallisirten Haemoglobin gethan habe, sehe ich mich genöthigt alle Angaben über farblose Blutkrystalle für Täuschungen, die auf Verwechselung mit andern Körpern, Verunreinigungen oder Zersetzungen beruhen, zu halten. Lehmann ist aus dem Grunde mit der Untersuchung dieser Krystalle nicht vom Flecke gekommen, weil er stets darauf ausging, farblose Krystalle zu erhalten und weil er nicht die Zersetzungen des Farbstoffes in ihrem Einfluss auf die Krystallisirbarkeit der Substanz beachtete.

Da nun die sämmtlichen von mir angegebenen Eigenschaften des Blutfarbstoffs, die sich bei der Spectraluntersuchung zeigen, den Blutkrystallen eigen sind und der Körper, welcher sie zeigt, von mir Haemoglobin zuerst genannt ist, so würde nach allge-

meinem Brauche erst dann die Bezeichnung Cruorin anwendbar
sein, wenn nachgewiesen würde, dass derjenige Körper, welchen ich
als Haemoglobin bezeichnet habe, ein Gemenge mehrerer Körper
sei; bis dahin wäre ich durchaus nicht einverstanden etwa die
Sache *in suspenso* zu lassen, denn ich halte einmal meine Untersuchungen für mehr werth als dass man sie wegen Zweifel, die ich
längst selbst durchgemacht habe und deren Unhaltbarkeit optische
und chemische Untersuchungen in Gemeinschaft zeigen, einfach bei
Seite schiebt.   Eine künstiche Darstellung von Blutfarbstoff aus
Haematin werde ich nicht versuchen, dagegen sind ausser mir
mehrere Practicanten in meinem Laboratorium mit der weiteren
Untersuchung des Haemoglobin beschäftigt.   Sobald ich Zeit
finde, werde ich meine Untersuchungen über diesen Körper, von
denen ich bis jetzt nur einige wichtige Resultate veröffentlicht
habe, im Zusammenhange ausführlich publiciren.

Hinsichtlich der Unterscheidung der Uebermangansäure
bedaure ich sehr Ihre berühmte Arbeit von 1852 in den Philosophical Transactions nicht in dieser Beziehung in der Erinnerung
gehabt zu haben; Sie haben in derselben bereits das Spectrum des
Uebermangansäure genau beschrieben.   Dagegen richtete sich
meine Notiz gegen eine erst 1858 erschienene Arbeit von H. Rose,
worin derselbe sich auf das phosphorsaure Manganoxyd stützt
und von diesem bis jetzt einzig in Lösung erhaltenen reinen
Manganoxydsalze sagen Sie nichts.   Dass meine Notiz nicht
überflüssig war, zeigt die Verbreitung, welche dieselbe gänzlich
ohne mein Zuthun erhalten hat.   Ich hoffe bald Gelegenheit zu
haben, in einer zu publicirenden Arbeit Ihre Priorität hinsichtlich
des Uebermangansäurespectrum zu wahren.

Einige neuere Mittheilungen über den Blutfarbstoff erlaube
ich mir Ihnen durch Post zu übersenden.

Mit grösster Hochachtung,
HOPPE-SEYLER.

The following undated memorandum, which appears to contain
a plan for experiments, has been found recently among Prof. Stokes'
papers.

BLOOD.

If scarlet haematoglobulin be reduced to purple by the iron
salt and then acidified by tartaric (suppose) acid [compare the
spectrum with scarlet haematoglobulin acidified and with haematin]

ammonia may *then* be added with little [try whether so, if *all* access of oxygen be cut off] precipitation. The spectrum is very marked, and differs much from that of purple haematoglobulin, ‖ ‖‖ instead of ‖‖‖‖. The substance readily absorbs oxygen, and then gives *very nearly* the spectrum of scarlet haematoglobulin; but on reduction again the former spectrum and not that of purple haematoglobulin appears.

[Try carbonic acid on a putrid solution shaken with air till the colouring matter is scarlet.]

Protosulphate of iron [try with a neutral tartrate] does not reduce scarlet haematoglobulin, but if a little tartaric acid be added, which gives a haematin-like spectrum, and then ammonia, the ‖ ‖‖ band substance, and not purple haematoglobulin, is produced, though there is besides some precipitation.

This looks as if scarlet haematoglobulin at the moment when it *tended* to split under the action of an acid was easily reduced, and thereby escaped splitting, being only modified.

Also as if in the actual splitting oxygen was given off (agrees with an experiment by Mayer, *Pogg. Ann.*).

The Presidential Address of Gen. Sabine on Nov. 30, 1864, contains a long account (*Roy. Soc. Proc.* XIII., pp. 505—510) of the work of Mr Darwin, in connexion with the award of the Copley Medal. The terms used in referring to *The Origin of Species* gave rise to vigorous protest, which is still remembered. The following paragraph of the Address (p. 508) is the one referred to in the letters printed below:

In his most recent work *On the Origin of Species*, although opinions may be divided or undecided with respect to its merits in some respects, all will allow that it contains a mass of observation bearing upon the habits, structure, affinities, and distribution of animals, perhaps unrivalled for interest, minuteness, and patience of observation.   Some amongst us may perhaps incline to accept the theory indicated by the title of this work, while others may perhaps incline to refuse, or at least to remit it to a future time, when increased knowledge shall afford stronger grounds for its ultimate acceptance or rejection.   Speaking generally and collectively, we have not included it in our award.   This on the one hand ; on the other hand, I believe that, both collectively and individually, we agree in regarding every real *bond fide* inquiry into the truths of nature as in itself essentially legitimate; and we also know that in the history of science it has happened more than once that hypotheses or theories, which have afterwards been found true or untrue, being entertained by men of powerful minds, have stimulated them to explore new paths of research, from which, to whatever issue they may ultimately have conducted, the explorer has meanwhile brought back rich and fresh spoils of knowledge.

Mr Francis Darwin, who has kindly given information on this subject, points out that in the *Life and Letters of Charles Darwin*, Vol. iii., p. 29, the quotation was made from *The Reader*, which has the words "we have expressly omitted it from the grounds of our award," instead of the words "we have not included it in our award," which appear in the authentic text quoted above.

A fragment of a MS. draft exists, apparently Gen. Sabine's, in which the words occur as in *The Reader*: but an addition to the first sentence of the extract quoted above is deleted, which ran as follows; "and they are distinguished by the same impartial judgment and the same extraordinary power of generalisation by which Mr Darwin's labours in every branch of Natural History are characterized."

The subject occurs also in the *Life and Letters of T. H. Huxley*, Edit. 1, Vol. i., p. 255, where a letter is given in which Prof. Huxley expresses his distrust of Gen. Sabine.

The replies from Huxley that are printed below have been taken from his own rough drafts, to which he must have attached some importance, as they were confided by him many years afterwards to the care of Mr F. Darwin.

LENSFIELD COTTAGE, CAMBRIDGE.

December 5th, 1864.

DEAR HUXLEY,

It seems that my ears were right after all, in spite of the concurrent testimony of three or four of you. The words "expressly excluded" do *not* occur in the address, so that Falconer need not have been so ready to attribute his not hearing them to his deafness, and I my not hearing them to the physical exertion consequent on continued loud reading after a bad cold. General Sabine has sent me both the page of his original MS. and the printed proof from which I read, and the words ran thus. After referring to the work on the origin of species, and saying that notwithstanding some difference of opinion "all will allow that it contains a mass of observation bearing upon the habits...of animals perhaps unrivalled for interest, minuteness, and patience of observation," and saying that some among us might be inclined to accept and others to refuse or postpone the acceptance of the theory contained in it, he adds, "Speaking generally and collectively, we have expressly omitted it from the grounds of our award." Now this is a plain matter of fact; the work was not, could not well have been, ignored before the Council. Darwin's supporters "expressly omitted" this work, about which there was, as we know, much difference of opinion, from among the grounds of the award. His claims were put forward with the distinct and avowed omission of this work.

In the case of a minor work it would be sufficient to pass it *sub silentio*. But that could not be done in the case of a work which has made so much stir as *The Origin of Species*. All the world would say the Medal had been given for that book, that it had been endorsed by the formal and solemn approval of the Royal Society.

Now there is a wide interval between putting a book in an *index expurgatorius* and declining to honour it with the Copley Medal of the Royal Society. Had no notice been taken of the express omission which as a matter of fact was made, any member of the Council would have a just cause of complaint, that the understanding on the strength of which alone, it may be, he voted for Darwin, namely that the Council should not be considered as committed in any way to that particular book, had been practically violated.

On the whole I submit:

1. That as a matter of fact the work on the Origin of Species was "expressly omitted" from the grounds on which the award of the Medal was asked to be made.

2. That good faith and truthfulness require that this omission should be stated.

3. That the words used imply this and no more, and neither were intended nor are calculated to have the effect of placing the work in a sort of *index expurgatorius*.

Yours very truly,

G. G. STOKES.

December 5th, 1864.

DEAR HUXLEY,

There was an expression in my last which is not perhaps rigorously exact. Instead of saying "declining to honour it with the Copley Medal of the Royal Society" I should have said simply "not honouring it with the Copley Medal of the Royal Society." For declining implies having been asked, and there was no asking in the present case.

Yours very truly,

G. G. STOKES.

MY DEAR STOKES,

I never had a clearer and more distinct impression in my life than that I heard the words " expressly excluded "; and at the risk of seeming presumptuous I would suggest that it is just possible that you may have inadvertently substituted the one word for the other, the sense of the two being for all practical purposes identical. But this is merely for your consideration; I quite admit that we are bound by the printed words, as they express what the writer meant to say, which is the only point of importance.

At the same time I disclaim all responsibility for the mistake, if there has been one. I stated openly and fairly what my own impression was, and if I erred it was for the President or yourself to correct me. The document was before you, and if any error became current through what I said, it really is not my fault.

However, I do not see that it makes any difference whether the words used were "expressly omitted" or "expressly excluded."

Not to fight about words let me briefly state my own position to you. As a Fellow of the Society I object to the following passage in the address :

"Speaking generally and collectively we have expressly omitted it from the grounds of our award"

on the grounds that

1. It clearly implies not only that the omission, but that the formal and public notification of that omission, was the result of a distinct determination of the Council as a body, arrived at after due consideration. And further that Darwin's friends accepted the medal for him, clearly understanding that such public notification would be made.

Now I cannot find any such record of such a determination of the Council; nor, since I know that both the proposer and seconder of Darwin were exceedingly astonished and annoyed when they became aware of (I won't say heard) the paragraph in question, can I conceive that any tacit understanding on the subject should have been entertained by them. Had I been a member of the Council myself I would not have accepted the medal for Darwin if the insertion of that paragraph had been a condition of the award.

2. But, even supposing that the paragraph in question had been inserted with the full knowledge and consent of Darwin's supporters, and as the result of an express resolution of the Council, I (in my capacity as a Fellow of the Society) nevertheless object to it and protest against it. I do so because I hold that the address to a medallist should state only the grounds upon which the medal is awarded, and should not indulge in critical remarks either favourable or unfavourable on any other subject. I can conceive no worse precedent to be established than one which should justify some future President, who had strongly opposed a medallist, in inserting of his own motion a derogatory paragraph in his address.

I think that Darwin's supporters had a right to base their claims on whatever grounds they pleased, and to omit anything they pleased : and if any member of the Council objected to vote for Darwin, unless his theory of the origin of species was not only *omitted* from the award, but stigmatised by the fact of that omission being publicly affirmed to be the "collective and general" act of the Council, I think he was bound to move that words to that effect should form a part of the resolution awarding the medal, and fairly and openly take the sense of the Council thereupon. If this course had been taken I should have no less objected to the result, but my remarks would have been directed against the Council and not against the President, who would (under these circumstances and these only, in my opinion) have been fully justified in saying what he did.

As I have repeatedly stated in public, I should be sorry to stand committed to the opinion that the truth of Darwin's hypothesis is demonstrated, and therefore it is not likely that I should have wished the Royal Society to place itself either directly or indirectly in the same position.

But, if it were really necessary to allude to Darwin's theory for the purpose of preventing the public from supposing that the Royal Society awarded the Copley Medal for it, I should have had no objection to its being done in a proper way after due consultation with those who were interested.

What I do protest against is that, without the knowledge and consent of Darwin's proposer and seconder, a phrase should have been inserted which compresses the maximum amount of offence into the handiest possible form for Darwin's opponents.

Then to sum up I object to the phrase quoted

1. Because it purports to be the act of the Council "collectively and generally," whereas so far as the evidence before me goes it was not so, but the act of the President expressing what he conceived to be the opinion of some of the Council.

2. Because, if it was the act of the Council, it is irregular and inexpedient to introduce negative considerations into an address which is properly concerned only with the positive merit of a person.

I may be quite wrong in my views, but I wish you to see at any rate that they are clear and definite and not based on any mere captious criticism.

Indeed, I may say that I never more anxiously pondered what it was my duty to do, than on the other evening, but (with one exception of which your letter twice reminds me) I should take the like course under like circumstances.

That exception is the phrase " index expurgatorius," which escaped me unwittingly, and for which I expressed my regret to the Treasurer the same evening.

I desired to say nothing that should be in the smallest degree wanting in due respect for either the person or the office of the President: so much so that I did not take the course which was open to me, by moving an amendment that "we should collectively and generally expressly omit" the passage in question from the printed address.

By so doing I should have dealt as disrespectfully with the President as, in my judgment, he will do with Darwin if any part of that passage is allowed to stand. I am glad to have had the opportunity of making these explanations, with which you can deal as you please.

I am, Ever yours very faithfully,

T. H. HUXLEY.

LENSFIELD COTTAGE, CAMBRIDGE.

December 7th, 1864.

MY DEAR HUXLEY,

Of course I cannot say that it is mathematically certain that in reading from a printed paper I did not through inadvertence read " excluded " for " omitted "; but it is *most violently improbable* that I should have made such a change, and the President, knowing as he must his own words, would be sure to have noticed it. It does not seem to me at all improbable that in your recollection of what was rapidly said you should have substituted for one word another which *you* deem of similar import.

However, you agree to accept " omitted," and the question arises whether this correctly describes what passed.

Now by the rules of the Royal Society the award of the Medal is as you know made by the Council alone. Therefore the " we " cannot mean the Royal Society at large, but the Council. The question then is, are the words " we have expressly omitted " a correct account of what passed?

The proposal of the Copley Medal to Darwin was as you know made by Falconer, who rested his claims on other grounds. The work on the Origin of Species was however too important to be passed *sub silentio*, and therefore he gave reasons for not including that among the grounds of the award. When the matter thus came formally before the Council, the President from the Chair used words of this general purport: " Then we are to understand that the work on the Origin of Species is not included among the grounds of the award of the Copley Medal to Mr Darwin." This was assented to, and the Council considered and voted on Mr Darwin's claims accordingly.

Now there are two modes of omission, by silence or by express mention. If the naming of a work by the proposer of a medal and saying why he did not include it, followed by a formal statement made from the Chair and assented to, that that work was not included among Mr Darwin's claims for the Medal, be not an " express " omission, I don't know what is.

But what are Mr Darwin and his friends to understand by this " express " omission? That the Council have passed any censure on the book? No such thing; but only that, while the Council were not prepared, or at least were not asked, to reward it with the Copley Medal, it was too important to be passed *sub*

silentio, and *had* to be mentioned in order to guard the Council
from being committed to it

Yours very truly,

G. G. STOKES.

MY DEAR STOKES,

I have never imagined that the "we" referred to
anybody but the Council.

I thought that Darwin was proposed for the medal by Busk*
and not, as you say, by Falconer, but that is a matter of no
moment.

Most unquestionably I do *not* think that an assent by the
proposer (whomsoever) to such words as "Then we are to understand that the work on the origin of species is not included among
the grounds of the award of the Copley Medal to Mr Darwin,"
from the President, affords a justification for the mention in the
address of the passage complained of, or any allusion whatever to
that work.

Even if the informal and general assent of Darwin's proposer
to such a proposition could be held to bind the Council, and could
be interpreted as the "general and collective" determination of
the whole body (a monstrous supposition), all that it could justify
would be silence respecting the book.

How can an agreement to say nothing about a book justify
one of the parties to that agreement, in telling all the world that
it "contains many observations...unrivalled for interest, minuteness, &c."? See address.

If that which you have mentioned occurred, and I gladly take
your authority for it, the case is worse than ever in my appre-
hension.

To this letter Prof. Stokes replied (Dec. 8), admitting that "in
my last I made some confusion between Darwin's proposer and
seconder which I beg you to conceive corrected": he then passed
on to a reminder connected with proof sheets which, as it turned
out, were being delayed on account of engraving.

* He was proposed by Busk, seconded by Falconer. The latter was unable to be
present at the first Council meeting for the award of medals, and therefore wrote to
Dr Sharpey the Secretary (see "More Letters of Charles Darwin," I. 252); he speaks
of the Origin as a "strong additional claim."

Huxley's reply on the latter subject winds up as follows :

" Now it is my turn to have a little chaff, as we have taken to that line. Don't you think that ' As Falconer is to Busk so is excluded to omitted ' is a good rule of three sum ? And that our worthy and usually very accurate Secretary, who could make the one mistake in writing, might (without too great risk) be supposed to have made the other in reading ?

" The above problem in proportion is of course of a private character."

July 16. 1871

My dear Stokes

I have sent you your

in a bandbox

please let me have my

back of the same

conveyance Yours
J H Huxley

Soon afterwards, in 1872, Prof. Stokes gained Huxley as a colleague in the office of Secretary of the Royal Society, and the relations of the two men became closer, as is illustrated by the characteristic autograph above reproduced, the more intimate association detracting in no way from their feelings of mutual confidence and regard. See *supra*, p. 48.

FOWEY, CORNWALL.
Sept. 26, 1884.

MY DEAR STOKES,

I have written to Mr White to go on with Gore's paper.

I shall be back in town next week and shall then read the article you have sent me (for which accept my thanks in advance) with great interest.

These problems* have a perennial attraction for me. I began life by thinking about them, and if I live long enough to extricate myself from the distractions by which I find my time most unsatisfactorily frittered away at present, I shall end it in the same occupation.

Ever yours very faithfully,
T. H. HUXLEY.

The opposition, started by Huxley, to the President of the Royal Society holding a seat in Parliament, has already been referred to (*supra*, p. 102). A more complete account, taken from *The Life and Letters of T. H. Huxley*, Vol. II. pp. 173–6, is in place here.

"Thus there is a good deal in his correspondence bearing on this matter. He writes on November 6 to Sir J. Hooker:—

I am extremely exercised in my mind about Stokes' going into Parliament (as a strong party man, moreover) while still P.R.S. I do not know what you may think about it, but to my mind it is utterly wrong—and degrading to the Society—by introducing politics into its affairs.

"And on the same day to Sir M. Foster:—

I think it is extremely improper for the President of the R.S. to accept a position as a party politician. As a Unionist I should vote for him if I had a vote for Cambridge University, but for all that I think it most lamentable that the Presidency of the Society should be dragged into party mud.

* Probably those of Natural Theology.

When I was President I refused to take the Presidency of the Sunday League, because of the division of opinion on the subject. Now we are being connected with the Victoria Institute, and sucked into the slough of politics.

These considerations weighed heavily with several, both of the older and the younger members of the Society; but the majority were indifferent to the dangers of the precedent. The Council could not discuss the matter; they waited in vain for an official announcement of his election from the President, while he, as it turned out, expected them to broach the subject.

Various proposals were discussed; but it seemed best that, as a preliminary to further action, an editorial article written by Huxley should be inserted in *Nature*, indicating what was felt by a section of the Society, and suggesting that resignation of one of the two officers was the right solution of the difficulty.

Finally, it seemed that perhaps, after all, a 'masterly inactivity' was the best line of action. Without risk of an authoritative decision of the Society 'the wrong way,' out of personal regard for the President, the question would be solved for him by actual experience of work in the House of Commons, where he would doubtless discover that he must 'renounce either science, or politics, or existence.'

This campaign, however, against a principle, was carried on without any personal feeling. The perfect simplicity of the President's attitude would have disarmed the hottest opponent, and indeed Huxley took occasion to write him the following letter, in reference to which he writes to Dr Foster:—'I hate doing things in the dark and could not stand it any longer.'

Dec. 1, 1887.

MY DEAR STOKES,

When we met in the hall of the Athenæum on Monday evening I was on the point of speaking to you on a somewhat delicate topic; namely, my responsibility for the leading article on the Presidency of R.S. and politics which appeared a fortnight ago in *Nature*. But I was restrained by the reflection that I had no right to say anything about the matter without the consent of the Editor of *Nature*. I have obtained that consent, and I take the earliest opportunity of availing myself of my freedom.

I should have greatly preferred to sign the article, and its anonymity is due to nothing but my strong desire to avoid the introduction of any personal irrelevancies into the discussion of a very grave question of principle.

I may add that as you are quite certain to vote in the way that I think right on the only political questions which greatly interest me, my action has not been, and cannot be, in any way affected by political feeling.

And as there is no one of whom I have a higher opinion as a man of science—no one whom I should be more glad to serve under, and to support year after year in the Chair of the Society, and no one for whom I entertain feelings of more sincere friendship—I trust you will believe that, if there is a word in the article which appears inconsistent with these feelings, it is there by oversight, and is sincerely regretted.

During the thirty odd years we have known one another, we have often had stout battles without loss of mutual kindness. My chief object in troubling you with this letter is to express the hope that, whatever happens, this state of things may continue.

I am, yours very faithfully,

T. H. HUXLEY.

P.S.—I am still of opinion that it is better that my authorship should not be officially recognised, but you are, of course, free to use the information I have given you in any way you may think fit.

"To this the President returned a very frank and friendly reply; saying he had never dreamed of any incompatibility existing between the two offices, and urging that the Presidency ought not to constrain a man to give up his ordinary duties as a citizen. He concludes:—

And now I have stated my case as it appears to myself; let me assure you that nothing that has passed tends at all to diminish my friendship towards you. My wife heard last night that the article was yours and told me so. I rather thought it must have been written by some hot Gladstonian. It seems, however, that her informant was right. She wishes me to tell you that she replied to her informant that she felt quite sure that if you wrote it, it was because you thought it.

"To which Huxley replied:—

I am much obliged for your letter, which is just such as I felt sure you would write.

Pray thank Mrs Stokes for her kind message. I am very grateful for her confidence in my uprightness of intention.

We must agree to differ.

It may be needful for me and those who agree with me to place our opinions on record; but you may depend upon it that nothing will be done which can suggest any lack of friendship or respect for our President."

HODESLEA, STAVELEY ROAD, EASTBOURNE.
Oct. 26, 1891.

MY DEAR STOKES,

Very many thanks for the copy of *Natural Theology* which has just reached me. I see it will need careful reading. I hope the report I saw the other day that you were not going to stand for the University again may be true. Scientific razors are too good for chopping parliamentary blocks!

Ever yours very faithfully,

T. H. HUXLEY.

From time to time letters passed between Charles Darwin and Prof. Stokes on philosophical points or on scientific business. The following letters will serve as illustrations, that from Prof. Stokes having been kindly supplied by Mr F. Darwin.

6, QUEEN ANNE ST, LONDON, W.
Thursday*.

MY DEAR PROF. STOKES,

Absence from home has prevented me from sooner thanking you most sincerely for the trouble which you have so kindly taken for me. I was rather crazy with curiosity to know what the chances were. I believe your way of stating the problem is rather better for me. I think I understand your two letters. The second way of calculating the case is much the best for me. I have made a copy for myself of your MS. sentence and have altered the few words and figures which are necessary. I cannot suppose that I have made any blunder; so if I do NOT receive your sentence back, I shall understand that it is right.

With my sincere thanks, pray believe me,

Yours truly obliged,

CH. DARWIN.

DOWN, BROMLEY, KENT, S.E.
Feb. 18*th* [before 1870].

MY DEAR SIR,

I have ventured to send my son to you to obtain a little information for me, on one point, if in your power to give it, and by this means you will be saved the trouble of answering this note. Have you ever attended to feathers, and can you tell me

---

* Mr F. Darwin points out that this letter, which he regards as characteristic, must be dated before 1868, as it refers to a passage in *Variation of Animals and Plants*, Ed. i. 1868, Vol. II. p. 5. Prof. Stokes had supplied a calculation as to the chance of transmission of abnormalities in man.

whether the splendid colours of the eye of a Peacock's tail depends on colouring matter, or on reflection ?   If on the latter, as appears the case, I much want to know, whether any change of structure, —as the distance of a film, or the distance of fine lines or points from each other—gradually, but perhaps not equally, increasing or diminishing, would account for the series of colours, which surrounds the eye, and passes into the general tint of the barbs at the circumference of the feather.   Will you be so kind as to look at the feather, and tell my son anything you can ?   Pray forgive me for troubling you and believe me,

My dear Sir, yours sincerely,

CH. DARWIN.

Feb. 28*th.*

I am very much obliged to you for your great kindness in writing to me at such length about the colours of the peacock's feathers.   As you say that you will look at it again, will you have the kindness to attend to one point, namely, whether a gradual thickening or thinning by little steps from the centre to the circumference of the film of colouring matter would account for the zones of colour which occur ; or must there be zones of different kinds of colouring matter ?

With very sincere thanks, believe me,

My dear Sir, yours truly,

CHARLES DARWIN.

LENSFIELD COTTAGE, CAMBRIDGE.

20*th Dec.* 1875.

MY DEAR SIR,

You may remember that some years ago you asked my opinion as to the cause of the colours in peacocks' feathers. I made some experiments, or rather observations hardly deserving the name of experiments, about it.   I felt however that it was a matter hardly to be attacked without a thoroughly good microscope which I did not then possess.   I expressed a leaning to the opinion that the colours were reflection-colours connected with intense absorption, and similar accordingly to the reflection-colours seen when some of the aniline colouring matters are poured in solution on glass, and the solvent allowed to evaporate.   I felt at the time I think misgivings as to whether so much play of colour as is observed could thus be accounted for.

As I have helped to lead you wrong if you followed my guidance, it is but just that I should direct you to a right solution. Some little while ago I was with Mr Sorby, who wished to show me some other experiments, and I found that he has been studying birds' feathers, as to their colours, and had arrived at the solution of the cause of the play of colours in the feathers of peacocks, humming birds &c. I am not sure whether he has yet published his results. If he has, I have no doubt he will be happy to send you copies of his papers if he has not done so already*. If not, I dare say he would explain to you in a few words their general nature. Though I feel confident of his permission, I refrain as a matter of principle from communicating to a third person what has been told me by a friend of his unpublished researches. Besides it is due to him not to spoil the pleasure he would feel in communicating to you for the first time his own results.

I am, dear Sir, Yours sincerely,

G. G. STOKES.

CHAS. DARWIN, Esq., F.R.S.

METEOROLOGICAL COUNCIL.

In 1866 Prof. Stokes became one of the original members of the Meteorological Council, then and until recently constituted as a Committee of the Royal Society which administered the meteorological affairs of the United Kingdom by means of a Treasury Grant in Aid. The Council was in the following years a very active body. His letters and papers contain masses of material relating to its work. When new projects were under consideration, he was subject to constant reference from the Office on all points of difficulty: his conscientiousness on these matters led him to put aside his own investigations in order to discharge to the full the public duty that he had assumed. As illustrations may be mentioned, the organisation of a plan of systematic observation of the altitude of clouds in 1885; a very interesting correspondence on the observation of trains of waves at sea and the deductions that may be made relating to the distant storms that originated them; the earliest practical construction and

---

* In the Royal Society Catalogue there are several entries of papers about this time, on the *colouring matters* of algae, black feathers, etc., by Dr H. C. Sorby, but apparently no paper relating to feathers of peacocks or humming birds.

testing of harmonic analysers; the improvement of anemometers, in which he was deeply implicated through the researches of his father-in-law Dr Robinson; the measurement of solar radiation; and many other topics. Correspondence on all these subjects is given *infra*, Vol. II.

The following letter from Prof. H. J. S. Smith, of Oxford, who became the first Chairman, testifies to the importance that was attached from the beginning to Prof. Stokes' cooperation.

UNIVERSITY MUSEUM, OXFORD,
April 11, 1866.

DEAR PROFESSOR STOKES,

I am well aware that Dr Hooker has already addressed you on the subject of the proposed Meteorological Council, and that you have given strong and well considered reasons for being reluctant to join it.

I venture nevertheless, though with great diffidence, to write to you upon the same subject.

Perhaps you may have heard from Dr Hooker that he intends (as at present advised) to suggest to the Council of the R. S. that I should become the Chairman of the proposed Council (supposing the present negotiations to terminate successfully).

I believe it is contemplated that three of the present Committee—De la Rue, F. Galton, and Gen. Strachey—should serve on the new Council; and certainly a more fitting selection could not be made.

If I became Chairman there would be but one vacancy remaining. This, I am strongly convinced, should be filled by an eminent Physicist. That is to say, if by any means it be possible, we should have you.

I understand your main objection to be that you might be taken away from Cambridge more than you would think right. The meetings would be fortnightly (I am informed), not weekly: they could easily be placed so as to suit your convenience with regard to the R. S.; and I do not think they need prolong materially your absences from Cambridge. The points, on which your advice would be urgently needed, would not be points of administration or routine (with such men as the three I have mentioned on the Council, I think you might be effectually shielded from such work). Your opinion would be wanted on scientific questions, and especially as to the scientific enquiries which ought to be undertaken by the office, and as to the persons who ought to undertake them. Such advice you could give without quitting Cambridge. I might add that (if I were Chairman) I should think nothing of coming to Cambridge to obtain it, only I am afraid that the prospect of such visitations might have a deterrent effect on you.

You will notice that I take a strong personal interest in the matter, and this is but natural, because, although there is only a possibility of my being the Chairman, that possibility alone is enough to make me more anxious than I like to say to obtain such help as yours. There is no one in all England whose name would carry so much scientific weight as yours; and, what after all is of more importance, there is no one who could tell, as you could, what attempts are worth making, what problems are utterly hopeless, in a word what can be done, and what cannot, toward giving a sound scientific direction to the work of the office. With such advice as yours I should feel comparatively safe; without it I am quite sure I should be much adrift. It is well known in the scientific world (excuse so personal a remark) that you give invaluable advice very freely to scientific men who ask you for it. I think it a little hard that the Meteorological Council should not have this advantage, because it would offer you a certainly very inadequate remuneration in return for it.

One word with reference to myself. Dr Hooker, and the three gentlemen I have mentioned, have all signified to me at various times their wish that I should be the Chairman. I strongly feel the importance and the honourableness of the office, but I have the strongest doubts as to whether I am the right man for the place. I am *very* ignorant at present of meteorology proper; and I have many other scientific interests upon which I could easily spend my whole time. I can truly say that a place on the Meteorological Council is not an object of ambition to me: there are other things (of less importance, no doubt) in which I know I can do some little bits of work; whether I can make anything of meteorology, and of the Meteorological Office, I do not know.

Under these circumstances I shall feel it a real kindness, if you would exert an independent judgment for me in the matter; and interpose to prevent a great mistake from being made, in case you think (as it is very likely you may think) the President and my other friends are making one in thinking of me. What I have done is to place myself in the President's hands; but I am very anxious that his hands should be guided aright; and it would never occur to me to think myself passed over, if he should change his mind or if the Council of the R. S. should not agree with him.

I will end by saying that what I should like best would be to see you Chairman; and by repeating that, if I should be appointed Chairman, your presence or advice would make an enormous difference in my hopes of usefulness.

Believe me, to remain, Very faithfully yours,

H. J. S. SMITH.

Pray answer this long letter *vivâ voce* sometime when we meet, and don't trouble to send a written answer.

The following letter to Mr W. N. Shaw, of date 1879, doubtless relates to the Researches on Hygrometry which Mr Shaw carried out at the request of the Meteorological Council.

LENSFIELD COTTAGE, CAMBRIDGE,
9 *August*, 1879.

I fear that any additional information that I have to give you on the subject of my last letter may now be too late to be of any use. We had not very definitely discussed the programme of the experiments, and therefore anything I could say would be rather lacking in authority. Still I may perhaps be able to give you some better ideas as to what we are aiming at.

There appears to be a general feeling that the wet bulb thermometer is not a satisfactory instrument. Yet it is the one almost exclusively used for meteorological observations as to the hygrometric condition of the air, at least for such as are continuous.

One objection is that it is liable to fail in frost. But quite independently of that, it is doubtful how far its indications even above the freezing-point can be relied on. The most obvious cause of uncertainty is the presumable influence of a greater or less rapidity with which the air passes over the surface of the instrument. We have an instrument at the Meteorological Office in which a current of air is made by mechanical means to pass over the surface at a constant rate. A trial how far this may remove the causes of variation would naturally form one part of the research. The instrument would be lent you for the purpose of experiment.

As the chemical method, though laborious, is presumably the

most accurate, the natural course would seem to be that you should first practise yourself with it, so as to learn what precautions it might be found necessary to take. Then the form of hygrometer which appeared next in point of accuracy would naturally be tried. This perhaps would be Regnault's; or perhaps Dines's, which has never I believe been thoroughly examined, might be found as accurate, and would be easier to manipulate. This might prove to be sufficiently accurate to be referred to as a standard, at least in provisional experiments, in examining a hygrometer which is presumably less accurate, such as the wet compared with the dry bulb thermometer. If so, a great deal of time might be saved by the avoidance of the necessity of having recourse to a chemical examination at every step.

Then again the hair hygrometer, which is a good deal used on the Continent, though not in England, should be tried. One advantage which it claims to have is that of acting equally well above and below the freezing-point. How far are its indications trustworthy? If trustworthy, what is the relation between its indications and the condition of the air? In other words, how are we to deduce the vapour tension from its reading and that of the dry bulb thermometer, regarded as two independent variables?

I hope that what I have mentioned will serve to give you a general idea of the research we wish to have carried out. In the conducting of the research, of course a great deal must depend on the additional knowledge gained as it progresses.

I am, dear Sir, Yours faithfully,

G. G. STOKES.

In July 1880, in consequence of the Tay Bridge disaster, a Committee was appointed by the Board of Trade to consider and report upon the question of wind pressure on railway structures, the members being Sir W. G. Armstrong, W. H. Barlow, Sir J. Hawkshaw, Prof. Stokes, and Col. W. Yolland, all Fellows of the Royal Society. Their report is dated May 20, 1881, and was published as a blue-book. It contains a rider as follows:

" We, the undersigned, concur in the above Report so far as it goes, but we think the following clause should be added, viz. :—

The evidence before us does not enable us to judge as to the lateral extent of the extremely high pressures occasionally recorded by anemometers, and we think it desirable that experiments should be made to determine this question. If the lateral extent of exceptionally heavy gusts should prove to be very small, it would become a question whether some relaxation might not be permitted in the requirements of this Report.

W. G. ARMSTRONG.
G. G. STOKES.

The opinion has in fact been widely held that the recommendations of the report, which required a maximum pressure of 56 lbs. per square foot to be allowed for, with the factor of safety 4, were too stringent, and that they have involved much unnecessary expense in structures such as the Forth Bridge. The problem of wind pressure on structures has recently been taken up at the National Physical Laboratory, at the instance of the Institution of Mechanical Engineers.

FROM SIR W. G. (AFTERWARDS LORD) ARMSTRONG.

NEWCASTLE-ON-TYNE,
28 March, 1881.

MY DEAR PROFESSOR STOKES,

I have received your letter of the 27th.

If the design of the Forth Bridge embraces two central girders *close together*, I agree with you that they might be taken as one, but I rather think the width required in that structure involves all the girders being widely separated.

As a general conclusion I think we should go far enough in the interests of safety if we decided,

1st. That the maximum wind pressure should be taken at 45 lbs. per sq. foot.

2nd. That in estimating the pressure on *inner* girders, both their number and the extent of shelter they derive from each other must be taken into account (as to the value to be assigned to that shelter I for one should be guided by your opinion).

3rd. That having regard to the extreme rarity of maximum pressures the factor of safety should be taken at 3 instead of 4.

I think our decision as to the maximum pressure should be chiefly based upon the fact that 45 lbs. per foot over an extended surface would be more than sufficient to overturn an entire train, and as no entire train ever has been blown over on the most exposed viaducts, the presumption is that no such wind has ever operated on the entire length of a viaduct.

The bare possibility of a still higher wind would be sufficiently met by the factor of safety, coupled with the fact that such a hurricane would stop all traffic and exempt the bridge from the strain due to passing trains, which is provided for by our rules.

If there be not another meeting, or, being one, I should be unable to attend, I will notify these views to the Secretary, but shall not press them against the opinion of the majority.

Very truly yours,

W. G. ARMSTRONG.

2 *May*, 1881.

I have this morning seen Mr John Hawkshaw, who produced the wind pressure report in an amended form with your initials affixed. He also showed me your letter in which you expressed a somewhat reluctant concurrence in the clause relating to the pressure on the secondary girders, but I observed that you made no reference to the more vital question of the maximum pressure of wind and the factor of safety.

I have a strong opinion that we were exacting far too much in requiring a resistance to be provided equal to 56 lbs. per sq. foot × 4, but if the Committee are agreed upon this point I shall not stand out alone against the rest. Only I should like to know definitely whether you have accepted the terms of the report in regard to maximum pressure and factor of safety, as well as in regard to the question of the pressure on secondary girders.

I go to Newcastle to-night and shall be glad of a reply addressed to me there.

The following letter from Prof. W. H. Miller doubtless relates to Prof. Stokes' intention of repeating Holtzmann's experiments on diffraction of polarized light by smoke gratings, which gave different results from his own; cf. *supra*, p. 61, also *Math. and Phys. Papers*, II. p. 327.

7 SCROOPE TERRACE,

19 *Jan.* 1867.

MY DEAR STOKES,

Wheatstone sold his 'Beugungserscheinungen' apparatus, Russgitters and all, to Talbot, who Wheatstone believes would be ready to let you have the use of them. They must be easy to make with such a dividing engine as that which Darker has. The thermometer-scale dividing engine at Kew, with a very trifling addition, would, I believe, serve to construct a Russgitter admirably. That dividing engine carries the tracing-point; this will not answer your purpose, you must make the plate of glass moveable, at least such is my opinion; the tracing-point might be such as Nörremberg uses.

A rod AB of glass having the tracing-point attached to a bit of wood C, fastened to the middle of the glass tube; the glass tube slides in two Y's, D, E. The tube is pushed and pulled backwards and forwards by a bit of thin wire, one end of which is fastened with a bit of sealing-wax or cork to the middle of the tube inside. The requisite additions might be made to the dividing engine for a few shillings, if you could get it free from thermometer work for a day. I have tried Nörremberg's mode of tracing lines and find that it succeeds admirably.

The thermometer divider at Kew has a very long screw, parts of it may possibly not be good enough for an optical Gitter. The probability is however strong, that some part of it of sufficient length may be good enough for the purpose.

Very truly yours,

W. H. MILLER.

To Professor G. QUINCKE of Heidelberg.

Springfield, Tiverton.
5th Sept. 1867.

My dear Sir,

I have been indeed remiss in not writing to thank you for your interesting papers published in Poggendorff's *Annalen,* which I received from time to time. I was particularly interested in your beautiful experiments on the alteration of phase in metallic reflection, and in transmission through a metallic film. As to the paper on the spot in the centre of Newton's Rings, as seen beyond the angle of total internal reflexion, I do not at this moment distinctly recollect its contents, and being from home I have not the papers by me to refer to. I do not recollect why. I did not write to you to mention my paper on the subject: it may have been pressure of work; it may have been laziness or procrastination; it may have been dislike to those "reclamations" that......so indulge in, or all these combined. Be that as it may, as you have kindly expressed an interest in my paper, I have written to a lady whom I left in my house, directing her where I think she will find some copies of my paper, and requesting her if she found them to send you one by the book-post. If she cannot find them I hope to send you one when I return home about the beginning of October.

The only papers I have published on this subject, or subjects closely connected (so far as I at present recollect) are (1) a paper 'On the perfect blackness of the central spot of Newton's Rings,' *Cambridge* (or *Cambridge and Dublin*) *Mathematical Journal* (about 1848?)*, and a paper 'On the metallic reflection of light exhibited by certain non-metallic bodies,' or a title very similar to that in the *Philosophical Magazine*, I think about the same date†. In a Library you would probably find the references at once by turning to my name in Poggendorff's catalogue. Not having access to such books here I cannot give them exactly.

Many years ago I observed a very curious phenomenon relative to metallic reflection in the case of the central spot of Newton's Rings, when formed between metal and glass at an angle exceeding that of total internal reflection. I did not publish an account of it, hoping to be able to accompany the account of the phenomenon by a mathematical explanation. I made some trials to get a theory, which I found by no means easy, but I never tried my best at it, and other matters caused me to lay the subject aside. If I find that you have not anticipated me, I think I shall publish an account of my observations, which were merely qualitative, in the *Proceedings* of the Royal Society ‡, and leave the explanation to a future effort on the part of whomsoever it may please to take it up. The phenomenon is exceedingly curious as indicating the existence in the case of metals, if I may so speak, of an angle of incidence determined by a sine greater than unity, in the neighbourhood of which the properties as to reflection rapidly change.

You ask me as to my views respecting the theory of metallic reflection, and the exactness of the received formulae as representing the phenomenon.

In the first place I would observe that the priority as to giving the formulae undoubtedly belongs to MacCullagh. Cauchy's

---

* *Math. and Phys. Papers*, II. pp. 56–81 (1848) and 89–103 (1849).

† *loc. cit.* IV. pp. 18–21 (1852) and 38–49 (1853).

‡ *Brit. Assoc. Report*, 1876 ; *Math. and Phys. Papers*, vol. IV. pp. 361–4. The subject has at length been examined mathematically by Prof. R. C. Maclaurin, *Roy. Soc. Proc.*, 1905, who finds that the phenomena noticed by Prof. Stokes are involved in the ordinary theory of metallic reflexion, without any need to make a call upon Airy's property for diamond which is referred to below.

formulae are nothing but MacCullagh's, reproduced with such variation of form as any algebraist might make, in transforming an imaginary expression so as to adapt it to calculation.

As to the difference in the values of the refractive indices of metals given by MacCullagh and Cauchy, which Moigno in his *Répertoire d'Optique* makes such a work about, that is a mere question of words. The formulae are got by introducing into Fresnel's formulae, in lieu of the refractive index, an imaginary expression, suppose $\mu(\cos\psi + \sqrt{-1}\sin\psi)$, and then treating the formulae as Fresnel had done in the case of total internal reflection. The two constants μ, ψ optically define the metal. The expression "refractive index" in its ordinary signification has no meaning as applied to a metal, and we may give the name to any function in place of $\mu(\cos\psi + \sqrt{-1}\sin\psi)$, provided it satisfies the condition $f(\mu, 0) = \mu$; so that when the metallic properties of the medium are supposed to vanish continuously, and the medium to diverge into an ordinary transparent medium, the quantity which we are pleased to call refractive index shall agree with that previously so designated in the case of media such as glass, water, &c. Now MacCullagh chose to take one such function*, namely $[\mu/\cos\psi]$, as that to which he gave the name "refractive index," and Cauchy chose another, namely...... Perhaps on the whole Cauchy's nomenclature is to be preferred; for when, as in your experiments, light is incident perpendicularly on the surface of a metallic film, the constant...is that which regulates the change of phase, in the same manner as the constant μ the refractive index in the case of a transparent plate.

MacCullagh left his formulae wholly without physical explanation (I say *explanation*, not *interpretation*—the interpretation is obvious) and in this respect Cauchy certainly made a step in advance, in showing that they were what would result from introducing into Fresnel's formulae, supposed otherwise established, the modification arising from a very rapid "extinction" or intense absorption of light on the part of the medium. The strongest confirmation I know of the general truth of this view is to be found in some observations I made on the nature of light reflected by crystals of permanganate of potash, described in the paper in the *Phil. Mag.* to which I have referred.

* MacCullagh's *Collected Papers*, p. 132 (Jan. 1837).

The deduction of the formulae from those of Fresnel readily
follows from the principles laid down by O'Brien with reference
to total reflection, in a paper published in the *Camb. Trans.* To
Eisenlohr's investigation I object that he needlessly (needlessly,
if the only object be to deduce the metallic formulae from
Fresnel's formulae) introduces Cauchy's *pretended but erroneous*
principle of continuity. The principle as used by Cauchy is in
fact obtained by *putting together* two conclusions which respectively result from two *mutually exclusive* hypotheses as to the
state of things at the boundary of two media.

Airy's observations showed for diamond, and Jamin's observations have since shown for transparent substances with few
exceptions, that the results of Fresnel's formulae deviate sensibly
from the observed phenomena in the neighbourhood of the polarizing angle. We should therefore expect *à priori* that the results
of the metallic formulae deduced from Fresnel's should differ
sensibly from the observed phenomena, more especially as the
deviations in the case of transparent substances are, on the
whole, greatest in the case of dense media, and metals are remarkable for their high specific gravity. I feel curious to know
whether you have worked with the central spot of Newton's Rings
under the circumstances I have described. If so, you can hardly
fail to have seen the phenomenon.

If you favour me with a line, will you have the goodness to
tell me (in case you know, but don't trouble yourself to find out
if you don't) whether Professor Magnus is in Berlin, and to give
me his Berlin address? I suppose "Berlin" simply would find
him. I owe him a letter, I am ashamed to say how long, to
thank him for an interesting apparatus he sent me.

I am, dear Sir, Yours sincerely,

G. G. STOKES.

The following letter from Sir John Herschel acknowledges the
receipt of the fundamental memoir ' On the communication of
vibrations from a vibrating body to a surrounding gas,' *Phil.
Trans.* 1868, *Math. and Phys. Papers*, v. pp. 299—324, in which
the explanation is given of the stifling of the sound of a bell when
immersed in hydrogen, a problem on which Herschel had speculated
(Ency. Metrop. art. *Sound*) many years before.

COLLINGWOOD, *Dec.* 25, 1868.

MY DEAR SIR,

Many thanks for your papers. I wish I was enough *au fait* at these *tours d'analyse* to follow them, but the facts about the effects of the lateral movement in diminishing the intensity of sound are very curious. Your experiment with the tuning-fork is a capital illustration, taken in conjunction with the fact that when *no* obstacle is placed to prevent the lateral movement the sound is inaudible to an ear placed at 45°.

I wish geometers would agreé and call the function

$$\iiint \frac{dx\,dy\,dz \times \delta}{r}$$

the Integral Proximity of a geometrical solid or material mass to a point without it, and so be rid of the notion of force in what after all is a purely abstract geometrical conception.

With the best wishes of the season, believe me

Dear Sir, yours very truly,

J. F. W. HERSCHEL.

The following letter from M. Jacobi, of St Petersburg, refers to his proposed attendance at the meeting of the British Association held at Exeter in 1869 under Prof. Stokes' Presidency. The determination of electric standards, in which the British Association has taken the leading part, is treated as regards its early stages from the German point of view in Werner von Siemens' volume of *Reminiscences*.

ST PÉTERSBOURG, 16–28 *Juin*, 1869.

MONSIEUR,—J'ai l'honneur de vous faire part que j'ai l'intention d'assister cette année à la réunion de l'Association britannique pour l'Avancement des Sciences à Exeter. J'espère pouvoir à cette occasion rénouer d'anciens rapports avec quelques illustres savants de votre pays, et de leur soumettre quelques questions qui se discutent plus aisément par un échange verbal d'opinions que par la parole écrite ou imprimée.

Surtout ai-je l'intention de me mettre en rapport avec le Comité "des étalons électriques" qui existe auprès de l'Association. Il s'agira de discuter l'opportunité ou plutôt la nécessité indispensable, de n'adopter qu'un seul "standard" pour mesurer la résistance des circuits électriques. Il est à regretter que déjà dès le début, c. a. d. dès qu'on avait commencé à reconnaître la nécessité de convenir sur certains "standards" pour la mesure des phénomènes du courant électrique, que dès le début, une diversité s'est produite à cet égard qui ne doit pas se perpétuer et contre laquelle il est de mon devoir de protester. En effet, cette diversité compromet essentiellement le bût que j'avais en vue en donnant en 1846 l'initiative à l'introduction d'une unité de mesure pour la résistance des circuits et en publiant en 1857 mon mémoire "sur la nécessité d'exprimer la force des courants électriques et la résistance des circuits en unités unanimement et généralement adoptées" (voir Bulletin de la Classe Physico-Mathématique de l'Académie Impériale des Sciences de St Pétersbourg T. XVI. No. 6 and 7, p. 81). Du reste, il ne sera pas difficile de parvenir à un accommodement pourvu qu'on mette de côte tout amour propre national ou personnel. Vous m'obligerez beaucoup, Monsieur, en employant vos bons offices, pour engager les membres du Comité en question, de se réunir à Exeter le plus complètement possible, pour que je puisse soumettre mes propositions à leur jugement éclairé.

J'ai l'honneur de vous envoyer sous bande quelques exemplaires d'un rapport "sur la confection des étalons protypes" des poids et mesures métriques, présenté à l'Académie Imp. des Sciences de St Pétersbourg dans sa séance du 20 May (1° Juin) par la Commission nommée ad hoc. L'Académie ayant adopté les conclusions de ce rapport, m'a chargé de faire valoir à l'occasion de ma visite à Exeter, auprès de l'Association britannique, les principes établis dans ce rapport, dont je vous prie de vouloir bien faire distribuer les exemplaires aux membres de l'Association qui s'interessent à l'adoption universelle du système métrique.

Malgré que j'ai visité votre pays à trois différentes reprises, le Devonshire est restée pour moi une "terra incognita." Je ne doute pas que les Officiers de l'Association n'aient pourvu aux moyens d'indiquer aux étrangers la meilleure route à suivre de Londres à Exeter pour pouvoir visiter au passage quelques remarquables points du pays. Il m'importerait d'autant plus d'avoir quelques renseignements à cet égard, qu'au départ, mon séjour à Londres sera probablement de trop courte durée pour pouvoir ne procurer ces renseignements moi-même.

Veuillez, Monsieur, agréer l'assurance de ma considération très-distinguée.

M. H. DE JACOBI,
Membre de l'Académie Impériale des
Sciences de St Pétersbourg.

MR BASHFORTH'S BALLISTIC INVESTIGATIONS.

In Dec. 1868 the War Office appointed a Committee to inquire into the ballistic experiments, now universally recognised as fundamental and classical in the Science of Artillery, that were conducted by the Rev. F. Bashforth * with his new chronograph. The following letter†, addressed to the Astronomer Royal, Prof. J. C. Adams, Prof. Stokes, and Capt. (now Sir) A. Noble, will explain the circumstances.

"The Professor of Applied Mathematics to the advanced class of artillery officers, Woolwich, Rev. F. Bashforth, has reported to the Secretary of State for War the result of a series of experiments which he has conducted at the public expense, but with a chronograph of his own invention and construction, for the purpose of determining the law of the resistance of the air to the motion of a projectile. Mr Bashforth has further represented that the results obtained are of high scientific importance, involving mathematical questions of considerable difficulty, and has asked that they may be referred to mathematicians of eminence for an opinion on their precision and value. The Secretary of State having acceded with pleasure to this request, I am directed to ask the favour that you will act as one of the referees, the others applied to being the gentlemen named in the margin.

Mr Cardwell desires to be informed by an independent scientific committee of reference (1), Whether it is now to be considered as proved that the resistance of the air varies practically as the cube of the velocity of the shot for all the velocities in use in gunnery, ranging from 300 to 1,900 feet per second, or for what range of velocities; and if not true, whether it is nearer the truth than any equally simple law before propounded.

(2) Whether this law of resistance is to be regarded as a new one, the discovery of which is due to Mr Bashforth.

(3) Whether the instrument devised and perfected by that gentleman for recording successive small intervals of time is susceptible of general employment at schools of instruction in gunnery.

(4) Whether any means of solving the same problem with equal precision existed before.

Upon the answers to these questions will depend in some degree the credit due to Professor Bashforth for his public and

---

* Mr Bashforth was second wrangler at Cambridge in the year when Prof. Adams was senior, and had been a fellow of St John's College, of which he is now an Honorary Fellow.

† See the official "Reports on Experiments made with the Bashforth Chronograph to determine the Resistance of the Air to the motion of Projectiles, 1865–1870"; Eyre and Spottiswoode, 1870.

private exertions in solving this important problem; and it is his desire, as well as that of the Secretary of State, that his claims to the scientific distinction of having been the first philosopher to discover, or at least to prove, the true law of resistance so long sought for, shall rest on the unbiassed decision of competent mathematicians.

The Secretary of State would wish the referees to agree to the tenor of a general answer to each of the above questions, but will be happy to receive separate reports, if desired, or opinions on the general question from other points of view."

On March 11 following the Astronomer Royal sent the following letter retiring from the Committee :

" My difficulties are much increased by the wide geographical separation from my colleagues.

Under these circumstances I should be glad if you would sanction my retiring from the Commission. I think that if it were understood that Professor Stokes is the leader, he would with advantage (particularly from his proximity of residence to Professor Adams) enter upon the philosophical points, which are nearly related to his favourite studies, and there would be good prospect of a result beneficial to science."

The report of the Committee, of which a first long draft exists in the handwriting of Prof. Stokes and Prof. Adams, was finally sent in on April 18, 1870, and occupies pp. 155—161 of the volume quoted above. The following extracts give the tenour of the report:

"Between October 1867 and May 1868, Professor Bashforth made a most valuable series of experiments with elongated shot of different diameters, the charges being varied so as to give a much greater range of velocities than in his earlier experiments. The object of these experiments was to determine whether the resistance of the air might be assumed to be proportional to the cube of the velocity for all practical velocities of the shot, and also whether it varies exactly as the square of the diameter.

The results showed that while for moderate variations of velocity, such as those met with in the same round, the resistance may be taken to vary as the cube of the velocity, yet for considerable variations of velocity this law no longer holds good; and that if for convenience the resistance be still represented by an expression of the form cv^3, the co-efficient c must be taken to vary continuously with the velocity. The experiments showed the values of the co-efficient c of resistance, corresponding to values of the velocity ranging from 850 to 1,600 feet per second. The co-efficient is found to attain a maximum value for a velocity of about 1,200 feet per second. Professor Hélie's value of the coefficient of resistance is found to be true only for velocities in the neighbourhood of 950 feet per second.

Between the beginning of May 1868 and the beginning of the following November, Professor Bashforth carried out similar experiments with spherical shot, in which the velocities observed varied from 1,000 to 2,100 feet per second. The co-efficient of resistance is of course very different in the case of spherical shot from the value corresponding to the same velocity in elongated shot.

In this case also the co-efficient appears to attain a maximum value for a velocity of about 1,200 feet per second, but it falls off very considerably for the higher velocities. The resistance for the same velocity was found to vary very accurately as the square of the diameter, both for spherical and elongated shot *.

We consider that these experiments of Professor Bashforth are admirably planned and that the results obtained are very valuable.

In conclusion we would beg to recommend that for the purpose of rendering our experimental knowledge of the resistance of the air more complete, an additional series of experiments should be made with Mr Bashforth's chronograph, on spherical projectiles moving with smaller velocities than those which occur in his previous experiments. The results would greatly add to the theoretical interest and value of those which have been already obtained."

BRITISH ASSOCIATION.

Prof. Stokes was President of the British Association at the Exeter meeting in 1869. It had been hoped to obtain for that occasion some prominent scientific man connected with the West of England: the Association were, however, disappointed in securing either Prof. J. C. Adams or Mr Fox Talbot, and in response to an intimation that "the unanimous opinion of the Council indicated yourself as the Physicist to be applied to in the event, which has now occurred, of the two local men of science being both of them disinclined to accept" he placed himself at their disposal, and devoted himself with his usual energy to the task. For very many years he was a regular attendant at the meetings of the Association, and there were few active Committees on the Physical side whose work was not influenced by him.

* It may be remarked that the method of dimensions, as applied in similar cases by Stokes (*Math. and Phys. Papers*, Vol. v. p. 106, 1881) and by Rayleigh (*Phil. Mag.* 1905 (1), p. 494), shows that for bodies of the same shape and of linear dimension l, the total resistance must vary as $l^2 \rho v^2 f(v/v_0)$, where ρ is the density and v_0 the velocity of sound in the gas. This is on the hypothesis that dimensional considerations apply in cases of turbulent motion, of which the result in the text is thus a confirmation; cf. under Osborne Reynolds, *infra*.

On a previous occasion, Cambridge 1862, when he was President of the Mathematical and Physical section, his friend Prof. W. H. Miller directing the Chemical section, he delivered what is probably the shortest introductory address on record in the following words:—

"It has been customary for some years, in opening the business of the Section, for the President to say a few words respecting the object of our meetings. In this Section, more perhaps than in any other, we have frequently to deal with subjects of a very abstract character, which in many cases can be mastered only by patient study, at leisure, of what has been written. The question may not unnaturally be asked—If investigations of this kind can best be followed by quiet study in one's own room, what is the use of bringing them forward in a sectional meeting at all? I believe that good may be done by public mention, in a meeting like the present, of even somewhat abstract investigations; but whether good is thus done, or the audience are merely wearied to no purpose, depends upon the judiciousness of the person by whom the investigation is brought forward. It must be remembered that minute details cannot be followed in an exposition *vivâ voce*; they must be studied at leisure; and the aim of an author should be to present the broad leading ideas of his research, and the principal conclusions at which he has arrived, clearly and briefly before the Section. It is then possible to discuss the subject-matter; to offer suggestions of new lines of experiment, or new combinations of ideas; and such discussions and suggestions, it seems to me, are among the most important business of a meeting such as this. Any one who has worked in concert with another zealously engaged in the same research must have felt the benefit arising from the mutual interchange of ideas between two different minds. Suggestions struck out by one call up new trains of thought and fructify in the mind of another; whereas they might have remained barren and unfruitful in the mind of the original suggester. The benefit of cooperation is by no means confined to the carrying out, according to a preconcerted plan, of a research involving labour rather than invention; it is felt in a most delightful form in the prosecution of original investigations. In a meeting like the present, we have the benefit of the mutual suggestions, not of two, but of many persons, whose minds are directed to the same object. The number of papers already in the hands of your Secretaries shows that there will be no lack of matter in this Section: the difficulty will rather, I apprehend, be to get through the business before us in the time prescribed. On this account the Section will, I hope, bear with me if I should sometimes feel myself compelled, in justice to the authors of papers which are placed later on our lists, to cut short discussions which otherwise might have been further prolonged with some interest."

This parsimony was in fact demanded by the necessities of the occasion. Among the papers contributed by eminent men to the Proceedings of the section were:—Capillary Attraction by Mr F. Bashforth, On the Differential Equations of Dynamics by Prof. Boole, On a Geometrical Mechanism by Dr Booth, two papers on Twisted Curves by Mr Cayley, On Quaternions by Sir W. Rowan Hamilton, On Pedal Surfaces by Dr Hirst, On the Forms of Waves by Prof. Macquorn Rankine; astronomical papers by Challis, Hennessy, Lassell, R. Main, Pogson, Spottiswoode, Claudet; on Rotatory Polarization in Crystals by A. des Cloizeaux; papers by Croll, Esselbach, C. Tomlinson, J. Ball, F. Galton, J. H. Gladstone, J. Glaisher, E. J. Lowe, R. Mallet, G. J. Symons, James Thomson, and many others.

But the absence of an address was amply atoned for by the presentation by Prof. Stokes of his Report on Double Refraction, which for so long a period remained the authoritative exposition of the progress of the Dynamics of Physical Optics and of the state of the problems then demanding solution. In the same year Cayley presented his second Report on Special Problems of Dynamics. The previous Cambridge Meeting in 1845, at which he expounded his explanation of the Aberration of Light by an irrotationally moving aether, had imposed on Mr Stokes the task of producing his other important Report on Recent Researches in Hydrodynamics, which was published the following year, as also Mr R. L. Ellis' Report on Recent Progress in Analysis.

At the meeting at Exeter in 1869, Prof. Sylvester presided over the Mathematical and Physical Section, and delivered the well-known eloquent address on the claims and privileges of Abstract Mathematics. Among readers of papers in the section were W. K. Clifford on the Theory of Distance, R. B. Hayward on the Equations of Dynamics, F. W. Newman (three papers), Rankine, W. H. L. Russell, R. Main (three papers), Neumayer, T. R. Robinson, Tait on Comets, C. Brooke, J. H. Gladstone, Janssen, Jellett on Optical Chemistry, Morren, G. J. Stoney, Magnus on Emission, Absorption, and Reflexion of Obscure Heat*, Symons, J. Glaisher, B. Stewart, J. P. Gassiot, J. W. Strutt, and many others.

---

* In this paper the priority of Balfour Stewart is remarked, " who, ten years ago, and several years before Kirchhoff and Bunsen had propounded their theory, published a paper in which he developed nearly the same ideas for heat as these philosophers did for light."

IMPROVEMENT OF TELESCOPES.

For a long series of years Prof. Stokes devoted much time to advising and assisting Rev. W. Vernon Harcourt, who had begun as early as 1834 a series of experimental researches devoted to the improvement of glass for optical purposes. The history of Prof. Stokes' collaboration was sketched by him at the British Association meeting at Edinburgh in 1872*. It seems that, at the Cambridge meeting in 1862, Mr Harcourt placed some of his prisms in Prof. Stokes' hands, with a view to his tracing out any connexion there might be between their fluorescent properties and their chemical constitution, which was of course exactly known. Prof. Stokes was able incidentally to determine their dispersive powers with more accuracy than had seemed possible. "This inquiry being in furtherance of the original object of the experiments, seemed far more important than that as to fluorescence, and caused Mr Harcourt to resume his experiments with the liveliest interest, an interest which he kept up to the last." Then, referring to the advantages arising from co-operation†, he goes on to say, "I do not think either of us working separately could have obtained the results we arrived at by working together." It is difficult to resist the impression, from reading this notice of his friend's work and his own, and from turning over the voluminous correspondence to which it gave rise, that the imperfect appliances at Mr Harcourt's disposal and the consequent poor quality and striation of the glass produced, probably prevented a great step in advance in the provision of optical appliances, which was not to be achieved until the subject was resumed by Abbe and his coadjutors, with organised practical co-operation on an industrial scale, at Jena many years afterwards.

A methodical notebook exists, containing a great mass of entries of determinations relating to compensation of dispersions of glasses and solutions made by Prof. Stokes during 1864—6. The following extracts will give an indication of the contents:

Experiments on influence of different substances on irrationality of dispersion.

Method. Two hollow prisms of 45° were procured from Ladd. A standard solution of one salt was prepared, and then the strength of other solutions was adjusted so that they should

* *Math. and Phys. Papers,* Vol. IV. pp. 339–343.
† He refers to the remarks on this subject on p. 198, *supra.*

achromatize the standard when the two opposing prisms were placed each at the minimum deviation.

Experiments commenced 10th Feb. 1866. Light from sky near sun. Standard solution used, 1 part by weight of common salt to 5 parts of water. Broad slit and collimating lens on plank, standard placed in hollow prism A.

1. CaCl solution: slightly green-refracting to standard; secondary spectrum just perceptible.

2. NaC̈: green: decidedly more so than last. N.B. Had to be heated to cause enough to dissolve, and after some time a portion began to crystallize.

3. HCl: green: more so apparently than the last. N.B. Had to be diluted to perhaps 3 times the volume to bring down the dispersion so as to match the standard.

4. NaS̈ dispersion too weak even when almost saturated.

13th Feb. 1866. In trying NO$_5$ diluted with 2 or 3 volumes of water the cement gave way. Therefore to-day I re-cemented with bees' wax, tying on the sides for security, and prepared several additional solutions. The following are the results of the comparisons:—...

Observations with Compensating Prisms.

Commenced for good Dec. 6th, 1864. At first the slit was used without any collimating lens or lens in connexion with the prisms; the prisms placed at as large a distance as the room would conveniently allow, and the secondary spectrum viewed through an achromatic telescope, care being taken that the pencil on the object-glass was as nearly centrical as might be. Afterwards a 44 inch focus lens was used at 6 or 8 feet distance from the slit, the prisms were placed close to the lens, and the spectrum viewed through a detached eye-piece (my lowest) with its focus at the conjugate focus. This was only tried for a few observations. Then the same lens was used as a collimator, and the achromatic telescope, with its deepest eye-piece unless the contrary is stated, was used for viewing the spectrum.

The first column gives the prism used and its angle, the next the composition of the compensating compound prism. The flint glass angles are written above, the crown-glass angles below. The unit is the angle ordered (1° 20′) but in the calculation the measured angles are employed. The two flint glass prisms having angles as ordered of $13\frac{1}{2}$ units each, and as measured of 1044', 1039', are denoted the first by 14 the second by 13. The angles used negatively are denoted by their numbers with a − written above. The order of the angles is from without inwards, *i.e.* towards the prism examined. Unless the contrary be stated the compensating prism was placed next the eye. The prism examined was placed at the minimum deviation, and the compound prism was so built up that the achromatism effected by it should be found near its minimum deviation.......

MINERAL DEPARTMENT, BRITISH MUSEUM,
Jan. 25, 1875.

MY DEAR PROF. STOKES,

Before Christmas I put in your hand a little paper of Professor Abbe's, of Jena, treating of a method of determining the refraction of various kinds of bodies.

But I do not think I explained to you—as I thought it would be explained in his letter accompanied by a translation by my friend Lettsom—that he wants your help in getting access to what you have published about the kinds of glass that are best fitted for correcting lenses in the matter of colour. If you should happen to have a copy of your report on Mr Harcourt's glass experiments, or could in any other form put the Professor at Jena on the track of what has been done, I am confident he would much appreciate your doing so.

Believe me to be very truly yours,
NEVIL STORY MASKELYNE.

TO PROF. E. WIEDEMANN.

LENSFIELD COTTAGE, CAMBRIDGE,
22 *Feb.*, 1879.

DEAR SIR,

On my return from London last night I found the printed notice of my paper*, which is possibly still in proof, and which I return with one or two remarks.

It is not so much that the images given by a pair of imperfect prisms are better than the spectral images that they show, that

---

* Perhaps for the *Beiblätter*. The paper is in *Roy. Soc. Proc.* XXVII. June, 1878, reprinted in *Math. and Phys. Papers*, v. pp. 40–51. The method was invented for dealing with Mr Harcourt's imperfect experimental glasses.

renders a pretty good measure of the ratio of their dispersions possible, as that any one-sidedness of distribution of colour arising from the prisms not achromatising each other is capable of being perceived even in somewhat imperfect images.

Even if you can get prisms of thoroughly good glass, the ordinary method requires extreme accuracy in the angular measurements to get the ratio of the dispersions with sufficient accuracy. The convenience of my method as applied to good glasses, such as those that would be used for an achromatic object-glass, is that it dispenses with any very great accuracy in the angular measurements that are required.

I have marked a passage in pencil with the number 2 because there are of course two secondary colours, a green and a purple, and not one alone. However the two are mentioned almost immediately afterwards, so that it does not signify.

I have to thank you for two papers received lately. The one on the measurement of the temperature of the electric discharge in gases interesting me particularly. It is, however, to my mind by no means conclusive, for this reason. The calorimeter, as I understand, was composed of glass and transparent liquids. Therefore the luminous rays would pass freely through it, and the same would probably be true for a portion of the ultra-red rays. Now it is conceivable, and I think very probable, that during the brief time that the gas was glowing a good portion of the heat passed off in the shape of radiation sufficiently refrangible to pass through the calorimeter, and consequently to give no indication of its existence.

If my supposition is right, it would be interesting to repeat the experiment, guarding against this source of error.

I am, dear Sir, yours sincerely,

G. G. STOKES.

At the meeting of the British Association at Belfast, in 1874, Prof. Stokes exhibited a telescope of $2\frac{1}{2}$ inches aperture and 21 inches focal length, the triple objective of which had been fashioned by Howard Grubb, and demonstrated the absence in it of secondary as well as primary dispersion; though of course the striation of the glasses, arising merely from imperfect manufacture, prevented the performance of the telescope from being good in other respects.

At the Bristol meeting in the following year he reported that, having secured the collaboration of Dr John Hopkinson, then connected with Messrs Chance, the glass manufacturers of Birmingham, he had tested whether the theoretical indications which pointed to a mixture of tartaric acid being effective in silicic as well as in phosphatic glass were confirmed, and had obtained a negative result.

These notes have been reprinted in *Math. and Phys. Papers*, Vol. IV. pp. 356—360; taken along with the previous report of 1871 they form all that was published on the subject.

From this time, and previously, Prof. Stokes, in association at first with Dr Romney Robinson, took much interest in the great enterprises in the construction of telescopes that were building up a world-wide reputation of the optical firm of Thomas Grubb and Son, of Dublin; and a voluminous correspondence with the members of the firm was kept up, lasting from 1869 to 1887. One of the problems to be settled at this period was the delicate question whether the impossibility of obtaining homogeneous discs, free from striation and internal strain, for the great Vienna Telescope, should lead to the abandonment of its construction.

Judging by the masses of minute numerical computations relating to the design of objectives that exist among his papers, Prof. Stokes must have experienced much delight in contributing his insight and experience to forwarding the great task of providing the material means for the continued triumphal advance of astronomical science.

When the question of a new telescope for photographic purposes came before the Cambridge Observatory Syndicate in 1892, and the experiment of trying a new plan was decided upon, it came as a surprise on some of the junior members of the Syndicate to notice the quiet and magisterial way in which Prof. Stokes virtually took charge of the whole problem, no step being thought of without his written report or reasoned opinion, always delivered with deference but with an air of familiarity and confidence, of which the origin was not then known to some of his colleagues.

To Mr THOMAS GRUBB, F.R.S.

CAMBRIDGE, *April* 5, 1871.

MY DEAR SIR,

A *truly* spherical speculum of a mile focus, or even considerably less, would do perfectly well for testing an object-glass as to spherical aberration. The aberration in a truly spherical surface of such radius would be quite insensible. It is the deviation of the surface from what it *professes* to be that we have to fear, a deviation of such character as to make the radius of curvature at the edge slightly different from what it is at the centre.

With a speculum of 15 in. aperture and a mile f. l., the distance between the surface near the edge and the tangent plane at the middle point is 0·0002220 in. (= t say). But the aberration is nothing like what would be produced by a surface-error of the same amount. If t' be the distance at the edge between the surface of the speculum, assumed to be an ellipsoid of revolution, and its sphere of curvature at the vertex, and if t' have such a value as to produce the same aberration that is occasioned by the spherical form in a surface of 2 miles radius, I find

$$t' = 0\text{·}00000000000003889 \text{ in.},$$

which shows how enormously telling a surface-error of this kind is compared with what results from the surface being ground to a slightly wrong curvature*.

I don't suppose it possible to construct a surface *à priori* sufficiently true to be trusted. If a speculum be used, the plan will be to grind it plane or flatly curved, test it, and repolish till the test is satisfied. The difficulty is to find a suitable test, the point to be tested not being whether the surface be plane or gently curved (that is of little consequence) but whether it is *uniformly* plane or curved, so as to give reflected images free from spherical aberration.

If the mirror be concave it might be tested (theoretically at least) by placing a luminous point at or nearly at the centre of curvature, and examining whether the reflected image was free

---

* For a speculum, the difference of optical paths, to which the spherical aberration is proportional, is twice the distance from the edge of the speculum to the osculating aplanatic spheroid corresponding to the actual conjugate foci. Change of curvature uniform all over merely alters the focus.

from spherical aberration. But on account of the great distance
involved, this would involve great steadiness of air, and independently of that would probably be very inconvenient.

The speculum, whether plane or gently concave or convex,
might be tested by means of a speculum known to be parabolic,
or an object-glass known to be free from spherical aberration, the
speculum or objective used for testing being of a size comparable
with the speculum to be tested.

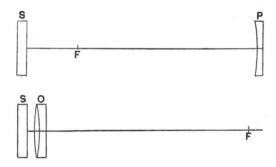

For this purpose arrange as in the figures, S being the plane
or nearly plane speculum to be tested, P the parabolic speculum,
O the object-glass if a refractor be used. Place a luminous point
nearly in the focus of the parabolic mirror, or the objective as the
case may be, and examine the image reflected first from P, then
from S, then from P again, or else refracted through O, reflected
upon S, and refracted again through O. The image and object
will of course nearly coincide. If the surface of S be true, the
image ought to be free from spherical aberration. If it prove
faulty, it can, I presume, be corrected by repolishing.

When once a true surface has been obtained, it may be used
for testing an objective just as the objective was used for
testing it.

From rough trials I made at Armagh, I think the most convenient luminous point would be obtained by means of a small
right-angled prism combined with a glass bead or microscopic
objective of short focus, reflecting light coming through a hole
illuminated by the sun or a lamp. The return image i would be
seen by the aid of a small obliquity. With the parabolic mirror
P a small reflector would be required to throw the return image
sideways for examination.

If S be convex, of true surface, and of f. l. F, an objective
made right by means of it would be right for an artificial star at
distance $2F$, and if this be as good as infinite, in its effect on
spherical aberration, the objective will be right for a star. If S
be concave, of f. l. F, the error of an objective made right by means
of it would be the same in magnitude as, but of opposite sign to,
the error of an objective made right for an artificial star at
distance $2F$, which may be disregarded under the same conditions
as before.

The surface of the speculum S would require to be as true as,
but not truer than, that of an ordinary speculum. The difference
between a sphere and a paraboloid for a f. l. anything like what you
propose would be quite insensible. If you could reckon on grind-
ing the surface spherical the thing is done ; if not, the only greater
difficulty attending S than attending an ordinary speculum consists, I presume, in the greater difficulty of testing. The speculum
S would of course require to be supported in the experiment
of testing it, or when tested using it for testing, as carefully as
that of a telescope, though as the imperfection of image produced
by a slight flexure is altogether different in character from
spherical aberration, a *slight* flexure would probably not much
interfere with its use.

The same method *might* be used for testing the correction for
chromatic aberration, but the simple test I gave you depending
on the observation of the secondary green, which should be about
midway between green and yellow, would probably be found better
for that purpose.

By rights the luminous point should be placed at such a
distance from the objective that its image should be at the same
distance, but it may be moved a little nearer to or further from
the objective without making any sensible difference.

The mercury or speculum plan has the advantage of not
requiring calculation, but perhaps after all the best and simplest
plan would be to turn the objective on to a test object (such as a
watch dial) at a considerable though finite distance, cover the
objective by a screen leaving exposed a central circle, focus, then
cover all but an annulus and focus again, and compare the shifting
of the eye-piece with the *calculated* difference of foci, supposing
the object-glass were right for a star.

I don't believe, however, that this would be a very sharp test.

CORRESPONDENCE WITH PROF. NEWCOMB.

WASHINGTON, *October* 16, 1872.

DEAR SIR,

    I have been much interested in your paper read before the British Association on achromatic combinations of glass, but am disappointed to perceive, from the abstract in *Nature*, that it can hardly be employed in large glasses, owing to the depth of curves required.

    As I am much interested in this subject, will you excuse the liberty I take in inquiring whether you know anything of a combination proposed by Schröder of Hamburg, of which one of the components is a magnesium glass?

Yours very respectfully,

SIMON NEWCOMB.

CAMBRIDGE, 28*th October*, 1872.

DEAR SIR,

    It is true that if we aim at destroying *completely* the secondary dispersion by means of the glasses which the late Mr Vernon Harcourt and I have investigated, rather severe curvatures have to be encountered, but if we consent to a triple objective in lieu of the ordinary double, we shall not require stronger curvatures than in the ordinary double. To get a power of the combination = 2, we require with a crown-flint double the following powers approximately,

crown + 5, flint − 3,

whereas in the Harcourt glasses we should require, in order to get the same combination-power 2, separate powers of + 7, − 5, respectively, very nearly. But if we consent to use a triple, employing two lenses made of the substitute for crown glass instead of one only of greater power, we may arrange the powers

as   + 5, − 5, + 2,

or as   + 4, − 5, + 3,

so that the highest power of any lens is not greater than with the ordinary crown-flint double. Or if we demand that the double form be retained and the curvatures not increased, the demand may be met if we consent to an increase of focal length of say from $11\frac{1}{2}$ to 15 apertures. I should, however, rather be in favour of using a triple, for with modern improvements I don't suppose the centring would occasion much difficulty. The cost of the objective no doubt would be $\frac{2}{3}$ or so that of an ordinary double, even if the substitute for crown glass could be made as cheaply as crown glass itself, which it could not; but the cost of mounting, depending on the focal length, would be the same.

I have worked out a number of triple forms, in which the spherical and chromatic aberrations are destroyed, consistently with the condition that the adjacent surfaces shall fit, so as to permit of cementing.

The ordinary crown-flint double, when made to fit, has approximately the curvatures

$$+1, +1; -1, 0,$$

or say $$+6, +6; -6, 0.$$

The triple fitting form, made from the new glasses, which I am disposed to prefer, has approximately the curvatures

$$+4, +6; -6, -3; +3, +2.$$

Messrs Chance, of Birmingham, are engaged in an experiment on the preparation of titanic glass on a convenient scale.

For practical purposes it would be best, I imagine, not to aim at the *complete* destruction of the irrationality, but leave a little, which would permit of having a greater difference between the substitute for crown and the low flint (or substitute for flint) as regards dispersive power, which would give lower curvature. Yet even if we do aim at complete destruction, the curvature with a triple would, as I have remarked, not be stronger than with an ordinary double.

M. Schroeder was so good as to send me a prism of the magnesian glass. According to my measurements the glass went a little, but only a VERY little, way towards meeting flint glass in point of irrationality.

Yours respectfully,

G. G. STOKES.

CELESTIAL SPECTROSCOPY.

About this time (1867) the application of the telescope to the spectroscopy of the Heavens on a large scale was initiated by Dr (now Sir W.) Huggins, for whom suitable instrumental equipment as regards the telescope was procured by the Royal Society, being placed in his hands on loan when it had been found that he was reluctant to become the absolute owner of it. The succeeding letter, of earlier date, from Sir John Herschel, relates to an eclipse expedition in which his son, now Colonel J. Herschel, F.R.S., took a prominent part.

To Dr (now Sir W.) HUGGINS.

LENSFIELD COTTAGE, CAMBRIDGE,
Dec. 8/68.

My dear Sir,

With reference to a proposal likely before long to come before the Council of the Royal Society, I should be glad to know if you approve of the following statement, and have anything to add to it.

What is the advantage to be gained by a large aperture in a telescope designed to be used in the spectroscopic observation of the celestial bodies?

(1) In the case of the stars aperture is everything. The total quantity of light we have to work by varies as the aperture. With a larger aperture, not only would the spectra of the brighter stars admit of being studied with greater precision and detail, but a field of research would be thrown open in the case of smaller stars which at present can hardly be attacked. It would thus too become possible to study separately and with precision the components of a double star. It would be very interesting to compare, and that in many instances, the components of such stars.

(2) In the case of nebulæ we have to deal with objects of finite size, and consequently have to consider the intrinsic brightness of the image. As this is not altered by increasing aperture and focal length in the same proportion, it might seem at first sight as if aperture were a matter of indifference. But practically the breadth of many of those objects is so small that the spectrum is most inconveniently narrow. The apparent breadth can only be increased either by increasing the magnifying power of the little telescope belonging to the spectroscope or by increasing the focal length of the principal telescope, and in both cases with loss of intensity *unless* (in the 2nd case) the increase of aperture keep pace with the increase of focal length.

If the nebula be so small as to make it desirable to use the cylindrical lens, it may for the present purpose be regarded as a star.

The nucleus of a comet may for the present purpose be treated as a star or a nebula, according to its angular breadth, but if the cylindrical lens be used the spectrum of the nucleus will be mixed with that of the comet.

Yours very truly,

G. G. STOKES.

COLLINGWOOD,

May 5/67.

MY DEAR SIR,

I have no experience in spectroscopy as applied to lights feebler than those of phosphorescent bodies, and from the trials I have made on these I should have felt disposed to believe it impossible to discern lines in spectra formed by lights much feebler. As regards the effect of large aperture in a telescope as a means of concentrating light, I confess myself at a loss to reconcile what seems to be the result of Mr Huggins's and Father Secchi's experience on Nebulæ with the undoubted optical proposition that the *intrinsic* illumination of a surface cannot be increased by *any* application of telescopic power. I mean that the luminosity of the focal image (what the American astronomers call its "albedo") is necessarily somewhat inferior to that of the object. And as it is a narrow slip of that image whose light has to be analysed, I cannot see in what manner telescopic vision, when the object has a sensible superficial or angular magnitude, *can* be of any use at all.

With the image of a star it is otherwise. There it is not the intrinsic albedo but the *total light* which is analysed, and that may be increased telescopically to any extent.

The red flames however have a very sensible angular breadth—quite that of any useful slit—and I confess I should be very much disposed to try a long tube *without an object-glass*, closed by a screen with a narrow slit so as to be directed at a tangent to the sun's disc wherever the flames happened to be, and to examine that slit spectroscopically by whatever means and appliances should have been found available in other cases.

The same mode of observation would, I am inclined to think, be available for comparing spectroscopically the light of the extreme border of the limb of the sun itself with that of the

14—2

central portion (without waiting for an eclipse)—choosing the time for observation that when the sun is on the meridian. In this case the slit should be the very narrowest scratch which could be ruled on a black-varnished glass, or something *as* narrow formed by metal edges, approached by a fine screw. For the flames it would require to be wider.

Possibly, too, the red flames may after all not be such *very* feeble lights; they may have many hundred or thousand times more "albedo" than a candle flame. If so they would assuredly give observable spectra in this way*. If not, I confess my own inability to perceive how a telescope (*quoad* telescope, i.e. otherwise than as an instrument for securing steady vision and exact direction) can help the observation. If it does, it does; but in a mode to me at present incomprehensible†.

I will write to the Astronomer Royal about it, however, as I have no prospect of being able to see him at Greenwich, and let you know what he thinks of the question unless he should prefer to communicate with you direct. Meanwhile I remain,

Dear Sir, Yours very truly,

J. F. W. HERSCHEL.

CORRESPONDENCE WITH SIR G. B. AIRY.

The close relations between Sir George Stokes and the Greenwich Observatory, through his activity in the organisation and reduction of geodetic surveys by pendulums and by triangulation, as well as in the practical optical improvement of telescopes—not to mention close agreement in scientific tastes—naturally led to extensive correspondence with Sir George Airy. To this connexion Stokes' office as Secretary of the Royal Society, and his position for a long series of years as one of the most active members of the Board of Visitors of the Observatory, contributed much: while in the two years of Airy's Presidency of the Royal Society, the novelty to him of the questions that arose, coming on the top of his strenuous and multifarious activities at the Observatory, and of advancing age, naturally made him lean

---

* The discovery that they do so, under high dispersion so as to dilute the background of sky-light, was made independently by Janssen and by Lockyer in connexion with this eclipse.

† See the preceding letter.

extensively on Stokes' help and counsel. Long habit of official correspondence has perhaps imparted a somewhat formal air to his letters to Stokes, and most of them are strictly confined to business; but a respect for exalted abilities and attainments, which was mutual, shines through them all. A valuable scientific correspondence is reprinted in the second volume.

The strong views held by Sir George Airy, in favour of strenuous effort to maintain the mathematical course at Cambridge in as close touch as possible with the requirements of mechanical and physical science, led him about 1868 to devote some of his few spare moments to promoting practically this cause. His contribution took the form of a textbook of the Theory of Sound. In the course of his preparations for it he came across Sir George Stokes' classical investigation of the circumstances which determine the communication of the motion from the vibrator to the atmosphere. In the attempt to find an illustrative case for a spherical vibrator that would work out in small compass, he hit upon the sectorial harmonic of type xy. His thirty years' masterful preoccupation with the business of the national Observatory, on whose history he has left so great a mark, had however left him rusty on such questions of analysis, which indeed had been first effectively grappled with by Stokes himself; he was for example puzzled at adapting the solution to the inside of a hollow bell without introducing infinities at the centre. A detailed correspondence with Stokes ensued, alternating between admiration for the lucid demonstrations of his master in physical analysis, as he called him, and further requests to put things in a way such that one not conversant with them could directly grasp their essence.

GREENWICH, *Feb.* 22, 1868.

MY DEAR SIR—I received last evening your letter of 20th, in which you advert to the present feeling in the University regarding the Smith's Prizes, and request the expression of any strong opinion on the subject that I may entertain.

I will allude to their presumed origin, and to the advantage of maintaining them.

I am not acquainted with the history of the foundation of the Prizes, but I remark that Dr Smith, the author of the *Harmonics* and the *Optics*, was eminently the promoter of Applied Mathematics in his day. Looking at this, and at the title (derived,

I believe, from his will) which they have always borne, I scarcely doubt that they were partly intended as a corrective to a spirit of too exclusively pure mathematics.

In receiving my opinion on the advantage of keeping up the institution in something like the original form—(in which, if I am not mistaken, a separate examination is expressly required by Dr Smith's will, and is made the condition of a grant to the Plumian Professor)—I must ask the representatives of the University to accept my commendations of the merits of the University system, and, for their sake, to give some indulgence to my remarks on what I think to be defects. I do say then that I regard the University as the noblest school of mathematical education in the world. And, in spite of the injury that is done by a too mechanical use of analysis, it is the most exact. There is no accuracy of teaching like that of Cambridge: and in the applied sciences that are really taken up there, of which I would specially mention Astronomy, the education is complete and excellent. But with this, there is an excessively strong tendency to those mathematical subjects which are pursued in the closet, without the effort of looking into the scientific world to see what is wanted there; and common circumstances produce a common effect on many men, till it becomes a peculiarity of the University.

The question then arises, how can this (considered as a fault) be corrected?

I have often wished that we could have in our examiners an element foreign to the University. But, looking to the extreme difficulty of this, I am very glad to accept the next available aid: namely, the assistance of Professors of the University, whose position makes them responsible to the world, and whose pursuits are connected with the science exterior to the University.

And this we have in the Smith's Prize Institution. The conditions to which I have alluded distinctly require a separate examination, but they do not distinctly require a morning's examination by papers instead of an Essay, &c., &c.

You mention as one of the causes of unfriendly feeling to the Smith's Prize Institution the idea that its examination is a sort of rival to the Senate House Examination. Certainly it is so, and I hope it ever will be so. Its function would cease if it were not so. The mere £50 for two prizes is a trifle: it is nothing in

comparison with the prizes of fellowships which are waiting for
the 1st and 2nd wranglers: the whole Foundation would be
nugatory if the examination were not independent, and therefore
necessarily a rival. As to giving the management of its small
pecuniary rewards to the Senate House examiners, and thus
ending the Institution entirely; or as to mixing the examiners
with the Senate House examiners and thus swamping them; in
either case, the Foundation will have perished.

The Smith's Prize Examination must, in my opinion, retain its
strictly independent character.

I do not think that there is any positive objection to the
Official Examiners or Trustees having unofficial Assessors if they
think fit.

And I see no objection to raising the question on the substitution of an Essay for a Paper Examination. I do not think it
easy to arrange: possibly the experience of the Adams' Essay may
serve as guide.

I heartily wish that I could assist you with any useful
suggestion on the last mentioned point, or any related to it.
But at present I do not see my way clearly to any change.
And, believing that the Prizes have been very useful, and that
the terms of the Institution, unaltered, contain the elements
of great utility, I am well content to let things remain as
they are.

I am, my dear Sir, Yours very truly,

G. B. AIRY.

February 24, 1868.

MY DEAR SIR—The periodical business of Monday compels me
to answer your letter of 23rd rather briefly.

1. Would there be advantage in expressing a hope that the
Smith's Prize Examiners will frame their questions *in concert*?
(Perhaps they do.)

2. I like your ultimate suggestion, division of the questions
into groups and requiring something in each.

3. And I suppose the limitation to the number of questions
to be answered is the only way of avoiding what is undoubtedly
a serious defect in the Senate House examination. On the degree
of limitation, you will judge much better than I. The understanding (in whatever way conveyed) of the importance attached
to accuracy and completeness will be very valuable.

4. The selection of the more choice candidates, the common discussion of their papers, and the casting vote to the previous Senate House decision, are all excellent. I am delighted at the earnestness with which you are going into this, and am sure that good will come of it. The Smith's Prize Examination will rise in credit (I do not care about the other questions), and it will become as heretofore the dominant examination. I am, my dear Sir, yours very truly,

THE WHITE HOUSE, CROOM'S HILL,
GREENWICH PARK, S.E.
Feb. 14, 1882.

MY DEAR SIR—Thank you for your orderly explanation of the steps which lead to the interpretation of dn.

I agree with you most completely as to the advantage—to a speculative mathematician—of fairly throwing down the reins and seeing which way the animals will pull them. But all, intellectually, thus gained, is mere dreams; if anything real can be extracted from them, well and good. But till that extraction is made, they are entitled to no higher privileges than the idlest dreams.

I do not in the slightest degree object to the devotion of time to these speculations or to the publication of them in the *Transactions* of the Mathematical Society. But the introduction of them into Academical Examination is, in my opinion, a very serious error. Examination has two objects: discrimination of merits in Academical pursuits, and guidance of future students; and for both these, the matter of which I have spoken is, as I believe, not only unfit but injurious.

I am, my dear Sir, yours very truly,

Feb. 9, 1883.

MY DEAR SIR—I am greatly obliged by the kind recollection of me by yourself, and also by the others of the Smith Prize Board, in sending me the copy of the Examination Papers (including the Dissertation paper, which I suppose I may attribute mainly to you). I very much value the maintenance of this connexion with the University.

I am pleased with this examination. I think that, generally, its tone is healthy; sounder in character than some past examinations, in parts which I will not further specify, but which you will remember.

But I find that mathematics go ahead of me (at least pure mathematics). I am not so clear on 'quantics' as I ought to be, though I know that Spottiswoode and Cayley value them highly.

Yours, my dear Sir, most truly,

June 2, 1886.

MY DEAR SIR—We are startled with the report that you cannot be present at the Visitation of the Royal Observatory. Cannot this be in some way altered?

Yours, my dear Sir, most truly,

G. B. AIRY.

OBSERVATORY,
March 29, 1887.

MY DEAR STOKES,

I send with this the papers I have relating to Hydro-statics, Optics, Sound and Heat, though some of these are very meagrely represented*. I find the little note-book you let me have very interesting.

You will also be doubtless glad!! to see such elaborate Smith's Prize papers. I have not yet been able to look over them, and it appears to me not a very easy job to do so. The papers appear to be of quite a different and superior stamp to those of last year, so far as I can judge.

We are just going to start, and have settled to go to Dawlish.

Yours very truly,

J. C. ADAMS.

FRIENDSHIP WITH JAMES CLERK MAXWELL.

The origin of the life-long intimacy between Sir George Stokes and Clerk Maxwell may be gleaned from the earlier of the scientific letters printed in Vol. II. It would appear that in his younger days Maxwell considered Stokes as a sympathetic master, to whom he could unreservedly report the successes and ambitions of his scientific efforts. In later years, after 1871, when Maxwell returned to Cambridge as first Cavendish Professor of Experimental Physics, the deep vein of moral seriousness in both their characters,

* Perhaps of the Portsmouth Collection of Newton MSS. of which the catalogue was published in 1888.

and a common interest in theological study, drew them still closer together. Their fields of scientific investigation were strikingly separate; though each of them was attentive and very appreciative to the work of the other, their efforts ran on different lines. With Maxwell the scientific imagination was everything: Stokes carried caution to excess. Maxwell revelled in the construction and dissection of mental and material models and images of the activities of the molecules which form the basis of matter: Stokes' published investigations are mainly of the precise and formal kind, guided by the properties and symmetries of matter in bulk, in which the notion of a molecule need hardly enter. Kelvin stands perhaps halfway between them. In the main features of his activity he could rightly be described, as he has himself insisted, as a pupil of Stokes; but along with this practical quality there worked a constructive imagination which doubtless, next to Faraday, formed the main inspiration of Maxwell, though it stopped short of the bold tentative flights into the unknown which in Maxwell's work turned out to be so successful.

When failing health overtook Clerk Maxwell and his wife, their relations with Sir George Stokes and his family were drawn closer still. In letters between Stokes and Kelvin the anxious solicitude of both in the matter of Maxwell's last illness comes out very strongly. On his death the arrangement of his affairs, both civil and scientific, devolved on Sir George Stokes' hands.

To Major W. A. ROSS, R.A.

CAMBRIDGE, 18 *May*, 1872.

My dear Sir,

I return the proofs of your book. I have not entered much into modern speculations about atomicity, and am not therefore competent to give a good opinion about its usefulness. My notion is that in chemistry there is a tendency at present to introduce to the student speculations at too early a stage, possibly facilitating a little his progress just at first starting, but going far to vitiate his philosophic intelligence and to prevent him from taking an impartial view of the evidence which may be adduced in favour of such speculations—leading him in fact to accept as dogma what he ought to accept, if he accepts it at all, on account of the simplicity with which it explains and groups together the known facts.

I am not sure whether you mean to use the polygons merely as illustrations or actually suppose that such is the shape of the molecules. If the latter I cannot go with you. We are far, I think, from knowing even as a probability what a molecule is ; but spectral phenomena prove to my mind that it is not a little complex even in the case of the so-called elementary substances.

Yours sincerely,

G. G. STOKES.

From Notes for a Report by Prof. Stokes (1888).

I confess that what the author regards as manifest seems to me very probably not true. What we know of isomeric compounds precludes the supposition that their chemical properties are, so to speak, functions of their molecular weights. And if this be not true of compounds, what right have we to assume that it is true of the molecular or even of the atomic weights of the so-called elementary bodies, which for anything we know to the contrary may be compounds of bodies yet more elementary which have never been isolated ?

But whether this assumption be true or not, it is of course open to any one to try whether he can discover, be it only as an empirical law, a relation between a series of numbers and the successive atomic weights of the elements. This is what the author has set himself to try.

How far there is evidence that the result has a real foundation in nature, or how far it is merely a playing with numbers, and getting out of a formula what had first been put into it, is a question on which it is hard to form a decided opinion without working a good deal, as the author has done, at the actual numerical trials. As far as I can judge by merely reading the paper, I confess the evidence for a natural foundation appears to me not to be strong. It does not of course follow that the arrangement is without value, were it only, as the author suggests, for purposes of chemical instruction, if it be held to be artificial rather than natural ; but of course a very different estimate would be placed upon its importance according as it was or was not supposed to have a foundation in nature.

ROYAL SOCIETY.

The amount of time and care that Prof. Stokes devoted, for a quarter of a century, to the well-being of the Royal Society has been amply recognized by the limited number of persons who are in a position to know the circumstances. The following letters and extracts may be of interest as illustrations.

To Prof. SYLVESTER.

R. S., 22 *April* 1872.

As to my doctrine of indebtedness in the matter of papers, I did not mean, as you seem to imagine, to speak of the balance of a debtor and creditor account, but only of the entry on one side. I cannot speak of the feelings of others, but I *can* speak of my own when I say that I do feel under obligation to the Societies that publish, and the referees that examine my papers. I don't mean to say but that a good paper is also an acceptable present to a Society. I look on it as one of those instances of mutual benefit that bind human Societies together, and I prefer leaving it in that courteous shape to going on to strike a balance.

When you remember that 100 papers or so per Session pass in one way or another through my hands, perhaps you will excuse my not having all the details of references at my fingers' ends.

If in this case there has been a minute ruffling of the previously smooth course of a friendly correspondence, it will at least have the advantage of preventing me from again forgetting your request.

P.S. On reading your letter again I see I misunderstood one par. by connecting the words "by some slight expression of regret or" with the passive verb immediately preceding instead of (as was intended) a passive verb much higher up in the sentence which was a long one. This altered the sense in a manner which (as I may freely say now that I perceive I misunderstood you) made it not a little offensive, and was the cause of my writing the last sentence in my letter.

I had *not* forgotten your *general* unwillingness to act as referee, but I had not remembered, nor do I now remember, nor with your letter open before me am I sure you assert, that your former unwillingness to act as referee amounted to a request that in *no case whatsoever* should a paper be sent you. I can hardly imagine that if it did I should have failed to make a mental note of it, but if it did I certainly owe you an apology for my forgetfulness. I shall understand it so for the future.

OBSERVATORY, ARMAGH,
4/8/74.
MY DEAR HUXLEY,

I have heard from Ray Lankester about his illustrations. I am afraid we are treating him as battledores do a shuttlecock.

If the snails knew they were to be honoured with TEN PLATES in the *Phil. Trans.* wouldn't they cock up their horns?

FROM SIR J. D. HOOKER, PRES. R.S.

Dec. 7/78.
MY DEAR STOKES,

I should long ere this have thanked you for the valuable hints as to the Address, but I have been occupied with Kew arrears, and also been away from home.

Your criticism on the crystalline nature of starch is most important, and I thank you for putting it so clearly to me: there is however a difference between starch being crystalline in structure, and what Nägeli contends for, which is, that the particles (molecules he calls them but not chemical molecules) which enter into the composition of starch have a crystalline character. The latter view is, as I am assured by Dyer (who is well up in the literature of the subject), universally adopted in Germany.

I must confess I cannot follow the physiologists' reasoning, the whole subject of the structure of the cell-wall as given in Sachs' *Lehrbuch* appears to me to be very speculative. I think it is a pity that the word 'molecule' was introduced at all, and that Nägeli's splendid physiological researches were accompanied with explanations that may be disputed by physicists. But I must give the state of the Science as accepted by physiologists, and cannot do this without an allusion to Nägeli and Strassburger's crystalline theory of the particles in cell-walls, starch grains, &c.

I have entered a caution by adding after believed—"by most

physiologists "—which correctly I believe represents the opinion
of the many German botanical schools, and of the rising Cambridge one under Mr Vines I suppose.

Lastly your criticism on my treatment of the "Radiometer,
&c." is quite just and indeed obvious to me.—I have altered it.
The more accurate determination of the Solar parallax I classed
with *advances* appealing to seekers of knowledge for its own sake,
and as having suggested the field of research in the parallax of
Mars—perhaps rather lamely! Again thanking you for this and
for many acts of kindness and consideration during my Presidency

From the "Pall Mall Gazette."

CAMBRIDGE, 8 *July*, 1880.

SIR,

In your impression of the 6th instant, in an article
headed "Who speaks first?" occurs an "illustration" which I
am sure the writer would not have used if he had been fully
acquainted with the circumstances of the case.

In January last Mr Huggins mentioned to me that he had
found that the flame of burning hydrogen on being photographed
gave a whole series of lines in the ultra-violet region of the
spectrum, and a few days later he showed me one of the photographs. We had some conversation on the subject, and it was
suggested that the spectrum was really that of incandescent
water. I rather think, but my recollection is not certain about
this, that he expressed some doubts whether he had best publish
the result he had obtained at once, or wait till he should have
examined other flames. Be that as it may, he adopted the latter
course.

A few months later Professor Liveing mentioned to me in
conversation that Professor Dewar and he had observed a set of
lines in the ultra-violet part of the spectrum which were referable
to water or appeared to be so. I of course said nothing about
Mr Huggins's discovery, which had been communicated to me in
confidence; and having been thus made the confidant of both
parties, I carefully abstained from saying anything to either
which might have the effect of either accelerating or retarding
the publication of their results.

Those who examine the *Proceedings* of the Royal Society will
know that Professors Liveing and Dewar have for a considerable

time been engaged together in a careful examination of the spectra of chemical elements and of some combinations. The further examination of the spectrum of water with a view to eliminating possible impurities and mapping it in detail naturally fell in with their work. The subject was fully worked out in time to send a paper to the Royal Society which was read at the last meeting in June. It is true that in a paper read very shortly before, Professors Liveing and Dewar mentioned that they had been working at the spectrum of water, and promised shortly to send a paper on that subject; and it was doubtless this mention that induced Mr Huggins to communicate a paper on his own work on the same subject without waiting till the materials for a longer paper on the spectra of different flames were ready for publication. Mr Huggins's paper on the spectrum of water was read at the same meeting as that of Professors Liveing and Dewar. We have therefore here another example, of which the history of science affords so many, of the same discovery being made perfectly independently and nearly simultaneously by independent workers following independent lines of research.

I will only add that in making the above communication I am acting in accordance with the wishes of both parties.

I am Sir, your obedient servant,

G. G. STOKES.

To an Author.

Lensfield Cottage, Cambridge,
2 *May* 1882.

It is of course no part of my duty as Secretary of the Royal Society to revise, or attempt to revise, papers that the authors send in. I have plenty of work of my own without that. Had I simply confined myself to my duty there was an end of the matter.

But I thought it a pity that it should not be published, merely from what were deemed by the referees, and I must say I agree with them, faults of style. I thought therefore that I would try if I could not revise it so as to make it more acceptable, of course submitting it to you for your approval before it was presented as your paper. But I have always loads of things on hand waiting to be done, and so this matter, which had not got to be done by a particular day, got put off.

1 *Nov.* 1884.

DEAR SIR RICHARD [OWEN],

In your letter of Oct. 10 you mentioned an interchange of the numbers of the plates illustrating your papers on *Sceparnodon* and *Notiosaurus*, and suggested that if the Biological Secretary had compared the plates before the volume was sent to press the mistake would not have happened.

In reply I expressed my great concern at the mistake, and took the blame upon myself, as I am editor of the *Transactions*.

On going to the Royal Society this week I was very much pleased to find that there had been no such mistake as you supposed. The numbers were all right, and it was merely that the stitcher of your private copies had not followed them.

Hence there is no blame for the Secretaries to take upon themselves; for I need hardly say that it does not form part of the duties of the Secretary of the Royal Society, whether the Biological or the Physical Secretary, to look over the shoulders of a bookbinder's assistant as he is doing his work lest he should make some mistake. Yours very truly,

PROF. MINCHIN ON NATURAL RADIATION.

A letter from Prof. Minchin, entering on a train of ideas in relation to the law of natural radiation which has recently been revived with success, elicited an interesting reply bearing on the exciting cause of all radiation.

ROYAL INDIAN ENGINEERING COLLEGE,
COOPER'S HILL, STAINES.

17*th Nov.* 1878.

DEAR SIR,

Professor P. Martin Duncan has advised me to address you with reference to experiments on spectra, and has allowed me to use his name as an introduction to you.

Can you tell me whether any of those who have investigated the heat and light intensities in different parts of the spectrum have determined numerically the values of the ordinates of the curves of intensities corresponding to abscissæ measured along the spectrum?

It has just struck me that a close connection between the Kinetic Theory of Gases and the theory of ethereal vibrations in the spectrum might be verified mathematically by determining the equations of the above-mentioned curves of intensity from a large number of measurements of ordinates.

For, in the Kinetic Theory, it is proved that the number of molecules of the gas which are moving about with velocities $> x$ and $< x + dx$ is proportional to

$$x^2 e^{-x^2/a^2} . dx,$$

where α is the mean velocity of the molecules (the gas being supposed homogeneous).

Now if we construct the curve $y = x^2 e^{-x^2/a^2}$, its shape is somewhat as in the following figure—

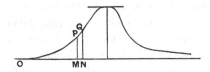

and this figure is extremely like the pictures of the curves of intensities of light and heat which one sees in the books. The maximum ordinate corresponds (as is *à priori* natural) to the mean velocity, and the number of molecules moving with velocities $> OM < ON$ is represented by the little rectangle $MNQP$.

There are many considerations connected with this curve of velocities which, I think, would throw light on the nature of the vibrations both in solid bodies and in gases. If we assume that x above has actually only values somewhere in the neighbourhood of α, it strikes me that the fact of *linear* spectra in gases might be explained.

I shall be much obliged if you will let me know whether your extensive work on this subject will furnish me with the required information, or direct me to some source whence I can obtain it. I am, dear Sir, yours truly,

GEO. M. MINCHIN.

L. C. C., 18 *Nov.*, 1878.

Dear Sir,

The ratios of the light and of the heat at different parts of the spectrum have certainly been determined numerically from observation. They were so determined for light by Fraunhofer. The results are given in his paper in the *Munich*

Transactions, in which he announced his discovery, or more exactly, re-discovery, of the fixed lines of the spectrum. See Vol. v. pp. 210 *et seq.* The intensity for heat also has been measured, among others by Tyndall.

I do not think the results can be applied in the way you propose. It appears to be one of the mysterious properties of the ether that it lets through solid matter without the excitement of luminous vibrations, or of vibrations of the same kind. For the production of such vibrations, it appears to be necessary that two portions of matter should be influencing each other's motion, be they two distinct molecules in what is spoken of as a collision, or be they portions of one and the same complex molecule. In the latter we must include the molecules, as we know them, of the so-called elementary bodies. A transparent gas requires an extremely high temperature to be luminous, a temperature far above that at which a solid metal would be brilliantly luminous. Yet according to theory the molecules of the former are moving about more freely than those of the latter. Take again a blow-pipe bead of say microcosmic salt. This will emit no sensible light when the platinum wire which holds it is nevertheless at a strong red or it may be at a white heat. It ought not to be so, were the motion of the molecules in the ether the only thing required for the production of light. Yours truly

G. G. STOKES.

CAVENDISH CHAIR; LORD RAYLEIGH.

On the death of Prof. Clerk Maxwell in the Autumn of 1879, the adequate continuation of the new departure which he had initiated and superintended, in the building and equipment of the Cavendish Laboratory, became a subject of anxious concern in the University of Cambridge. It was felt that every effort should be made to obtain the services of either Sir W. Thomson or Lord Rayleigh, as head of the Laboratory. At an early stage it had become known that the former would not be able to break up his life-long connexion with Glasgow: thus effort was concentrated on inducing Lord Rayleigh to offer himself to the suffrages of the Senate, to whom the selection belonged. As in the previous case of Maxwell, and in others, it devolved on Prof. Stokes to be the means of communication. The following letter was accompanied by a statement:—

"We the undersigned members of the Senate of the University of Cambridge, whose names are on the Electoral Roll, are of opinion that in the probable event of the re-establishment of the Professorship of Experimental Physics, which has been terminated by the lamented death of Professor Clerk Maxwell, it would tend greatly to the advance of Physical Science and to the advantage of the University that Lord Rayleigh should occupy the chair:"

signed by 125 members of the Senate, practically amounting to the whole of the official part of the University.

LENSFIELD COTTAGE, CAMBRIDGE.
18 November, 1879.

DEAR LORD RAYLEIGH,

When we were speaking last week about a successor to Maxwell, it never occurred to me to look on you as a possible occupant of the chair. Some of your friends however here have thought that you might be induced to accept it; and the enclosed memorial will show how acceptable your consent would be.

In the Grace establishing the Professorship, Feb. 9, 1871, it was enacted that the Professorship should terminate with the tenure of office of the Professor first elected, unless the University should decide by Grace of the Senate that the Professorship should be continued.

In the *Reporter* of to-day appears the following Grace which is to be offered to the Senate next Thursday:

"That the Professorship of Experimental Physics, established by Grace of the Senate, Feb. 9, 1871, be continued, subject to the regulations then enacted so far as they are now applicable."

The regulations laid down by Grace of the Senate on Feb. 9, 1871, declared:

"It shall be the principal duty of the Professor to teach and illustrate the laws of Heat, Electricity and Magnetism, to apply himself to the advancement of the knowledge of such subjects, and promote their study in the University."

They further required 18 weeks' residence and a course of lectures in each of two terms at least. I mention these as you may have left your copy of the *Ordinationes* at the office of the Commission, and might wish to refer to it. I do not think it worth while to mention the other regulations, which are of minor consequence.

15—2

I have now put you in possession of the principal facts of the case. I will only say in conclusion that I should be greatly pleased to hear that you were disposed to look favourably on the proposal, and the paper enclosed will show how widely that feeling is shared.

I remain, dear Lord Rayleigh, yours sincerely,

G. G. STOKES.

TERLING PLACE, WITHAM, ESSEX.
Dec. 1/79.

DEAR PROF. STOKES,

After much consideration I have made up my mind to stand for the Professorship of Experimental Physics, feeling unable to resist a request such as that which you forwarded to me*. I can only hope that my friends are right in their view as to my qualifications for the office.

Yours very sincerely,

RAYLEIGH.

Dec. 1/79.

DEAR PROF. STOKES,

Sidgwick looked at Lensfield House†, but we conclude from his report that there are not enough rooms in it, though the lawn is attractive. We had heard of the house in St Peter's Terrace, and think that it would suit.

This being the case, I think I ought no longer to retain the refusal of Lensfield House.

With thanks, Yours very sincerely,

RAYLEIGH.

The following letter from Prof. Ketteler, of Bonn, whose theoretical and experimental work has been in the forefront in advancing the theory of the dispersion of light, is of interest for the history of the progress of optical science.

* In a letter from Henry Sidgwick (*Life*, p. 342) to his sister, from Terling Place, May 29, 1879, a passage occurs: "As for other news, we are just now anxious about Rayleigh's coming to Cambridge.... He has not yet decided. It is rather a wrench to give up leisure and the comforts of a country house unless one is *quite* sure that one's duty to society requires it."

Lord Rayleigh held the chair, residing in Cambridge during term, for four years.

† Next to Lensfield Cottage; afterwards and still the residence of Dr and Mrs Laurence Humphry.

BONN der 16 *Juni* 1880.

SEHR GEEHRTER HERR!

Der Unterzeichnete erlaubte sich schon seit einer Reihe von Jahren, Ihnen dann und wann Separatabdrücke seiner optischen Arbeiten zugehen zu lassen. Diesmal giebt er sich die Ehre, die vier letzten von ihm publicirten Abhandlungen gleichzeitig zu übersenden. Zwei weitere kurze Aufsätze: "Constructionen zur anomalen Dispersion" und: "Theorie der chromatischen Polarisation einer senkrecht zur Axe geschliffenen dichroitischen Platte" sind unter der Presse.

Da ich Sie, geehrtester Herr, in der optischen Wissenschaft als hochstehende Autorität verehre, so wollen Sie es mir zu gute halten, wenn ich der heutigen Sendung einige Worte an Sie beifüge. Ich habe mich in meinen Arbeiten bemüht, die theoretische Optik auf Grundlage der Annahme des Zusammenschwingens der Aether- und Körpertheilchen umzubauen und insbesondere die Gesetze der Spiegelung und Brechung an und in den absorbirenden Mitteln zu erforschen. Meines Erachtens habe ich die Aufgabe für die isotropen Mittel ziemlich vollständig und für die anisotropen wenigstens ausreichend gelöst. Zu bearbeiten bliebe daher nur noch die Rotationspolarisation, indess wird eine Theorie derselben, wie ich meine, erst nach weiterer experimenteller Untersuchung absorbirender drehender Substanzen Aussicht auf Erfolg haben.

Am ausführlichsten verbreiten sich meine Arbeiten über Metall- und Total-Reflexion und über anomale Dispersion. Bezüglich der letzteren dürfte es mir gelungen sein, die früheren Ansichten Boussinesq's, Strutt's, Sellmeier's und Helmholtz's glücklich zu einem einheitlichen Ganzen zu verschmelzen. Die Integration der bezüglichen Bewegungsgleichungen giebt dann Beziehungen, die ich als das Haupt- und Grundgesetz der ganzen Dioptrik betrachte. Verbindet man dasselbe mit den von mir umgeformten Uebergangsbedingungen des Lichtes und unter Zuziehung des Incompressibilitätsprincips, so erhält man die Amplituden und Phasen nicht bloss der gespiegelten, sondern auch der gebrochenen Welle und damit insbesondere die Gesetze der sogenannten Metallreflexion, die übrigens mit den früheren, wohl als empirisch zu bezeichnenden Formeln Mac-Cullagh's und Cauchy's übereinstimmen. Wenn sich freilich in unerwarteter und überraschender Weise herausstellte, dass der Refractions- wie der Absorptionscoefficient der absorbirenden Mittel mit dem

Einfallswinkel veränderlich ist, so wird dieser Umstand leider gerade für die Aufnahme und Anerkennung der Theorie ein arges Hinderniss abgeben. Um so mehr musste mich daher die Wahrnehmung erfreuen, dass die von mir aufgestellten allgemeinen Formeln die Fresnel'schen Gesetze der Totalreflexion als einen speciellen Grenzfall umfassen, da ja eben der "streifende Strahl" derselben, wie längst thatsächlich anerkannt wird, nach Refraction und Extinction vom Incidenzwinkel abhängt.

Leider wird meinen Arbeiten zur Zeit noch die Anerkennung versagt; man behandelt sie wie so manche Eintagsproducte, die auf diesem Gebiete neuerdings namentlich mit Maxwell'schen Ideen herumphantasiren. Die meisten Celebritäten stehen zudem bekanntlich jetzt der Optik fern, und was insbesondere die Aufnahme Deutscher Arbeiten in Frankreich betrifft, so ist dieselbe seit 1870 noch immer sehr erschwert.

Vielleicht, hochgeehrter Herr, darf ich die Hoffnung hegen, dass namentlich Sie meinen theoretischen Resultaten voll und rückhaltlos zustimmen. Wäre das wirklich der Fall, so dürfte ich mir wohl die dringliche Bitte gestatten, Sie möchten gütigst bei etwa sich darbietendem Anlass, etwa in einer der gelehrten Körperschaften Ihres Landes, auf meine Arbeiten aufmerksam machen. Sollte es sich empfehlen, die Separatabdrücke derselben etwa der Royal Society vorzulegen, so würde ich gern erbötig sein, eine mögligst vollständige Zusammenstellung derselben einschliesslich der von mir herausgegebenen Werke: "Astronomische Undulationstheorie oder die Lehre von der Aberration des Lichtes. Bonn 1873" und: "Beobachtungen über die Farbenzerstreuung der Gase. Bonn 1865" zu Ihrer geneigten Disposition zu stellen.

Indem ich mir die Ehre erbitte, Ihnen auch meine bevorstehenden Publicationen übermitteln zu dürfen, verharre ich mit vorzüglicher Hochachtung als

Ew. Hochwohlgeboren, ganz ergebenster,

E. KETTELER, Professor.

PROF. OSBORNE REYNOLDS.

The following letters relate to the researches of Prof. O. Reynolds. In the earlier work on gas-theory, he and Prof. Clerk Maxwell proved to be engaged in the same field; though they benefited by each other's work, they were not able to come to an agreement on general principles.

FALLOWFIELD,
6 *Nov.* 1879.

MY DEAR SIR,

Thank you for your letter. Of course ever since I saw you I have been expecting the sad intelligence [of the death of Clerk Maxwell]. I was talking to Dr Joule on Tuesday night when he was almost cross with me for saying that there appeared to be no hope.

I quite agree with what you say as to the proper course to be taken with my paper or letter, whichever it may turn out to be.

When you write next will you kindly tell me if I am right in supposing that the matter in a preliminary communication in the *Proceedings* may be taken up again and more amply dealt with in a final communication?

I have been experimenting now for a long time on the actual motion of water—photography, vortex rings, &c.—and I have still more that I want to do before communicating the results. But last spring I arrived at what seems to me to be the theoretical explanation of the formation of eddies, and I should like to communicate this shortly without waiting till I have worked up all the rest. What I have done is to show that elements having rotational motion must, according to the dynamical theory, remain and accumulate in certain positions with respect to the solid surface against which the fluid is moving. The positions which the reasoning assigns for the accumulations agree with those at which the eddies and vortices are formed. I think I have thus solved the puzzle of the formation of vortex rings, and completed the proof that the resistance encountered by solids, other than skin resistance, is to be explained by the accumulation of eddies in their rear, as you suggested in 1843.

Believe me, yours sincerely,

OSBORNE REYNOLDS.

16 *March*, 1880.

DEAR PROFESSOR STOKES,

Thank you for your letter and the comments on my paper. I think that the right course is to publish my letter.

As I pointed out when I last wrote about it, I should like my letter to appear as having been written before Prof. Maxwell's death—it was written before I was aware of his extreme illness and on the moment of my first receiving his paper. This requires that the date should remain unaltered, and so the alteration you have made in the first sentence will not stand, for this makes it appear that the letter was written after Maxwell's death.

I think that the fact that Maxwell's paper has been published removes the objection which you have raised against this sentence as it originally stood.

I have nothing to urge against your second alteration, at all events if you were to add a line to your comments stating that Prof. Maxwell was Referee—which statement appears to me to be on all accounts desirable. However this I leave to your judgement.

April 25th, 1883.

MY DEAR DR STOKES,

Thank you very much for your letter and the criticisms which it contains. I am very sorry that I have so badly expressed my meaning.

I had no intention whatever of laying down the conditions of dynamical similarity, although I now see that Art. 6 not only bears this construction but really fails to express what I meant. I shall be glad to adopt any means by which this may be remedied.

The "evidence" referred to in that article is not evidence that (as I put it)

$$l\rho U/\mu = \text{(an arbitrary constant)}$$

represents a condition of dynamical similarity, but evidence that, as this constant varies, for some particular value the dynamical condition underwent a critical change, which might be a change from a condition of stability to one of instability. I have referred to two terms as though they were the only two terms in the equations; but this was an oversight and not intended.

Taking cylindrical motion parallel to the plane of xy, the equation of motion becomes on eliminating p

$$\frac{d\zeta}{dt} = -u\frac{d\zeta}{dx} - v\frac{d\zeta}{dy} + \frac{\mu}{\rho}\nabla^2\zeta.$$

What I meant was that the right-hand member of such equations consists of two sets of terms

$$u\frac{d\zeta}{dx} + v\frac{d\zeta}{dy} \propto \frac{U^2}{l^2},$$

$$\frac{\mu}{\rho}\nabla^2\zeta \propto \frac{\mu}{\rho}\frac{U}{l^3}.$$

Consequently if the effect of one of these sets of terms is for stability and the other for instability, or roughly if one is positive and the other negative, there must be some particular value of lpU/μ for which $d\zeta/dt$ will change sign; or, taking $d\zeta/dt$ to represent the eddying motion, any small eddy disturbance would diminish up to this particular value and then increase.

It was this evidence of a critical value for, what may well be called, the parameter of dynamical similarity, that I referred to as having been "overlooked,"—and the critical values, for I have shown there are two, both of which are determined with more or less exactness, which I considered as two new physical constants. Of course these constants are contained in the equations of motion, but it seems that they are none the less new since the integrations necessary for their determination have not yet been effected.

My first intention was to include the mathematical and physical investigation in my paper, and I wrote the introduction with that view; but finding my time running short I decided to include only the experimental, and in cutting down the reference to the mathematical portion I cut out rather too much.

Will you kindly suggest what course I should take?

The following statement was made by Sir George Stokes as President of the Royal Society in presenting a Royal Medal to Prof. Osborne Reynolds on Nov. 30, 1888.

"A Royal Medal has been awarded to Professor Osborne Reynolds for his investigations in mathematical and experimental physics, and on the application of scientific theory to engineering.

"Professor Reynolds was among the first to refer the repulsion exhibited in that remarkable instrument of Mr Crookes's, the radiometer, to a change in the molecular impact of the rarefied gas, consequent upon the slight change of temperature of the moveable body due to radiation incident upon it; and in an important paper published in the *Philosophical Transactions* for 1879, he deduced from theoretical considerations the conclusion that similar phenomena might be expected to be observed in bodies surrounded by a gas of comparatively large density, provided their surfaces were very small. He verified this anticipation by producing on silk fibres, surrounded by hydrogen at the atmospheric pressure, impulsions similar to those which in a high vacuum affect the relatively large disks of the radio-meter.

" In an important paper published in the *Philosophical Trans-actions* for 1883, he has given an account of an investigation, both theoretical and experimental, of the circumstances which determine whether the motion of water shall be direct or sinuous, or, in other words, regular and stable, or else eddying and unstable. The dimensions of the terms in the equations of motion of a fluid when viscosity is taken into account involve, as had been pointed out*, the conditions of dynamical similarity in geometrically similar systems in which the motion is regular; but when the motion becomes eddying it seemed no longer to be amenable to mathe-matical treatment. But Professor Reynolds has shown that the same conditions of similarity hold good, as to the average effect, even when the motion is of the eddying kind; and moreover that if in one system the motion is on the border between steady and eddying, in another system it will also be on the border, provided the system satisfies the above conditions of dynamical as well as geometrical similarity. This is a matter of great practical import-ance, because the resistance to the flow of water in channels and conduits usually depends mainly on the formation of eddies; and though we cannot determine mathematically the actual resistance, yet the application of the above proposition leads to a formula for the flow, in which there is a most material reduction in the number of constants for the determination of which we are obliged to have recourse to experiment.

" There are various other investigations of Professor Reynolds's which time would not allow me to enter into, and I therefore merely mention his investigation of the relation between rolling friction and the distortion produced by the rolling body on the surface on which it rests, that of the effect of the change of temperature with height above the surface of the ground on the audibility of sounds, and his explanation of the effect of lubrication as depending on the viscosity of the lubricant."

---

* By Prof. Stokes himself in 1850 in his Memoir on Pendulums; cf. *Math. and Phys. Papers*, Vol. III. p. 17. The subject was further developed with weighty applications to balloons (1873), winds, waves, and clouds (1886-90) by Helmholtz.

19, LADY BARN ROAD, FALLOWFIELD, MANCHESTER.

March 31, 1902.

DEAR SIR GEORGE,

Thank you very much for the third volume which I am very pleased to have, not only for itself but also on account of the donor. Your visit was a great pleasure to us, and we hope you were none the worse after your expeditions.

With our kind regards, Yours sincerely,

OSBORNE REYNOLDS.

VISCOSITY OF GASES.

The following letters relate to Prof. Stokes' reduction of Crookes' experiments on the rate of decay of oscillation of a mica plate in very high vacua. The presumption expressed in the first letter that nothing quantitative could come out of this mode of experimentation was on the face of it not unreasonable; but an examination of Prof. Stokes' masterly discussion* shows that in fact more elaborate appliances might really have been a hindrance rather than an aid. The important memoir by Kundt and Warburg was translated in *Phil. Mag.* 1875.

STRASSBURG IN ELSASS.

d. 26 *Marz*, 1881.

GEEHRTER HERR!

Aus einer der letzten Nummern der *Nature* ersehe ich, dass Sie sich für Untersuchungen über Reibung in den Gasen auch jetzt noch in hohem Grade interessiren, indem Sie aus den Versuchen des Hr Crookes die Reibungscoefficienten berechnen.

Ich erlaube mir daher Sie darauf aufmerksam zu machen, dass bereits 1875 vom Herrn Warburg und mir sehr ausführliche Untersuchungen über Reibung und Wärmeleitung verdünnter Gase angestellt sind. Dieselben sind nicht bloss viel umfassender als die des Herrn Crookes, indem dieselben bereits alle von letzterem gefundenen Resultate enthalten, sondern unterscheiden sich auch wesentlich dadurch von denen des Herrn Crookes, dass sie so ausgeführt sind, dass sie eine strenge Berechnung zulassen. Die Untersuchungen finden sich:

* *Phil. Trans.* 1881, *Math. and Phys. Papers*, Vol. v. pp. 100–116.

Poggendorffs *Annalen*, Bd. CLV. pp. 337—365, 525—550, CLVI. pp. 177—211, 1875.

Ich erlaube mir, Ihnen durch die Post Separatabzüge der betreffenden Abhandlungen zu zusenden.

Hochachtungsvoll ergebenst,

A. KUNDT,

Professor der Physik.

STRASSBURG IN ELSASS. 16/4. 81.

SEHR GEEHRTER HERR!

Gestatten Sie mir, Ihnen meinen besten Dank für Ihre ausführliche Mittheilung vom 29 Marz auszusprechen. Ich habe Ihr Schreiben auch Herrn Professor Warburg, welcher mit mir die Untersuchung über Reibung und Wärmeleitung ausführte, zugesandt.

Wenn die Versuche des Herrn Crookes auch nach einer andern Methode als die unsern ausgeführt sind, und wenn bei den Reibungsversuchen das Evacuiren der Apparate von Herrn Crookes auch etwas weiter getrieben ist, als von uns, so ist doch unzweifelhaft von uns—wie ein näherer Einblick in unsere Abhandlungen Ihnen gezeigt haben wird—die Abnahme der Reibung eines Gases mit Zunahme der Verdünnung nicht bloss experimentell nachgewiesen, sondern auch die Theorie der Erscheinung aus der Gastheorie, das *Auftreten der Gleitung*, etc. etc., bereits vor Jahren entwickelt. Und wie Sie selbst schreiben :

Consists the chief interest of Mr Crookes' experiments in tracing the falling off of the damping effects of the gas, when excessive exhaustions are reached.

Entschuldigen Sie, wenn ich für heute mich mit dieser kurzen Erwiederung begnüge.

Eine ausführliche Darlegung der Resultate unserer Untersuchungen, gegenüber denen des Herrn Crookes, würde ich lieber nur in Gemeinschaft mit Herrn Warburg geben, mit dem ich gemeinschaftlich die betreffenden Arbeiten ausführte. Letzterer ist aber seit einigen Jahren nicht mehr in Strassburg, und mithin ein gemeinschaftlicher Schreiben zur Zeit wenigstens nicht möglich.

Mit der Versicherung ausgezeichneter

Hochachtung ergebenst,

A. KUNDT.

LETTERS FROM DR JOHN RAE, F.R.S.

14 *April*, 1881.

I have much pleasure in replying to your note of yesterday's date on the subject of the Aurora *.

In 1846–7 I wintered at Repulse Bay and again in 1853–4 at the same place, on the west side of Hudson's Bay in latitude 66° 32′ North, longitude 86° 56′ W., variation of compass 62° 50′ west, dip or declination of needle 88° 12′.

On the last occasion I did not pay particular attention to the aurora as the displays were few and unimportant—but on the first visit (1846–7) I find recorded 39 appearances of the aurora from September to February inclusive, say six months—these displays were never fine and sometimes very faint, always of a pale yellow or straw-colour, generally in the form of an arch to the magnetic SOUTH and not at a great height; sometimes there were streaks or beams of the same colour described pointing towards the zenith, these were generally seen in a southern direction. The movements of both forms were seldom if ever varied.

The magnetic needle suspended by a filament of silk in a snow-hut was never visibly affected by these aurorae.

I may mention that there was water uncovered by ice, owing to rapid currents during the whole winter, in the direction of the magnetic south, about 30 or 40 miles from winter quarters.

At Fort Confidence, on the shores of Great Bear Lake, latitude 66° 54′ N., longitude 118° 49′ W., variation of compass 48° 30′ E., where I wintered with Sir John Richardson in 1848–9, the movements of the aurora were watched with great care and attention. We had a number of fine displays of the flashing, rapidly moving and variously coloured aurora, which you appropriately term "flickering," and it was these that acted so powerfully on the suspended magnetic needle. These aurorae were usually seen towards the *magnetic north*, and they were generally at a greater altitude than the others, sometimes as high as the zenith.

I may mention that both Indians and Eskimos assert that the bright, rapidly moving aurorae produce sound. I shall be happy to answer any other questions.

---

* Cf. Prof. Stokes' introductory lecture on Solar Physics, delivered at South Kensington, April 6, 1881, on behalf of the Committee on Solar Research, and reported in *Nature*, Vol. XXIV. pp. 595–8 and 613–8.

22nd October, 1881.

I duly received your letter of to-day's date on the subject of
Glaciers, and regret that I cannot answer either of the two
questions you ask.

Unfortunately the lands near to where I wintered in the
Arctic are neither high enough nor steep enough to have glaciers
of any kind, "*thick* or *thin*," and there was no opportunity of
studying the subject.

My own idea is that Arctic glaciers must have a downward
motion more or less rapid during the whole year, summer and
winter. I believe the alternation of heat and cold—or I should
rather say of temperature—would of itself cause motion, especially
near the upper surface.

We know that ice, two or three feet or more thick, contracts
very considerably in a few hours by a sudden fall of 15 or 20
degrees of temperature. I have found cracks in the ice on Lake
Winnipeg three or four feet wide formed by this cause—during a
single night—almost stopping our sledge journey; this gap soon
freezes up. Then the weather gets milder, the ice expands, and
with the new additional formation, is too large for the lake, and
is forced up into ridges.

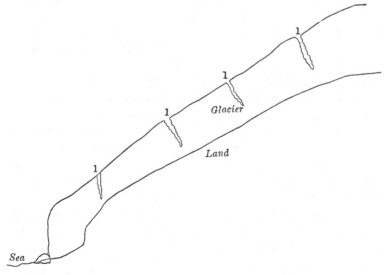

1 = Cracks formed by contraction caused by cold.

This process goes on, at every cold *snap** alternating with milder weather.

I find on looking over some of my journals that during the winter months—November to April inclusive—we had during each month two or more alternating short intervals of colder and milder weather, each of which visibly affected the ice on the lakes in the manner I have described.

1st November, 1881.

I thank you much for your letter, telling me of "Moseley's theory" of ice or glacier movement, of which I certainly had never heard.

There was a curious case of lake-ice movement (perhaps there are many instances of the same kind) observed in Canada, which may not have come under your notice.

A long, rather narrow and shallow lake interfered with the making a straight road or railway (I forget which). To obviate the difficulty a line of piles was driven into the mud across the lake, rather nearer one end than the other, on these a road was constructed.

Before the end of winter the posts were so much displaced or injured by ice movement that the road had to be given up. This damage took place before the breaking of the ice, and as there was little or no current, must have been caused by contraction and expansion, such as I mentioned in my former letter.

Some time before 1880, owing to increasing difficulty in writing, Prof. Stokes took to using a typewriter, with great advantage to his correspondents and much satisfaction to himself.

"I have now had this machine for several years. Some weeks ago I tried to run a race with myself, writing down a passage I knew by heart at first with the machine, and then with the pen. I found that with the machine I could get through a given amount of matter in about 70 per cent. of the time required for the pen. Besides, writing with the pen is laborious, whereas writing with the machine is rather an amusement than otherwise."

* "Cold-snap," an American term, meaning a rather sudden increase of cold.

To PROF. S. P. LANGLEY*.

LENSFIELD COTTAGE, CAMBRIDGE,
27 *April,* 1881.

DEAR SIR,

The importance to man of solar radiation is too well known to require me to dwell upon it. If there be reason to suspect that the amount of radiation is variable, an investigation of the nature and laws of its variation is not merely a question of deep scientific interest, but may prove to have very important practical bearings.

Now that which seems to me to be by far the most probable view of the nature of sun-spots leads to the expectation that when they abound radiation is more copious than on the average, and when they are absent less copious. I do not regard the spots themselves as the cause of increased radiation, rather the reverse, but as being a sequel of a condition of activity, of which they form the most easily observed manifestation, and which is of such a nature as to be accompanied by increased radiation.

Of course a change of radiation must tell upon the meteorology of the Earth. But the problem of the Earth's meteorology is one of such extreme complexity that I do not think that we can directly connect such things as temperature of the air, height of barometer, rainfall, and so forth, with solar radiation, in such a manner as to make out variations of the latter from variations of the former.

All, it seems to me, that we can do in this way is to fall back on the general principle that if there be a system, however complicated, which is acted on by a disturbing cause, and the latter be subject to a periodic variation, the disturbed system will show a variation of similar period. From observations on the disturbed system continued for a sufficient length of time

---

* The absolute measurements of solar radiation by Prof. Langley, whose death has recently been so great a loss to science, began in 1880 by the use of his then newly invented bolometer, and were continued for many years at both high and low levels; among the many marks of appreciation of the results, the award of the Rumford Medal of the Royal Society may here be mentioned. Cf. the memorial notice by C. G. Abbot, in *Astrophysical Journal,* May, 1906. The great importance attached to actinometry by Prof. Stokes, and the labours he undertook in order to evolve trustworthy instruments of observation, have already been noticed in connexion with the Meteorological Council.

we may infer the existence and length of period, and likewise the epoch, of a periodic disturbance, if such there be. From a periodic disturbance we may infer the existence of a periodic disturbing cause, of which the length of the period may be thus ascertained, but *not* the epoch. Hence, even if the periodicity of a change in the amount of solar radiation were more regular than we have reason to think that it is, a great while must elapse before we could make out in this way the existence and length of period of such a change, while we should remain in ignorance of the epoch.

The most hopeful way, it seems to me, for arriving at some knowledge of the variations, if any, in the amount of solar radiation, is by the employment of some form of actinometer. But for the employment of such an instrument it is of great importance to get rid as far as may be of the haze and dust which are contained in the lower portions of the atmosphere of the Earth,— in fact, to make the observations in as elevated a locality as reasonably may be. I think it very probable that in this way direct evidence may ultimately be obtained of a variation in the amount of solar radiation.

It is with a view to a study of solar radiation that I attach so much importance to a good elevation. But besides the study of direct radiation, the meteorological conditions at a high elevation are also of great interest, especially if observations at such a station are combined with others at a comparatively low level, taken at a place not far off in a lateral direction.

I am, dear Sir, yours sincerely,

G. G. STOKES.

To WARREN DE LA RUE, F.R.S.

30 *June*, 1881.

DEAR MR DE LA RUE*,

It seems to me that the only thing that can be done in a free balloon towards the measurement of atmospheric electricity—but it is a thing which there would be much interest in doing—is this:

There would be no difficulty in letting a rope of half a mile or so in length hang down from the balloon ; indeed that is often

---

* A long series of letters from Mr Warren de la Rue (1855–1885) exists.

done as a matter of aerial navigation. This rope might have belonging to it a thin copper wire covered with gutta-percha or india-rubber. This wire is connected above with one pair of quadrants, and with one coating of the Leyden-jar or other condenser, or with one pole of the battery of numerous small elements, according as a condenser or battery is used*. The other pair of quadrants is connected with the second coating of the condenser or second pole of the battery. The copper wire is connected below with some form of collector. Water would freeze; and besides, a supply of water, and the tube to lead it down to the bottom of the wire, would weigh too much. I think the best form of collector would be a brush of long, very fine wires. A spirit flame, fed by a supply below, might compromise the safety of the aeronaut; for he would probably want to draw up the rope, and then the flame might come in contact with escaping gas. Connect the car with the needle.

Or, what would probably be more convenient, connect the car and one coating or pole with one pair of quadrants, the other coating or pole with the other pair, and the insulated wire with the needle.

I have written to Sir William Thomson to ask him what he thought would be the best form of condenser collector.

The wire must not be used in thundery weather, and even in ordinary weather half-a-mile might be too great a length to begin with, for fear of any accident.

In 1872, the subject of the protection of the harbour of Madras occupied a great deal of his attention as a member of the Governmental Committee to consider the question, and led to much experimenting (see p. 32). A long and patient correspondence with the Engineer of the harbour that had been demolished by the waves, who was an old schoolfellow, still exists.

To W. PARKES, Esq., C.E.

1 VICTORIA BUILDINGS, HUNSTANTON.
4 *September*, 1882.

I got Stevenson's book on Harbours out of the University Library, and read it all with the exception of some details which did not concern me. I was much interested by it.

* In *this* arrangement the condenser or battery would have to be insulated.

I do not for a moment deny what you call the impulsive action of the moving water when its motion is checked or altered. But I do altogether deny that the existence of such an action entails the abandonment of the fundamental property of a fluid that the pressure, while varying from point to point, is at any point alike in all directions, a consequence of which is that the pressure is normal to the surface. Your ideas about an oblique action of the obliquely descending wave seemed to me to involve an abandonment of this fundamental principle.

What would actually take place if a wave were to descend obliquely in the manner you drew, is this:—the direction of motion would be rapidly deflected, and the motion itself a good deal checked. This would entail a considerable augmentation of pressure about the part of the block on which the water descended. This great increase of pressure is of brief duration, though not by any means so brief as those to which the term "impulse" is commonly applied. But this pressure remains normal to the surface, and consequently has no tendency to upset the block.

Had the block been so shaped as to present a sort of parapet on the harbour side, the case would have been altered. The checking of the horizontal motion would have produced a considerable pressure, increasing from the seaward side towards the side of the parapet—I speak it will be seen of an imaginary structure which an engineer would be careful *not* to make—and this pressure, acting horizontally against the side of the parapet, would tend to upset the block into the harbour. The little roughnesses of the block may act as such parapets, but they are too insignificant to be taken into account.

As to the additional wall you propose, why should it not be bonded to its neighbour in the manner you propose for the two existing walls? or if the necessity of providing for settlement should forbid this, then by chains, or at least bars formed of two or three links, in lieu of railway bars? Of course they would cost more, but I should suppose that the cost would be a very insignificant fraction of the cost of the repair as a whole.

I remain here till Saturday week.

To Sir John HAWKSHAW, C.E., F.R.S.

LENSFIELD COTTAGE, CAMBRIDGE.

4 *August*, 1882.

I enclose you, or rather send you on this sheet, a roughly copied drawing of the system I suggested for what you may call sliding bond. I should explain that the angles chosen and the proportion of length to breadth are not the result of any calculation, but were merely taken as representing something like what seemed suitable.

The breadth of a pier may take in any number, greater than one, of half breadths of a block. The figure below takes in five. On the next leaf I have inked in four only. The latter is most nearly analogous to Mr Parkes's wall.

Of course different patterns of keying might be suggested. The one I have drawn is perhaps the simplest and best.

The piles of blocks stand like basaltic columns, so that while bond is preserved, there is no interference with the principle of independent settlement, to which Mr Parkes seemed to attach so much importance.

22 *May*, 1882.

My dear Spottiswoode,

I have sent in a short paper* to the Royal Society containing a possible explanation of a photographic phenomenon suggested to me by an observation I made about solar phosphori, not containing anything new in principle, but in a form to suggest the photographic application, which was not suggested when I made the experiment, not at all a new one, in a pure spectrum.

I asked Captain Abney whether an explanation had been given of the photographic phenomenon, and he mentioned to me an explanation, which he told me was commonly received. From the language, partly involving a theory which I do not share, in which he expressed himself, I did not at first think the common explanation was likely to be correct. But when I stripped the explanation of the theoretical language, and ascended to what I suppose are the facts, it struck me as very likely to be the true explanation. Still, Captain Abney thinks my explanation a possible one, and proposes "after my paper is read" to examine the thing experimentally.

* *Math. and Phys. Papers*, vol. v. pp. 117—124, where Sir W. Abney's explanation is given.

ACTINOMETRY.

The Government Committee on Solar Physics was instituted in the early eighties, in connexion with the Solar Physics Observatory established at South Kensington, under the direction of Prof. Norman Lockyer, to promote insight and investigation into the problems connected with the nature and changes of the Solar Radiation, by which all physical activity on this Earth is ultimately determined. Prof. Stokes felt it to be his duty to accede to the request that he should become a member of this Committee: he became its Chairman, and the guiding spirit in its work, which occupied a great deal of his time for many years. One of the first problems taken up at his instigation was the improvement of methods of measuring and recording the direct intensity of solar radiation, and its variation from time to time. There is a long correspondence with Balfour Stewart from 1882 to 1888 on the construction and testing of actinometers designed by the latter. Stokes had a keen view of the importance of the subject, and was very sanguine that important results would come out of it[*]. His energy may be learned from the published Reports of the Solar Physics Committee, extracts from which are reproduced in *Math. and Phys. Papers*, Vol. v. pp. 125—139, and in Vol. ii. of the present collection.

In 1891-2 he was engaged in a prolonged correspondence with H. F. Blanford, F.R.S., the Meteorologist of the Government of India, who was then in this country conducting experiments with Violle's type of actinometer, with the object of testing its reliability for use by ordinary non-expert observers. These letters, which would amount to about 30 pages of print, evolve a sort of treatise on the objects to be aimed at, and the defects to be avoided, in obtaining a continuous measure of the total Solar activity as received at the Earth. Though Balfour Stewart's second type of instrument, fully analyzed as to its principles and performance by Sir George Stokes in a report to the British Association in 1872[†], had worked satisfactorily, and though electric differential methods of easier application have since been successfully elaborated by Ångström and others, these considerations must still repay attention.

[*] Cf. *supra*, p. 241.

[†] "On the best methods of measuring the direct intensity of Solar Radiation," *Math. and Phys. Papers*, vol. v. pp. 243—253.

A series of promising tests of the second Stewart actinometer was made by Prof. H. McLeod, F.R.S. at Coopers Hill and continued by Mr W. E. Wilson, F.R.S., in 1896, at his observatory at Daramona, Westmeath, with Stokes' active advice and cooperation. After a period of intermission, activity in the subject has now revived. The Royal Society has appointed a Committee, with Prof. Callendar as Chairman, to whom the income of their Gunning Fund has been assigned, to work out reliable self-recording methods of observations such as may be put into extensive operation in the meteorological services of Great Britain and India. The International Union for Solar Research, organized about the same time by Prof. Hale and having Prof. Schuster as Chairman of Committee, is also promoting determinations on the basis of Ångström's instrument.

22 *Jan.* 1884.

MY DEAR DR SCHUSTER,

I had intended to have written to you a few words relative to your letter in reply to my enquiries about the Eclipse Report.

I never took my theory of the possible effect of a difference of radiation from the sun, according to its state, as possibly occasioning electric discharges constituting the aurora, as being more than a more or less plausible conjecture. Still, it is not so easily to be disposed of by considering the smallness of any change in the total radiation as we receive it at the surface of the earth, or even on a high mountain. A mere increase of total radiation would have little effect. For no effect would be produced by radiations with respect to which the air was perfectly transparent. Now, it is very transparent, not merely with regard to the violet, but to rays of higher refrangibility for some considerable way beyond....

THEORY OF LUBRICATION.

The new light thrown on the principles of the lubrication of machinery by the results of the late Mr Beauchamp Tower's experiments, conducted about 1883–4 on behalf of the Institution of Civil Engineers, naturally claimed Prof. Stokes' interest and attention. In his Presidential Address to the British Association, Montreal, Aug. 1884 (*Scientific Papers*, Vol. ii. p. 344), Lord Rayleigh independently mentioned this subject. "We may, I believe, expect from Prof. Stokes a further elucidation of the processes involved." See also the investigation of Osborne Reynolds, *Phil. Trans.*, 1886. The letter which follows is the only record of Prof. Stokes' investigation that has been found.

To BEAUCHAMP TOWER, M. Inst. C.E.

21 *April*, 1884.

Dear Sir,

An investigation submitted to me by Mr Walter Browne, which I believe is going to be published, called my attention to your interesting report on friction experiments in the November number of the *Proceedings* of the Institution of Mechanical Engineers. I thought I would try what would be the results to which we should be conducted by the application of the known equations of motion of a viscous fluid, which in the present case are very much simplified, since the inertia of the lubricant does not sensibly come into account.

My investigations lead me to some numerical calculations which I have not worked out; nor have I time for it at present, as my lectures are close at hand. A good deal of light, however, seems to be thrown upon the results of the experiments—I mean, on the causes of what was observed. Thus, the considerable pressure in the film of oil under the brass comes out. It appears that the point of nearest approach of the surfaces is by no means directly under the point of bearing,—as I may call it, that is, the point where the line of the resultant weight cuts the surface,—but considerably in advance of that. Thus if we suppose the journal to be going round in the direction of the hands of a watch, whereas the point of bearing will be at the hour-figure XII, the point of nearest approach of the surfaces will be away towards the hour-figure III; I don't say so far away as III. Hence when the direction of rotation is changed, the point of nearest approach is shifted altogether to the other side of the hour-figure XII. Hence when the journal has been revolving for some time in one direction, so that the metals get worn to fit, so that they are no longer in contact here and there, but separated by a film of oil, and the friction thereby reduced,—on now reversing the rotation they come in contact in a new place, and have to wear themselves to fit before the friction is reduced to its normal amount.

My theory explains, not only the greater facility of lubricating a crank than a journal, which was mentioned and explained in the paper or discussion, but also the greater facility of lubricating the axle of the quadrant lever of a pump, which was mentioned but not explained.

I am disposed to think that your syphon lubricator leaves nothing to be desired in point of lubrication, and that the reason why a higher friction was obtained with it than with the oil bath was, not a defect in the system of lubrication, but the cutting away of the brass. For I am led to regard the side portions of the brass as very important. I have not however as yet worked this out, though I see the way to do it. The calculations however would take more time than I can give at present.

I thought it well to inform you of what I had been doing, in case you should be going on with your experiments. For in carrying out the experiments it is well to have any suggestions that theory may indicate.

There must have been considerable generation of heat in the film, and the journal probably got hotter than the oil in the bath. Might it not be well to get a pit drilled in the axis of the journal deep enough to hold the bulb and a little of the stem of a small thermometer?

I thought it well to communicate to you these preliminary indications of theory, as I cannot at present work the thing fully out on account of my lectures.

I am, dear Sir, Yours very truly,

G. G. STOKES.

FROM PROF. E. WIEDEMANN.

ERLANGEN.

18 *July*, 1887.

DEAR SIR,

Please accept first my best thanks for your so very kind letters.

I should like to observe that it is not a solution in glycerine itself that showed me phosphorescence, but one in gelatine added with glycerine. Experiments with gelatine and water gave no sure results, as the borders dried more quickly than the interior; and I was not sure if the observed phosphorescence did not come from the former. By an addition of glycerine this difficulty is avoided.

The experiments on the infra-red have been made by Becquerel, but he has not tried the same substances as I myself. These have from their anomalous dispersion a special interest.

I am now very busy at work about fluorescence and phosphorescence. As soon as I have got results I shall take the liberty of giving notice of them to you, who have laid the foundations for us younger workers in this chapter.

I am, dear Sir, yours most obedient,

E. WIEDEMANN.

31 *Dec.*, 1895.

DEAR PROFESSOR WIEDEMANN,

I write to thank you and your colleague Mr Schmidt for several most interesting papers on fluorescence, the last of them, on the fluorescence of the vapours of potassium and sodium, in some respects the most interesting of all, not only for its very great importance in relation to astrophysics, but also for its bearing on the theory of fluorescence. I wrote to Mr Lockyer to ask him if he had seen the last, saying that I regarded it as of great importance in relation to solar physics. He telegraphed to me, asking me for the reference, saying that he had looked into the *Annalen*, but could find no paper with that title. You were so good as to send me two copies, so I answered his letter, sending him one.

But the earlier papers you sent me are also of very great interest. I had often thought that perhaps something interesting might come out of the examination of a fluorescent substance in the gaseous form, if it could be obtained in such a form without the use of too high a temperature; but I feared that in most cases the substance would be decomposed, or that the temperature required to volatilise it might, even without decomposition, be unfavourable to fluorescence, as is the case, for instance, with a bead of microcosmic salt (phosphate of soda and ammonia) containing a little sesqui-oxide of uranium, which when taken out of the jet of flame from a blowpipe is not fluorescent, but becomes very strongly so on cooling. From accounts I had read of the behaviour of the substance, I thought that perhaps diethylo-phenylamine might give a chance, but it is not a substance to be obtained in commerce. I certainly had no idea that so many substances as you mention would show fluorescence in the gaseous condition. Nor should I ever have imagined that complex organic substances would have sustained an electric discharge without being immediately decomposed. Nor should I have imagined that such a chemically simple substance as potassium or sodium in the gaseous state would have shown fluorescence.

I need perhaps hardly say that Kirchhoff's law as to the equivalence of emission and absorption is only established for the case in which you have the same temperature all round, and matter all round sufficiently thick or opaque to prevent radiation from passing through it. It does not necessarily hold good when, as in experiments with fluorescence, the radiation comes in within a certain solid angle, but emission takes place in all directions.

P.S.—1895 is just out: a happy new year to you.

10 *Januar* 1896.

HOCHGEEHRTER HERR!

Haben Sie zunächst den allerherzlichsten Dank für Ihre so freundlichen Zeilen. Dass die Resultate gerade Ihr Interesse erregt haben, ist für Dr Schmidt und mich von hohem Wert.

Es scheint, als ob sich allgemein der Satz aufstellen lässt: Jeder Körper fluoresciert, sobald er nur hinreichend absorbiert und dafür gesorgt wird, dass die Dämpfung innerhalb seiner Moleküle hinlänglich klein ist.

Bei einer grossen Zahl anorganischer Substanzen ist uns der Nachweis der Fluorescenz gelungen, indem wir sie in festen Körpern lösten, so bei Mangansulfat. Einem Schüler von mir ist derselbe Nachweis für Eisensulfat, Nickelsulfat u.a.m. geglückt. Herr Dr Schmidt wird in einer demnächst erscheinenden Arbeit zeigen, dass man dasselbe Ziel bei festen Lösungen organischer Substanzen erreichen kann.

Durch diese Nachweise ist dann die Fluorescenz aus dem einer nur vereinzelt auftretenden Erscheinung zu einer ganz allgemeinen geworden, die nur durch sekundäre Umstände verhindert wird, unter gewöhnlichen Verhältnissen zu Tage zu treten.

Mir scheint überhaupt, als ob die Durchforschung der von mir als Luminescenz zusammengefassten Strahlungserscheinungen gegenüber der reinen Temperaturstrahlung immer mehr in den Vordergrund treten muss. Die letztere stellt einen idealen Grenzfall dar.

Versuche über die Fluorescenz des Na, K Dampfes, solche über Electro-Luminescenz an diesen Dämpfen, am Cd, Zn, Hg Dampf lehrten uns übrigens, dass selbst bei einatomigen Gasen der Leuchtmechanismus ein weit complicierterer ist, als man gewöhnlich annimmt. Auch unter dem Einfluss von Entladungen erhält man bei Natrium* und Kalium continuierliche Spectren in grün respl. rot. Beim Quecksilber treten 4 Spectren auf. Bei Zink und Cadmium scheinen ein Linien—ein Banden—und ein continuierliches vorhanden zu sein.

Wir können im Allgemeinen die Bedingungen für das Auftreten dieser Spectren aufstellen. Dem Banden-Spectrum entspricht stets relativ viel Dampf (der Druck beträgt aber nur wenige Millimeter) und mittelstarker Energie-Zufuhr.

Beim Hg Dampf scheint ausserdem bei bestimmter Art der Anregung der Leuchtenergie-Inhalt, d. h. der auf innere Bewegung enkommende Anteil der Energie, ein relativ grosser Bruchteil der gesammten Energie zu sein. Es steht dies in einem gewissen Widerspruch zu den Folgerungen der Kinetischen Gastheorie.

Eine Erklärungsmöglichkeit scheint mir in der Annahme zu liegen, dass die absorbierten Strahlen zunächst schnell gedämpfte Schwingungen im Molekül hervorrufen, die eventuell sekundär zu langsam gedämpften Veranlassung geben. Das Molekül wäre mit einem Lecher'schen Drahtsystem zur Erzeugung electrischer Resonanzschwingungen zu vergleichen.

Vielleicht darf ich noch erwähnen, dass uns unsere Untersuchungen auf einen noch wenig beachteten Spectraltypus geführt haben, der aus einem continuierlichen Band besteht, in dem einzelne Emmissions Minima hervortreten. Er findet sich vor Allem bei Verbindungsspectren, von denen wir infolge eines einfachen Kunstgriffes eine ganze Anzahl in Entladungsröhren untersuchen konnten. Ein junger Engländer aus Oxford, Herr Jones, soll einen Teil dieser Versuche weiterführen. Indem ich Ihre gütigen Neujahrswunsche auf das beste erwidere verbleibend

Hochachtungsvoll Ihr ergebenst,

E. WIEDEMANN.

* Cf. however the recent advances by R. W. Wood, *Phil. Mag.* 1905.

LENSFIELD COTTAGE, CAMBRIDGE.

27 *July*, 1887.

DEAR MR GRUBB,

I do not know whether you have had much to say to measures of length or weight of extreme precision, your measures being mostly angular. Perhaps, however, when you were in Paris for the Conference about Astronomical Photography you may have come across something about the *Comité Internationale des Poids et Mesures.*...

Now the point is this. A few years ago England joined the Comité, paying a lump sum of about £1500 in lieu of back expenses, and an annual subscription of £318 according to the tariff, a subscription which, in a few years, would be likely to be halved when the initial work of the establishment was done, and the maintenance would consequently become less expensive. For this England has the right of getting comparisons of standards made at the establishment with all the accuracy that can be secured, and of purchasing standard metres or kilogrammes. Now with a view to retrenchment the Government are, or had been, thinking of retiring. England would then be shut out from the use of the office, or rather establishment, for comparing such standards as might be required, or from purchasing verified standards.

Of course we have a Standards Office of our own, but it is in a makeshift sort of building, and the appliances, especially those for securing a constant temperature, are by no means such as can be supplied at the Breteuil establishment.

If from your own experience, or independently of your own experience, you should have formed a distinct opinion as to the desirableness or inutility of England's continuing to belong to the Comité, I should feel obliged if you would kindly let me have the benefit of it.

I am, &c.,

G. G. STOKES.

In the year 1888 Prof. Stokes officiated in a capacity then novel, as "independent reporter on experiments made in conformity with the order of the High Court of Justice" in the matter of the patent case, Edison *v.* Holland, involving the construction of filaments for incandescent electric lamps. After witnessing experiments conducted by both sides in laboratories selected by them, Prof. Stokes drew up a lengthy and detailed report for the Court, which was incorporated in the official documents of the suit.

DR ISAAC ROBERTS, F.R.S., ON PHOTOGRAPHY OF NEBULAE.

KENNESSEE, MAGHULL, NEAR LIVERPOOL.
15 *February*, 1889.

MY DEAR SIR,

As soon as the clouds will permit I will try a few experiments with the telescope in order to prove the correctness of your inference as to the cause of the diffraction effects by the plate holder and supports, and will then send you the proofs.

I am sending for your acceptance three enlarged photographs of nebulae in Orion, that you may see the effects of different exposures in the development of the nebulosity. The one dated 4th inst. with exposure of $3^h 25^m$ throws much new light upon the extensions in space of this nebula, and it is now seen that the three nebulae M 42 and 43 and L 1180 either form one vast nebula or that separate nebulae are overlapping each other. I inferred by a photograph which I took in Nov. 1886 (see paper in parcel) that this extension would probably be shown to exist, and the February photo supplies the evidence that it is so.

Yours very truly,

ISAAC ROBERTS.

To DR WILLIAM POLE, F.R.S.

COURTOWN, NEAR GOREY, IRELAND.
Aug. 22nd, 1889.

MY DEAR SIR,

*   *   *   *

It is a curious question how far to you a colour is more and more wanting in illumination, according as to us it is more and more red. I think your colour-top experiments will enable a normal-eyed person to decide, by trying whether your matches, while to a normal-eyed person of very different colours, are yet matches to us as regards intensity, *i.e.* illumination. It will be difficult for us to judge on account of the differences of colour. I am sending for a colour-top to try, as I did not bring mine here with me.

It would be well to put in the proof from whom you got the colour-top (Mr Bryson, of Edinburgh?); because the value of your colour-top experiments depends in some measure upon their being reproducible, and we could not trust to the identity of coloured papers obtained from different sources.

There can be no doubt whatsoever that blue and yellow (*i.e.* what to us are blue and yellow) make white, *i.e.* corresponding to a given very good blue, there can be found a very good yellow which, mixed with it in proper proportion, will make white. Of course it is not every blue and yellow which by their mixture will give white; the nearest approach to white will be either a very dilute red or a very dilute green, according to the original hues. But I cannot agree with you in your inference (Art. 16) that if this be so there must be four primary colours; or rather I should agree with you or not according to the sense in which the expression "primary colours" is understood. If you use the term in Maxwell's sense, then I think your conclusion invalid, and for this reason, that you *assume* blue and yellow to be two of the primaries, whereas we have no means of knowing what two at least of the three primaries are; for the phenomena of colour blindness render it very highly probable that one of the primaries is the red, which to you is invisible; that is, that the fullest red of that hue which we can produce is the primary red *plus* white.

But if you mean by primary colours standard colours selected to a certain extent (as I believe) conventionally, then I agree with you that we require four, in order to describe with sufficient readiness all other colours as mixtures of them.

What you say about the apparent change of hue of pinks (in Art. 19 H) renders an additional experiment highly desirable. The result got by laying on thinly on white paper a pigment, which thickly laid on gives a crimson, is not strictly the mixture of the crimson with white. The following experiment would show whether the observed result is anything more than the common, almost universal, phenomenon of a change of hue with a change in the thickness of the coloured medium looked through.

Having chosen a crimson pigment which when laid on more and more thinly gives a pink (to you almost neutral), passing through the strict neutral from dull yellow to dull blue, paint the colour full (at least fully enough to give a dirty yellow), and then see whether the hue changes to dull blue by what is strictly mixture with white. The mixture may be obtained by combining a crimson and white disk on the colour-top, successively diminishing the crimson sector, or by reflection at and transmission through a plate of glass, varying the inclination, &c.

The result, whatever it be, of thus strictly mixing crimson with white ought to be mentioned.

I am, dear Sir, yours very truly,

G. G. STOKES.

L. C. C., 4 *Feb.* 1892.

Dear Captain Abney,

If in normal vision there are three fundamental sensations of colour, X, Y, Z, and if in a colour-blind person one of these, Z, is wanting, while the remaining sensations X, Y are the same as normal, and if we could draw curves representing for each sensation the distribution in the spectrum, we might at first sight be disposed to say, since the two sensations which the C.B. has are identical with two out of the three which are normal, the curves for X and Y in the C.B. would agree with the curves for the same sensations in the N.E. person. ...But I think *a priori* considerations show that most probably this is not true, unless as a sort of rough approximation. Light must be absorbed in order to produce the sensation of vision, and the most probable explanation of the mode of stimulation of the nerves of vision is that there are three different substances, having different modes of absorption in passing along the spectrum, the result of the absorption of light by which is to stimulate the nerves which excite the three fundamental sensations respectively. Now as the light travels on in the place where the absorption takes place, all three substances absorb it, but in different proportions, according to the part of the spectrum. Call the substances x, y, z. In one place (P) in the spectrum z might absorb feebly or not at all, while at another place (Q) z might absorb powerfully. At P then the removal of z would leave the absorptions by x and y the same as before, but at Q the removal of z would leave more light for x and y to absorb, hence the ratio of the ordinates at P and Q for the X curve, and similarly for the Y curve, might very well be altered by the removal of z. Hence it is probable that neither the X curve nor the Y curve would be the same (considering only the ratio of the ordinates for different parts of the spectrum) for C.B. as for N.E. This may easily be illustrated by conceiving how light would be absorbed by a mixture of an ammoniacal solution of a copper salt with chromate of ammonia, and taking the total absorption by each constituent of the mixture as the analogue of the stimulating effect of light for one kind of nerves, and then considering how the curves of absorption for either constituent would be affected by the removal of the other constituent.

We have no right whatever to assume, nor is it probably true, that the same luminosity is produced by the expenditure of the same amount of radiant energy of given refrangibility, whether it be expended in the stimulation of sensation X or Y or Z. Yours, &c.

After considerable hesitation Prof. Stokes accepted in 1890 the office of Gifford Lecturer on Natural Theology in the University of Edinburgh for the period 1890—92. His scruples find expression in the preface to his published volumes of lectures, and are referred to in the following letter from the Principal of the University of Edinburgh.

13 *May*, 1890.

DEAR SIR GEORGE,

Thank you very much for your letter. I quite understood the matter as you say; but, as I mentioned to Prof. Tait, I was still anxious for such a definite acceptance as your letter now conveys.

Yours very truly,

W. MUIR.

*Leave-taking of Royal Society; from Presidential Address,
Nov.* 30, 1890.

The proceedings of to-day bring to an end my long tenure of office in the Royal Society, which has extended now over thirty-six years, during the last five of which I have held the honourable office of your President. I am deeply sensible of the kindness which I have always experienced from the Fellows, and of the indulgence with which they have overlooked my deficiencies, due in part to the pressure of other work. It cannot be without a strong feeling of regret that I come to the close of an official connexion with the Society that has now extended over full half my life. But I feel that it is time that I should make way for others, and that I should not wait for those infirmities which advancing years so often bring in their train; besides which there are personal reasons which led me to request the members of the Council not to vote for my nomination for re-election as your President.

And now it only remains to me, as virtually my last official act as your President, to perform the pleasing duty of delivering the medals, which the Society has to award, to the respective recipients of those honours.

Lighthouse Illumination.

The following letters refer to a Report, made by Sir G. Stokes to the Board of Trade, on some controversial matters dealt with by the Trinity House Report on Lighthouse Illuminants.

33 to 36 CAPEL STREET, DUBLIN.

27 *Feb.* 1891.

DEAR SIR,

I write to ask you to be so kind as to enable me to correct a mistake which has got into the minds of some of my friends with respect to your report on the Trinity House South Foreland experiments. You were asked by the Board of Trade to report whether the results arrived at by the Trinity House as stated in their report were justified by the record of the experiments contained therein. Inasmuch as in your report you advised that gas might be used with advantage in certain cases, and the Trinity House made no such recommendation, your report answered in the negative the question to which the Board of Trade asked you to reply; but you also stated that for specially important sea lights, the experiments showed that electricity offers the greatest advantage. This, as I understood, was merely quoting the words of the Trinity House, and in effect saying that the record of the experiments bore out their conclusions in this respect; but the notion has got abroad that you and your colleagues *yourselves* made experiments and tested the electric light against other lights in foggy weather, and as the result of these experiments gave your opinion that the electric light was the best light for lighthouses in foggy weather. I have pointed out to some of my friends that although I implored the Board of Trade to ask you to make these experiments they would not consent and you never made any such investigation, and that unless you had personally done so I was quite sure you would not make any pronouncement on the subject.

In the interests of truth and fair play will you kindly send me a line to say that I am right in this respect.

Yours truly,
JOHN R. WIGHAM.

I said that you were only giving your opinion as to how far the record of the Trinity House report justified its conclusion, and not giving any opinion as to the experiments themselves, and that my opinion (which of course has nothing to do with the matter) was that had you made such an investigation you would have pronounced against the electric light, my opinion being formed not only because the experiments at South Foreland were to my own knowledge incomplete and inadequate, and really pointed in that direction, but because my experience has been that the electric light in fog is much inferior to gas.

The extraordinarily foggy weather which has prevailed in London during the past winter has given remarkable evidence of the uselessness of the electric light in fog. You yourself have probably noticed that while during the dense fogs we have had, the electric lights were like farthing candles, the generally diffused glow of a number of gas-lights in shop windows and other places *did* give a certain amount of useful illumination.

J. R. W.

L. C. C., 28 *Feb.* 1891.

We made no experiments of our own on the relative efficiency of the electric light and gas in fog; we judged by what we found in the report, united of course with anything that we might happen to know of our own knowledge bearing on the subject.

I do not think that any conclusion in this respect can be drawn from their relative behaviour in a London fog, into which smoke so largely enters. In thinking of any possible future experiments to be economically carried on, it often struck me that while they might be near some town or village, for the sake of a gas supply, it was indispensable that the town should be quite a small one, so as not to furnish any amount of smoke worth mentioning.

In *haze* the electric light is far more cut down than that of gas or oil, as is mentioned in the report itself; but it does not appear that in dense fog there is any great difference, provided the fog be clean.

Investigations in Optical Chemistry.

9 FITZROY SQUARE, *Dec.* 2, 1857.

MY DEAR SIR,

Many thanks for your note.

You are quite right in reluctantly admitting the identity of the three substances in question by the deportment of the substances with ether alone. It is extraordinary how the presence of a small amount of impurity often alters the characters of a substance. Satisfactorily the question can be solved only by the preparation of the bodies in a state of purity, by their ultimate analysis and by a careful comparison of the characters of the bodies thus individualised.

This is almost more than a physicist would undertake I fear, and I would therefore advise you to bring the subject before the Chemical Society in the form which it has at present, and suggestively stating the optical observations which you have made and which will give very considerable assistance to any Chemist, actually engaged in the inquiry of these substances, or inclined to take up their study.

I am sorry that I am unable to take the matter in hand myself, but my leisure is very limited and the perfect chemical elucidations of the subject may require weeks and even months.

The sooner you send your notice to the Chemical Society the better*.

Yours very sincerely,

A. W. HOFMANN.

G. G. STOKES, Esq., F.R.S.

* There are three communications ' On the existence of a second crystallisable substance (Paviin) in the bark of the horse-chestnut,' *Q. J. Chem. Soc.*, 1859; 'Note on Paviin,' *Q. J. Chem. Soc.*, 1860; 'On the optical characters of Purpurine and Alizarine,' appended to a paper by Dr E. Schunck, *Q. J. Chem. Soc.*, 1860. See *Math. and Phys. Papers*, Vol. IV. pp. 112—126.

KERSAL, MANCHESTER.

8th August, 1891.

DEAR SIR GEORGE,

Accept my best thanks for your kind letter of the 3rd inst. Having obtained my MS. from the Royal Society I have suppressed the passage marked by you on p. 13 and substituted one, in which I state, as you suggest, that you had informed me that you had found phyllocyanin and phylloxanthin to be derived from blue and yellow chlorophyll respectively*. It is possible, I think, that the body with one band in the green is the first product derived from yellow chlorophyll, and that my phylloxanthin is formed by the further action of the acid on the first product.

I agree with you that the difficulty experienced in obtaining these substances in a state of purity arises from the presence of fat.

On referring to my paper I find that I do not speak of crystallising phylloxanthin, but say on the contrary that it is almost always amorphous even under the microscope. I believe if quite pure it would crystallise in brown needles. In my letter to you I used the word crystallise inadvertently.

I think you are right in supposing that the change seen when green leaves are treated with boiling water is due to the presence of a minute quantity of acid, which acts on the chlorophyll and causes decomposition.

I believe that by the prolonged action of hydrochloric acid phyllocyanin undergoes some change.

I am, yours very truly,

EDW. SCHUNCK.

* E. Schunck, "Contributions to the Chemistry of Chlorophyll, No. IV." *Roy. Soc. Proc.*, June 16, 1891, pp. 302—317; cf. p. 311. "In a written communication which Sir G. Stokes has kindly addressed to me, he informs me that he has satisfied himself that by decomposition with acids blue chlorophyll yields phyllocyanin, whereas yellow chlorophyll gives phylloxanthin. This interesting fact affords a striking confirmation of the views held by him regarding the complex nature of chlorophyll."

In the early period of fluorescence there was much correspondence between Prof. Stokes and Dr Schunck on the analysis by this means of organic bodies.

30 *July*, 1891.

DEAR LORD RAYLEIGH,

I have to thank you for your note of the 26th inst. informing me that my paper on chlorophyll, pt. 4, will be published in the *Proceedings* of the Royal Society, also for the Report of Sir G. Stokes and a letter from him accompanying it. Having read Sir G. Stokes' remarks and critique, I think it would be well for me to see the MS. again before it is sent to the printers, so that I may have an opportunity of making some alterations or additions, as the case may be, in accordance with the suggestions of Sir G. Stokes.

I think it would be desirable in the interests of science, if Sir G. Stokes would embody his remarks and suggestions, which are of very great value, in a "Note," or still better in a separate communication which should set forth his views, not only on my experiments, but on the subject in general, a subject on which his researches have thrown so much light.

I am, yours very truly,

EDW. SCHUNCK.

The following memorandum has been found among Sir G. Stokes' papers.

According to Dr MacNab, plants containing green chlorophyll grains, when placed in darkness, partial or complete, change colour from the destruction of the chlorophyll. Sachs says—*Text Book of Botany*, page 669—that in the leaves of rapidly growing angiosperms the absorption and disappearance of the chlorophyll takes place in a few days. If the temperature be high, he adds, cactus stems with slow growth and the...of solaginella remain green for months in the dark.

It is probably true also of conifers, as they have been kept in the dark during the winter months without injury, as at Berlin where they were protected by mats, &c.

Sachs says, page 665, "If the temperature is sufficiently high the green colouring substance is formed in the cotyledons of conifers, and in the leaves of ferns as well as under the influence of light....The formation of the green colouring matter in the cotyledons of angiosperms does require light, but in both cases it ceases in a low temperature."

THE following letters, kindly communicated through Prof. Sir G. H. Darwin, are part of a correspondence with Rev. Canon Russell, Geashill Rectory, King's County, who drew Prof. Stokes' attention to an appearance as of a luminous vapour rising from some flowers, which, after insolation, have been taken into the dark.

4, WINDSOR TERRACE, MALAHIDE.
25 Sept. 1891.

DEAR CANON RUSSELL,

The marigolds arrived this morning by the morning's post. It was a bright sunny morning, and I put them in a tumbler of water and placed it on a window-sill facing the South. About 12 or 1 the sky became overcast, and I took the tumbler into a room in which I had closed the shutters. I could not see anything of the nature of a luminous vapour, nor any moving lights. But I saw one thing which *perhaps* is what you refer to, though it is only a conjecture that it may. When the room was nearly but not quite dark, I saw certain parts of the flower appearing as if they were faintly luminous. This was not seen when the room was perfectly dark, nor when I went first into the room. In that case it was not seen till the eyes had got sensitive by remaining in darkness for a little time.

I do not know whether what I saw is the thing you describe, but the origin of what I saw is plain enough. It depends on the enormously different sensitiveness of the eye to rays of high and low refrangibility. Suppose you start with high and low rays of about the same intensity as far as the eye can judge, and then reduce them both enormously, but in the same proportion. The high rays will be seen long, long after the low ones cease to be visible. In order that the eye should have full sensitiveness for these rays of high refrangibility, it must be given some rest in darkness or near darkness, after it has been exposed to light of common intensity.

The yellow and orange petals absorb the more refrangible part of the spectrum. But some petals are paler, approaching somewhat to white. Hence, when the flower as a whole is viewed in a room nearly but not quite dark, the orange and full yellow petals are not seen. But the petals or parts of petals which approach to white reflect no inconsiderable portion of the blue rays. Hence these are seen by the sensitive eye, though the other flowers, or the rest of the flower, appear dark.

The evening light absorbs the rays of high refrangibility, more so a great deal (in proportion to the rays of low refrangibility) than the rays of the sun when low, or than the rays of a lamp or candle. Hence the faint evening light would be specially favourable for the exhibition of the phenomenon if it really be what I have suggested.

I always felt a difficulty about the phenomenon not being seen in perfect darkness, supposing its origin to be a luminous vapour.

If from my description you have reason to think that what I have described is not the same thing that you saw, it would be well not to quote my letter, as it would be only misleading. But whether what I saw *is* the same as what you mentioned, I do not know.

Yours sincerely,

G. G. STOKES.

27 *Sept.* 1891.

I got your letter to-night, and I will write a few lines in reply before going to bed.

I thought that you would hardly be deceived by such a very simple thing as that which I mentioned....

Your inability to see it in perfect darkness seems fatal to my theory, unless it can be explained in this way:—the luminosity is presumably from the description very faint, and in perfect darkness there is nothing by which you can guide the axes of the eyes as to where to look. Thus not merely is the point of the field, where the luminosity would come, unknown, but the luminosity would be reduced to half intensity. For the chances are that the axes of the two eyes converge to a nearer or more distant point, so that the luminous spot, if it were bright enough, would be seen double, as the images would fall on non-corresponding points of the two retinas. Thus instead of one faint image you would have two, in different parts of the apparent field, and the intensity would be divided between them, and might then become too faint to be seen at all.

IN the period 1881–4 there was considerable correspondence with Mr D. E. Hughes in relation to the experimental papers on magnetism which he was then communicating to the Royal Society.

From 1876 to 1884 there is a correspondence with Sir J. Conroy relating to his investigations on metallic reflexion and the absorption of quasi-metallic substances*.

In 1891 there was correspondence with Dr O. Wiener regarding his experiments on stationary light waves, and their bearing on the direction of the vibration in polarized light. Among other things, experimenting by means of fluorescence was suggested, which was carried out independently at a later period by Nernst and Drude at Göttingen.

To this time also belongs a correspondence with Dr V. Schumann on the discovery by the latter of a further ultra-violet expansion of the spectra of gases, to which air is not transparent.

In these and other cases, only the letters to Prof. Stokes are available.

The following letter from Prof. Arthur Smithells, F.R.S., is included here by his permission. During the same year (1892), to which he refers, Prof. Stokes was much interested in the spectroscopic investigations carried on by his colleague Prof. Liveing, at Cambridge.

THE UNIVERSITY, LEEDS.

July 6, 1905.

DEAR LARMOR,

Certainly you may publish any of Stokes' letters to me. I have been intending for long past to re-read his letters and select some of the more interesting ones for you. I will do so within the next few days. Meanwhile it may perhaps be of service to you if I tell you the following facts.

In 1892, *a propos* of some publications of mine about hydrocarbon flames, a correspondence took place between Sir G. Stokes and Prof. Armstrong, and this was subsequently published by Prof. Armstrong in the *Proceedings* of the Chemical Society†. In one of the letters Stokes had suggested that the much disputed "Swan spectrum" of hydrocarbon flames was, in view of my

---

* For a theoretical discussion, see R. C. Maclaurin, *Roy. Soc. Proc.*, 1906.

† *Math. and Phys. Papers*, Vol. v. p. 235.

experiments, probably due to carbon monoxide. I was at that time but poorly versed in the literature of spectrum analysis, but I had got the idea that gaseous spectra were much more likely to be due to disturbances occurring during the formation and decomposition of molecules than to purely thermal actions, and I had concluded that the Swan spectrum was due to the formation of carbon monoxide. Although I did not know Stokes personally at that time, I ventured to write and submit to him my views on the genesis of spectra. I received an immediate reply, and a long correspondence ensued, which lasted in fact almost to the time of his death. I also had a considerable number of long talks with him on the subject. He was of opinion that chemical combination was *a* source of spectra, inasmuch as the new-born molecule (as he was fond of calling it) would be in a state of special agitation; but he firmly believed that heat alone could produce gaseous spectra *. When in 1892 E. Pringsheim sought to prove that all line spectra in flames were due to chemical action, he read the papers with eagerness, but he repudiated the conclusions with equal warmth. I endeavoured to defend the Pringsheim view, which accorded with my own, and I made a great many experiments, many of them suggested by Stokes, to test the question. I tried a long time with sodium to see if the vapour gave the *D* absorption when chemical action was excluded. He took great interest in this and continually made suggestions. At the same time he interested himself in all the other work we were doing about the spectra of compounds (compounds of copper and gold) and especially about the Swan spectrum. Indeed I found it difficult to keep up my end of the correspondence.

Stokes considered the fact that most gases cannot be made to glow thermally as merely the reciprocal of the fact that they do not absorb appreciably, and he suggested to me to use iodine vapour and nitrogen peroxide, both of which gave a positive result. So far as iodine is concerned this had been previously observed by Salet. I kept Stokes very fully informed of the work on the Swan spectrum, and his advice was as valuable as it was profuse. He considered the evidence connecting this spectrum

[* His view seems however to have fluctuated: see *supra*, p. 94, in which he probably means that purely thermal collisions, such as occur among the molecules of perfect gases, are too soft to excite radiation to a sensible degree; and that gases are absorbers and therefore also radiators, of intensity depending on the surrounding temperature in the sense of the general theory of exchanges (the thin Fraunhofer lines having a different origin) only when they are imperfect.]

with carbon monoxide to be in the end quite convincing, and in one of his last letters he says he is sure I shall live to see the day when my view (which was more justly his view) will be generally accepted.

My relations with Stokes constitute one of the great experiences of my life. He seemed to me both intellectually and morally to tower above other men, and to be free of all pettiness of mind or character. He was willing to lend his time and his talents without stint to anyone whom he deemed an honest worker, and the only thing he seemed to exact in return was that there should be no fuss or effusiveness of acknowledgement. In the union of greatness with modesty I can only name Bunsen as his equal. Though in conversation he gave little sign of having any interest but science, I had many opportunities of discerning more human feelings; and of the extent and manner of his personal kindness to me I should find it difficult to speak adequately. What Stokes did for his generation can hardly be estimated. As one of a multitude I should like gratefully and affectionately to lay a stone upon his cairn.

Yours sincerely,
ARTHUR SMITHELLS.

The following partial draft of a letter does not reveal the name of his correspondent. It is addressed from Prof. Tait's house.

8, BELGRAVE CRESCENT, EDINBURGH.
19 July, 1893.

MY DEAR SIR,

I will begin a letter to you now, and expect to continue it from time to time. I will say nothing at present about your solution, but confine myself to some general considerations relating to *jets*. I will confine myself to the case you have taken up— that of motion (steady) in two dimensions without external forces acting. I will also suppose, as you have done, that the jet is symmetrical with respect to its axis. Naturally some of the ideas will have a wider application.

July 20. I did not get far yesterday. The same identical problem frequently presents itself in totally different physical applications. In our problem, whether we think of the stream as walled in part and free in part or as walled the whole way, the stream function has two constant values at the boundary and at the axis; and we may figure to the mind a conducting solid of the same form as the fluid, and having constant permanent temperatures at the boundary and at the axis. The intermediate stream lines may then be replaced by isothermals. It will be convenient to think of an infinite number of stream lines so spaced that the flux across one interval between line and line is the same as that for another, and consequently at a distance in the jet the lines are equidistant and parallel. Call the fluid between two consecutive lines a filament. In the free part of the boundary (in the case in which some part *is* free, *i.e.* in the case of a jet) the external filament may be of uniform thickness from *minus* infinity (*i.e.* infinitely far in the jet) to the mouth ; for the pressure being constant, the velocity must be also constant. Now if there be a gentle swelling, as you have drawn in one of your figures, the filaments in the swollen part must be thickened, as we see still more readily if we think of the stream lines as isothermals, which violates the condition of the external filament being of uniform thickness throughout the free part. And even if we supposed that the external filament preserved its uniform thickness, and that it was only as we went inwards that the thickness increased, still that would not do. For it is easy to prove that at the free boundary dp/dn, n being measured along the normal, equals DV^2/R, D being the density, V the whole velocity, and R the radius of curvature, the pressure decreasing inwards. But this would demand that the filaments should get thinner as we go in, whereas they get thicker. And if we say that the first filament is of normal thickness, that is, the same as at infinity, then as we go in they at first get thinner, then after passing through the normal thickness get thicker than normal well before we reach the axis, that evidently would imply such a *bizarre* alteration of thickness as could only occur near the mouth, not at a little distance, where the motion would have got regular. In the same way it may be shown that we cannot have a long contraction followed by widening out again.

These general considerations will I think suffice to show the general character of jets.

First, suppose that the vessel has an orifice like the end of a trumpet inverted, that is, with the broader end directed inwards, and such that the tangent at the mouth has still a very considerable inclination to the axis. . Or suppose that the...

AFTER giving up active administration at the Royal Society, on retirement from the office of President in 1890, Sir George Stokes continued to give his assistance for two years as Member of Council and Vice-President. After his final withdrawal, the Council was not slow to adopt on the first opportunity the obvious course of proposing him for the Copley Medal, the highest award that the Society confers. The congenial duty fell to his successor in the Presidency, and lifelong friend, Lord Kelvin, on the next anniversary, Nov. 30, 1903, to present to him the Medal, after adverting to his scientific achievements in the following terms *.

"In presenting the Copley Medal to Sir George Stokes I feel that no 'statement of claim' is needed. Nevertheless, it is interesting to recall to memory something of the great work that he has done in mathematical and physical science. Fifty-two years ago he took up the subject of fluid motion, with mathematical power amply capable to advance on the lines of Lagrange, Fourier, Cauchy, Poisson, in the splendid nineteenth century 'physical mathematics,' invented and founded by those great men; and with a wholly original genius for discovery in properties of real matter, which enhanced the superlative beauty of the mathematical problems by fresh views deep into the constitution of matter.

"In the purely mathematical part of his hydro-dynamical subject, he advanced from the 'infinitely small' waves of Cauchy and Poisson to deep-sea waves of such considerable steepnesses and lengths and heights as are seen in nature—on water 500 fathoms deep, or more—after a severe gale far away from land; and he has shown how to carry on his mode of solution right up to breaking waves at sea and tidal bores in shallow water.

"His enunciations and solutions for motion of viscous fluids, —rich with applications to natural phenomena, the distance of audibility of sound, the suspension of clouds in the air, the subsidence of ripples on a pond and of waves on the ocean after the cessation of wind; and rich in aids to scientific investigation, as in the theory of the pendulum in air—have added to hydro-dynamics a previously unknown province, in which the exceeding difficulty of the mathematical work deserves every capable effort

* Roy. Soc. Proc., Vol. LIV. pp. 389—391.

to advance it, on account of the vast practical and scientific importance of the issues.

"His 'instability' of the motion of a viscous fluid, in the neighbourhood of a solid, gives the key to the scientific mystery of turbulent motion in practical hydraulics, and in wind blowing over solid earth; the hearing of sounds in the direction of the wind which are inaudible at equal distances in the contrary direction; the flow of water in rivers, culverts, and water-supply pipes; the 'skin-resistance' of ships, the scientific consideration of which has done so much to make 22 knots a proper speed for travelling by sea.

"Of true dynamical science in all these subjects Stokes' early work was the beginning. It also first gave true views as to that very important practical subject, the rigidity, and the resistance against compression, of solids; views which would be false if a majority of votes in the scientific world of 1893 could decide between truth and error.

"In optics and the undulatory theory of light, Stokes has been the teacher and guide of his contemporaries. His Report to the British Association in 1862, 'On Double Refraction,' showed with perfect accuracy and clearness the outstanding difficulties, but called special attention to the door which Green had opened for escape from them. That Report has given the starting impulse and essential information for nearly all that has been done for the subject in England since its publication.

"By his own experimental and mathematical work on the polarisation of light by a grating, and of the light of the blue sky; by his experimental investigation of 'epipolic dispersion' (or 'fluorescence'); and, perhaps more than all, by his accurate *measuring* work, from which he drew an exceedingly rigorous verification of the accuracy of Huygens' geometrical construction for the double refraction of Iceland spar, Sir George Stokes has done much to make the Undulatory Theory of Light sure and strong as it is—a codification of laws divined by Huygens and Fresnel. But he has done more than this. He has not merely left to mathematicians and speculative physicists a desperate problem to find the dynamical explanation of those laws. He has given (perhaps only in conversation) what seems to me certainly the true clue to the dynamics of the Undulatory Theory of Light, by pointing out that we must look not merely, or not at all, to change of shape of the portion of ether within a wave-length in the motion constituting light, but also, or altogether, to its absolute rotation*, for explaining the efficient force."

The final years of the century were marked by the announcement of far-reaching discoveries in Physics in which Sir George Stokes naturally showed deep interest, especially as he had closely pondered over the phenomena of the electric discharge in vacuum tubes for many years, in connexion with Sir W. Crookes' investigations. Selections from a long correspondence on this subject will appear below. On the announcement of the discovery of the Röntgen rays Stokes was not long in making up his mind that the hypothesis of longitudinal aethereal vibrations, which had been broached for their explanation, must be excluded. His well-known explanation, that they differ from ordinary light in being excited by pulses entirely irregular, like 'hedge-fire' on a target as distinct from regular volleys, was unfolded to the Cambridge Philosophical Society on Nov. 9, 1896†. It was further elaborated in the Wilde Lecture delivered to the Literary and Philosophical Society of Manchester on July 2, 1897; and Prof. Lamb has recently recalled (in conversation) the force and enthusiasm of his *extempore* exposition, the published report having been drawn up subsequently from shorthand notes taken during the lecture. The theory indeed raises questions regarding the degree of regularity, if any, that is to be assigned to ordinary light such as steady temperature-radiation, which cannot yet be regarded as entirely settled. The distinctness and force of Sir G. Stokes' convictions on these matters is illustrated in the following note‡ communicated to the French Academy, June 26, 1897, which was overlooked in the collection of his scientific papers.

"Les *Comptes rendus* du 5 juillet 1897 contiennent (p. 17), une Note où M. G. de Metz décrit une expérience dont le résultat, suivant lui, ne peut être expliqué que par l'une ou l'autre de ces deux hypothèses: ou dans un vide extrême les rayons X sont capables de déviation magnétique, ou les rayons cathodiques

---

* The most telling justification of this outcome of MacCullagh's ideas, by means of a model of a rotational aether, described first in *Comptes Rendus*, 1889, is due to Lord Kelvin himself.

† *Math. and Phys. Papers*, Vol. v. p. 256 and p. 258.

‡ *Comptes Rendus*, Vol. 125, pp. 216—218.

peuvent traverser la paroi en verre d'un tube de Crookes. Je
ne crois pas qu'aucun des termes de cette alternative contienne
l'explication exacte, et je demande à l'Académie la permission de
lui soumettre ce que j'estime être la théorie vraie du phénomène.

" Tout tend à prouver que les rayons X sont une agitation de
l'éther, et l'on peut regarder aujourd'hui comme pratiquement
établi que cette agitation est transversale. Si ces rayons sont une
agitation de l'éther, supposer qu'ils sont capables de déviation
magnétique prête le flanc à de très grandes difficultés théoriques ;
je ne sache pas d'ailleurs qu'une telle déviation ait été expéri-
mentalement démontrée dans aucun cas. Quant aux soi-disant
rayons cathodiques, il me paraît absolument évident que ce ne
sont pas du tout de vrais rayons, mais bien des courants de
molécules chargées d'électricité, projetés par la cathode. Il y
aurait sans doute une grande difficulté dans cette manière de
voir si nous étions obligés de supposer ces molécules capables
de passer à travers la paroi en verre d'un tube de Crookes,
d'autant plus que, Crookes lui-même l'a montré il y a déjà
longtemps*, les rayons cathodiques sont arrêtés par une mince
pellicule de verre, de quartz ou de mica. Mais il n'est nullement
nécessaire d'avoir recours à cette supposition pour expliquer les
résultats obtenus par M. de Metz. Il me semble évident que les
phénomènes qui se présentent dans les hauts degrés de vide sont
de la nature de ceux qui ont été étudiés par MM. Spottiswoode
et Moulton sous le nom de *relief-effect†*. Les deux masses d'air
extrêmement raréfié, situées respectivement dans le tube de
Crookes et dans le tube cylindrique, constituent les deux arma-
tures d'une bouteille de Leyde, dont le diélectrique est formé
par la portion de la paroi du tube de Crookes, limitée au contour
du tube cylindrique. A chaque décharge de la bobine d'induction,
un torrent de molécules électrisées négativement est projeté
contre l'anti-cathode ou la première surface du diélectrique,
laquelle communique sa charge, ou une bonne partie de sa
charge, soit directement à l'anode, soit, en premier lieu, à quelque autre partie de la surface interne du tube de Crookes.
Toute charge momentanée de la première surface du diélectrique
agit inductivement sur le contenu du tube cylindrique, et produit
réciproquement une décharge entre la seconde surface du diélec-

* Crookes, *Philosophical Transactions for* 1879, p. 150.
† Spottiswoode et Moulton, *Philosophical Transactions for* 1879, p. 177.

trique et le cylindre d'aluminium relié à la terre; et, dans cette phase de décharge réciproque, où la seconde surface agit comme cathode, les molécules sont projetées de cette seconde surface exactement comme de la cathode du tube de Crookes, et elles affectent de même un écran au platinocyanure de baryum.

"Bien que, comme j'en suis pleinement convaincu et comme, j'imagine, le pensent la plupart des physiciens, les rayons cathodiques et les rayons X soient de nature complètement différente, ils sont également capables d'affecter une plaque photographique ou d'exciter la fluorescence d'un écran couvert de platinocyanure de baryum. Cela admis, les résultats obtenus par M. de Metz trouveront une explication très simple. Lorsque l'air à l'intérieur du tube cylindrique était à la pression atmosphérique ou seulement à un degré de vide modéré, la fluorescence observée sur l'écran était due aux rayons X. Car, ainsi que Lenard* l'a montré, les rayons cathodiques, à supposer qu'ils existent, seraient promptement arrêtés par l'air et ne pourraient par conséquent atteindre l'écran. En conséquence, les rayons produisant la fluorescence étaient trouvés insensibles à l'aimant. D'autre part, à un vide élevé, les rayons cathodiques, constitués par les molécules que projetait la surface rendue cathode par induction, étaient capables d'atteindre l'écran; et, comme ils étaient à même d'exciter une fluorescence beaucoup plus intense que les rayons X, l'effet observé était principalement dû aux rayons cathodiques; et, par conséquent, les rayons excitants étaient trouvés susceptibles de déviation par l'aimant.

"En présentant cette explication, je tiens à me garder contre la pensée qui pourrait m'être attribuée d'expliquer de la même façon l'apparition de rayons cathodiques venant de la seconde surface d'une plaque d'aluminium dont la première surface reçoit des rayons cathodiques. Dans ce cas, le processus est probablement plus direct et présente, je suis porté à le penser, quelque analogie avec l'électrolyse."

In 1898 Mr A. Hilger, the optician, essayed the difficult task of the construction of Prof. Michelson's echelon transmission grating, then recently invented, and there is a series of his letters to Prof. Stokes asking for directions and reporting progress. In the second letter on Sept. 24 he writes, "I cannot tell you

* Lenard, *Wiedemann's Annalen*, Vol. 51, p. 225; 1894.

how much I am obliged to you for the trouble you have taken in this matter. I had written at least half a dozen letters but without results, even some without reply." Since then Mr Hilger and his successors have occupied a foremost place in the construction of these delicate instruments, and most of the powerful echelons now in use in Europe and even in America have been produced by them.

THE pleasure felt by Prof. Stokes in beautiful natural surroundings moved him to public action on various occasions. When a suggestion was made to abolish the rivulet which runs down by the side of the footpath in Trumpington Street, he was active in dissent. At the Hills Road end of Lensfield Road a beautiful poplar tree, now, unfortunately, docked from age, stood out in the footway : a proposal to remove it called forth a protest in the form of a circular of which the following constituted the preliminary portion of the draft :

" A scheme has for some time been in motion pleading for certain improvements (as they are called) in this Road to which it would be well to draw the attention of the inhabitants.

" In common as I know with many others, I have often been struck with the beauty of this Road in walking from Downing Terrace towards the Hills Road. The eye rests on a beautiful line of plantation on the right, while the grounds of Downing lie on the left. Short as the road is, I cannot but regard it as one of the ornaments of the Town. The fine poplar tree at the corner is also a highly ornamental object as seen from St Andrew's Street and elsewhere...."

CAMBRIDGE JUBILEE CELEBRATION.

For fifty years Sir George Gabriel Stokes had held the appointment of Lucasian Professor of Mathematics, and the event was officially celebrated by the University in the first week in June, 1899. The feelings which prompted the University of Cambridge to organize this signal mark of respect are set forth in an article contributed by Prof. J. J. Thomson to the *Daily Chronicle* at the beginning of the celebration.

" It is to real students of science, and to them perhaps alone, that Stokes's work is known. When difficulties have to be overcome, when the subject seems obscure and uncertain, they turn

to Stokes's writings, certain that if he has dealt with the subject
at all it will have become clearer and more definite by passing
through his mind.......It is impossible in a brief sketch to give
any adequate idea of Stokes's contributions to science. His earliest
researches were on hydrodynamics, in which subject he opened up
entirely new regions. In optics he has—to quote the words used
by Lord Kelvin when presenting him with the Copley Medal, the
highest honor British science has to bestow—been the teacher
and guide of his contemporaries.......His papers, besides being
remarkable for the value of the results they obtain, are models
of scientific exposition, 'perfect in form and unassailable in
accuracy.'

"In Stokes there is that rare but effective combination of the
highest mathematical powers and the greatest experimental skill,
and some of his greatest triumphs have been due to this combination. With simple apparatus (his experiments were mainly
done at a time when physical laboratories hardly existed in this
country) he has solved some of the most crucial questions in
physical optics, and the simplicity and success of the experiments
with which he illustrates his lectures are a source of unfailing
delight to his audience. In this combination of qualities he
resembles his great predecessor Newton, between whose career
and his there are many striking coincidences.......

"No notice of Sir George's life could leave unmentioned the
unsparing help and encouragement he has throughout his long
career given to younger men; many most important researches
made by others have owed their inception to his suggestions, and
their completion to his encouragement.

"We may well congratulate ourselves on the prosperity of
Natural Philosophy in this country, as within three years we
have celebrated the jubilees of two such men as Lord Kelvin
and Sir George Stokes."

Distinguished men of science, and representatives of Universities and other learned bodies at home and abroad, to the
number of about one hundred and fifty, participated in the
celebration, which commenced with a conversazione in the Fitzwilliam Museum on the Thursday evening.

On the Thursday afternoon the annual Rede Lecture had
been delivered in the Senate House by M. Alfred Cornu, of the
Institute of France, who announced as his subject "La Théorie

des Ondes Lumineuses: son influence sur la Physique moderne."
After alluding to the marvellous material results that had arisen
in modern times from patient and profound study of the laws
of natural phenomena, M. Cornu introduced his discourse as
follows.

"De là une série de questions qui s'imposent à l'attention de
tous. A quelle occasion le goût de la Philosophie naturelle, si
chère aux philosophes de l'Antiquité, abandonnée pendant des
siècles, a-t-il pu renaître et se développer? Quelles ont été les
phases de son développement? Comment ont apparu ces notions
nouvelles qui ont si profondément modifié nos idées sur le mécanisme des forces de la Nature? Enfin, quelle est la voie féconde
qui, insensiblement, nous conduit à d'admirables généralisations,
conformément au plan grandiose entrevu par les fondateurs de la
Physique moderne?

Telles sont les questions que je me propose, comme physicien,
d'examiner devant vous : c'est un sujet un peu abstrait, je dirai
même un peu sévère ; mais nul autre ne m'a paru plus digne
d'attirer votre attention, à la fête que l'Université de Cambridge
célèbre aujourd'hui, pour honorer le cinquantenaire du professorat
de Sir George-Gabriel Stokes, qui, dans sa belle carrière, a précisément touché d'une main magistrale aux problèmes les plus
profitables à l'avancement de la Philosophie naturelle.

Ce sujet d'autant mieux a sa place ici qu'en citant les noms
des grands esprits qui ont le plus fait pour la Science, nous
trouverons ceux qui honorent le plus l'Université de Cambridge,
ses professeurs ou ses élèves, Sir Isaac Newton, Thomas Young,
George Green, Sir George Airy, Lord Kelvin, Clerk Maxwell, Lord
Rayleigh ; et le souvenir de gloire qui se perpétue à travers les
siècles jusqu'au temps présent rehaussera l'éclat de cette belle
cérémonie."

After a masterly sketch of the development of optical science
from Descartes and Newton onward to Stokes and Maxwell and
Hertz, he concluded with these words,

"Mais, je m'arrête, Messieurs ; je sens que j'ai assumé une
tâche trop lourde en essayant de vous énumérer toutes les richesses
que les ondes à vibrations transversales concentrent aujourd'hui
dans nos mains.

J'ai dit, en commençant, que l'Optique me paraissait être la Science directrice de la Physique moderne.

Si quelque doute a pu s'élever dans votre esprit, j'espère que cette impression s'est effacée pour faire place à un sentiment de surprise et d'admiration en voyant tout ce que l'étude de la lumière a apporté d'idées nouvelles sur le mécanisme des forces de la Nature.

Elle a ramené insensiblement à la conception cartésienne d'un milieu unique remplissant l'espace, siège des phénomènes électriques, magnétiques et lumineux ; elle laisse entrevoir que ce milieu est le dépositaire de l'énergie répandue dans le monde matériel, le véhicule nécessaire de toutes les forces, l'origine même de la gravitation universelle.

Voilà l'œuvre accomplie par l'Optique ; c'est peut-être la plus grande chose du siècle !

L'étude des propriétés des ondes envisagées sous tous leurs aspects est donc, à l'heure actuelle, la voie véritablement féconde. C'est celle qu'a suivie, dans sa double carrière de géomètre et de physicien, Sir George Stokes, à qui nous allons rendre un hommage si touchant et si mérité. Tous ses beaux travaux, aussi bien en Hydrodynamique qu'en Optique théorique ou expérimentale, se rapportent précisément aux transformations que les divers milieux font subir aux ondes qui les traversent. Dans les phénomènes variés qu'il a découverts ou analysés, mouvement des fluides, diffraction, interférences, fluorescence, rayons Röntgen, l'idée directrice que je vous signale est toujours visible, et c'est ce qui fait l'harmonieuse unité de la vie scientifique de Sir George Stokes.

Que l'Université de Cambridge soit fière de sa chaire Lucasienne de Physique mathématique, car, depuis Sir Isaac Newton jusqu'à Sir George Stokes, elle contribue pour une part glorieuse aux progrès de la Philosophie naturelle."

Those who had the privilege of attending will not forget the grace and charm of M. Cornu's prelection. Their recollection will be tinged with profound regret that so soon afterwards, in April, 1902, his brilliant career was fated to come to an end ; the melancholy privilege was reserved for Sir George Stokes to formally move the Cambridge Philosophical Society to record their feelings of deep loss on the death of an Honorary Member closely bound to Cambridge by many ties.

The reception was held by the Vice-Chancellor on behalf of the University at the Fitzwilliam Museum, in the evening. A few minutes after ten o'clock the ceremony of presenting marble busts of Sir George Stokes, the work of Mr Hamo Thornycroft, R.A., was proceeded with in the central gallery. A temporary platform had been provided, on which seats were taken by the Vice-Chancellor, Prof. Sir George Stokes, Lord Kelvin, and the Rev. C. H. Prior, who represented Pembroke College.

Lord Kelvin, addressing the company, said* that they had met for a celebration of a great man and of Natural Philosophy in the University of Cambridge—Natural Philosophy in the broadest sense of the term, of which foundations had been laid by Sir George Stokes that would render the nineteenth century memorable in future centuries. Sir George Stokes commenced as an undergraduate in Pembroke College: his first experimental work was made when he was a junior Fellow. He (Lord Kelvin) well remembered that in Pembroke College there were no physical laboratories; and the first physical laboratory in European Universities, he believed—certainly in these islands—was in Sir George Stokes' rooms, which he occupied as a junior Fellow about the years 1840 to 1843. Sir George there commenced experiments on the properties of matter. If he were asked what was Sir George Stokes's province, he would say, Natural Philosophy in the broadest and widest sense of the term. Sir George Stokes had not interpreted dynamical theory in any narrow sense. If they considered the condition of Natural Philosophy in 1840, and in the present year, they might form some idea of how vast had been the results that had accrued in his life's work. Lord Kelvin pointed out that Sir George Stokes had the courage and spirit to attack subjects absolutely at the limits of the range of the mathematics of that time. Sound, light, elasticity, as mathematical problems on the one hand and properties of matter on the other, were the studies of Sir George Stokes; and the results of his efforts had been of splendid benefit to the world of science. Elasticity of solids Sir George Stokes put on its right footing. When he reflected on his own early progress, he was led to recall the great kindness shown to himself, and the great value which his inter-

---

* For the speeches, which were not printed in the official Order of Proceedings (*University Reporter*, or Introduction to special volume of *Camb. Phil. Soc. Memoirs*), acknowledgment is due to the reports in the *Cambridge Chronicle*.

course with Sir George Stokes had been to him through life. Whenever a mathematical difficulty occurred he used to say to himself, "Ask Stokes what he thinks of it." He got an answer, if an answer was possible; he was told, at all events, if it was unanswerable. He felt that in his undergraduate days, and he felt it more now. Lord Kelvin alluded to the valuable work of Sir George Stokes with respect to the suspension of clouds, based as he said on irrefragable mathematics. To Sir George Stokes belonged the distinction of having foreseen, in lectures if not in print, the grand principles of spectrum analysis. His published papers contained but a small part of the work he had done for science. All workers in science in the University of Cambridge, all the communicators of papers to the Royal Society during the thirty years he was Secretary and the five years he was President, would agree with him in saying that Sir George Stokes had published in his own name but a very small part of the good he had done to the world. How many beginners in researches had been helped by Stokes: he had counselled them how to obtain more knowledge, and he had helped them to secure results. Consider then the debt of gratitude due to him, not only for what he had published himself but for what he had done for others. His published papers and the discoveries contained therein had produced a monument more enduring than marble. But, still, they would like to have a marble monument, a tangible and visible sign for the men who knew Sir George Stokes and the work he had done, and for future ages; and, therefore, he felt it a great privilege indeed to be allowed to unveil the two busts, one designed by the subscribers for the University of Cambridge and the other for Pembroke College. He had the honour to ask the Vice-Chancellor to accept one of the busts for the University of Cambridge, and to ask Mr Prior to accept the other bust for Pembroke College. He was no judge of the art of sculpture, but he believed any work of Mr Thornycroft was certain to be a masterpiece of art. In that respect he did not venture to express an opinion, but he did say that each bust was an excellent likeness of Sir George Stokes, and if it was like Sir George Stokes it was something like one of the best things that existed in the world.

The Vice-Chancellor, Dr A. Hill, Master of Downing College, expressed thanks on behalf of the University, and Mr Prior on behalf of Pembroke College.

Brilliant weather favoured the proceedings on the Friday, when the formal celebration of Sir George Gabriel Stokes' jubilee as Lucasian Professor passed off with great dignity and success.

The Vice-Chancellor presided at a Congregation' in the Senate House at eleven o'clock, when the representatives of learned Societies in all parts of the world presented formal Addresses of congratulation to Sir George Stokes. Abbreviated contents of some of them are printed as an Appendix.

Sir George Stokes, on rising to respond, received an ovation. He quietly expressed his thanks for the great honour they had done him in coming to that celebration. It was an event which he never could have expected. But when the proposal was made to him by the Council of the Senate, he thought, although he felt unworthy of so great an honour, that it was his duty to accept the invitation. As he looked back on his long life, he must say, in spite of the flattering addresses that had been presented to him, that he ought to have worked a good deal harder. But perhaps it was because he had not done so, that he was there before them that day in excellent health. Perhaps it would be well that those who worked at physical subjects should avoid overworking. He assured them he felt deeply grateful for the manner in which they had honoured him.

The Vice-Chancellor entertained upwards of four hundred guests at luncheon in Downing College. At a quarter to three another Congregation was held in the Senate House, which was again filled in every part. The Chancellor, the Duke of Devonshire, entered the building with Sir George Stokes; and in the procession were the Vice-Chancellor accompanied by the Registrary, Dr Jebb and Sir John Gorst, representatives of the University in Parliament, the Bishop of Ely, the Bishop of Bristol, Lord Lister, Lord Kelvin, Lord Rayleigh, the Heads of Houses, Doctors, Professors, Proctors, and other University officers. Then followed those delegates from the various countries who had been selected as recipients of Honorary Degrees to mark the occasion: Marie Alfred Cornu, Jean Gaston Darboux, Albert Abraham Michelson, Magnus Gustav Mittag-Leffler, Georg Hermann Quincke, Woldemar Voigt.

The following Address, as approved by the Senate, and sealed with the University seal, was read by the Public Orator, Dr Sandys, and presented to Sir George Gabriel Stokes by the Chancellor.

Baronetto insigni
Georgio Gabrieli Stokes
Iuris et Scientiarum Doctori
Regiae Societatis quondam Praesidi
Scientiae Mathematicae per annos quinquaginta inter
Cantabrigienses Professori
S. P. D.
Universitas Cantabrigiensis.

Quod per annos quinquaginta inter nosmet ipsos Professoris munus tam praeclare ornavisti, et tibi, vir venerabilis, et nobis ipsis vehementer gratulamur. Iuvat vitam tam longam, tam serenam, tot studiorum fructibus maturis felicem, tot tantisque honoribus illustrem, tanta morum modestia et benignitate insignem, hodie paulisper contemplari. Anno eodem, quo Regina nostra Victoria insularum nostrarum solio et sceptro potita est, ipse eodem aetatis anno Newtoni nostri Universitatem iuvenis petisti, Newtoni cathedram postea per decem lustra ornaturus, Newtoni exemplum et in Senatu Britannico et in Societate Regia ante oculos habiturus, Newtoni vestigia in scientiarum terminis proferendis pressurus et ingenii tanti imaginem etiam nostro in saeculo praesentem redditurus. Olim studiorum mathematicorum e certamine laurea prima reportata, postea (ne plura commemoremus) primum aquae et immotae et turbatae rationes, quae hydrostatica et hydrodynamica nominantur, subtilissime examinasti; deinde vel aquae vel aëris fluctibus corporum motus paulatim tardatos minutissime perpendisti; lucis denique leges obscuras ingenii tui lumine luculenter illustrasti. Idem etiam scientiae mathematicae in puro quodam caelo diu vixisti, atque hominum e controversiis procul remotus, sapientiae quasi in templo quodam sereno per vitam totam securus habitasti. In posterum autem famam diuturnam tibi propterea praesertim auguramur, quod, in inventis tuis pervulgandis perquam cautus et consideratus, nihil praeproperum, nihil immaturum, nihil temporis cursu postea obsolefactum, sed omnia matura et perfecta, omnia omnibus numeris absoluta, protulisti. Talia propter merita non modo in insulis nostris doctrinae sedes septem te doctorem honoris causa nominaverunt, sed etiam exterae gentes honoribus eximiis certatim cumulaverunt. Hodie eodem doctoris titulo studiorum tuorum socios nonnullos exteris e gentibus ad nos advectos, et ipsorum et tuum in honorem, velut exempli causa, libenter ornamus. In perpetuum denique observantiae nostrae et reverentiae testimonium, in honorem alumni diu a nobis dilecti et ab aliis nomismate honorifico non uno donati, ipsi nomisma novum cudendum curavimus. In honore nostro novo in te primum conferendo, inter vitae ante actae gratulationes, tibi omnia prospera etiam in posterum exoptamus. Vale.

Datum in Senaculo
mensis Iunii die secundo
A. S. MDCCCXCIX.

LS

The Chancellor then, in the name of the University, said he had the pleasure and the honour of presenting to Sir George Stokes a gold medal, struck in commemoration of the distinguished services which, during the period of fifty years, he had

rendered to the University, especially in the advancement of those sciences in the study of which, since the days of his illustrious predecessor Newton, Cambridge had held an honourable preeminence. The Duke next announced that M. Cornu and M. Becquerel had requested permission to present a medal of the highest scientific renown.

M. Cornu, in the name of the Physical Section of the Institute of France, presented to Sir George Stokes the Arago medal, which had been ordered to be struck for this occasion, as a testimony of the high estimation in which the Academy held his valuable and important work, and to perpetuate the memory of his jubilee.

Sir George Stokes said he was aware that his own University were going to present him with a medal, and he need not express the deep gratitude he felt for that distinguishing mark of their approbation. He wished he were more worthy of it. But he was not aware until M. Cornu rose that he was about to be honoured also by the French Institute with the medal just placed in his hands. He requested M. Cornu to convey to his *confrères* his acute sense of the honour which the Academy of Sciences had conferred on him.

The ceremony of conferring the honorary Degrees then followed, the recipients being introduced to the Chancellor by the Public Orator in Latin speeches describing their achievements.

At the close of these proceedings there was a garden party at Pembroke College; and in the evening the University entertained the delegates at dinner in the Hall of Trinity College, under the presidency of the Chancellor.

The congratulations of the Berlin Academy took the form of the election of Sir George Stokes as a Foreign Member, as the following letter testifies.

KÖNIGLICHE AKADEMIE DER WISSENSCHAFTEN.

BERLIN, *den* 29 *Mai* 1899.

HOCHGEEHRTER HERR,

Herr Kohlrausch, der den Wunsch hatte, Ihnen unser Diplom als auswärtiges Mitglied, die höchste Ehrenbezeugung unserer Akademie, persönlich zu überreichen, ist im letzten Augenblick durch Krankheit zurückgehalten worden. Er wird seinen Besuch später ausführen. Wir aber fügen unserem Diplom die allerherzlichsten Glückwünsche zu Ihrem Jubiläum zu.

In ausgezeichneter Hochachtung,

DIELS,

vorsitzender Secretär
der Königlich Preussischen Akademie
der Wissenschaften.

It was natural that the Cambridge Philosophical Society should mark their gratitude to their distinguished Fellow, whose work had been one of the great ornaments of their *Transactions*, in some special manner. A proposal was set on foot to prepare a

special volume of Memoirs, which should be presented to Prof.
Stokes on the occasion of his Jubilee, and should also form

Vol. XVIII of their *Transactions*.   Accordingly a magnificent
quarto volume of nearly 500 pages, with plates, containing 22

memoirs, by Fellows of the Society and some of the Honorary Members in other countries, was published in 1900, with a portrait frontispiece here reproduced, and with the following dedication:

IN HONOREM
GEORGII GABRIELIS STOKES
PHYSICAM ET MATHEMATICAM
APUD CANTABRIGIENSES
IAM QUINQUAGINTA ANNOS PROFITENTIS

Two of the following letters convey congratulations from some of Sir George Stokes' distinguished friends in the French Academy of Sciences on his election as one of the eight Foreign Associates, which it had been hoped to arrange to coincide with his Jubilee. After them are printed letters conveying congratulations, from M. Becquerel's father and from M. Cornu, on the earlier occasion of his election as Correspondent in 1879.

PARIS, *le 26 février* 1900.
TRÈS HONORÉ CONFRÈRE,

J'avais appris quelques jours avant votre élection, par Mr Poincaré, qui avait passé quelques jours à Londres, le malheur qui vous a frappé, il y a deux mois*.

Nous avons tous été très attristés de votre chagrin, et nous espérons que le témoignage d'estime et d'admiration, que notre Académie vous a donné, adoucira un peu l'amertume de votre grande et juste tristesse.

Absent de Paris pendant quelques jours, à cause d'un deuil de famille, je n'ai pas pu répondre plus tôt à votre si aimable lettre, ni assister hier à la Séance de l'Académie. Mais je suis sûr que, suivant l'usage traditionnel, le Bureau de l'Académie vous aura officiellement annoncé votre élection par une lettre des Secrétaires perpétuels.

Mr Joseph Bertrand est en le moment fort malade et sa signature n'aura peut-être pas figuré au bas de la lettre: nous sommes extrêmement inquiet de sa santé. Ses forces déclinent chaque jour. On craint que ce soit l'effet d'un cancer du foie.

Veuillez agréer, je vous prie, illustre et très honoré Confrère, l'expression de mon respectueux dévouement.

A. CORNU.

* The death of Lady Stokes.

Paris, 9, rue de Grenelle,

le 1er *mai* 1901.

Très honoré Confrère,

Je vous suis bien reconnaissant, au milieu de votre douleur personnelle, d'avoir pris la peine de m'adresser une lettre aussi émue à l'occasion de la mort de mon pauvre frère. Vous avez bien voulu rappeler qu'aux fêtes de votre Jubilé c'est moi qui ai eu l'honneur de vous offrir au nom de l'Académie des Sciences la Médaille Arago, témoignage bien sincère de l'admiration que nos confrères portent à vos travaux.

Je vois que vous êtes bien entouré d'affections, puisque vous êtes allé vivre au milieu de la famille de Madame Humphry votre fille, dans une résidence contigue avec celle que vous avez habitée pendant de longues années avec la compagne que vous pleurez. Ce doit être pour vous une consolation bien douce.

J'ai tardé quelques jours à répondre à votre lettre, ayant été un peu détourné de mes occupations ordinaires par les tristes circonstances que je viens de traverser: j'ai d'ailleurs attendu d'avoir pu faire au Secrétariat de l'Institut les rectifications d'adresse que vous avez bien voulu me communiquer.

Veuillez agréer, je vous prie, très honoré Confrère, l'expression de mes sentiments de respect et d'admiration.

A. CORNU.

Paris, 20 *février*, 1900.

Cher et vénéré Confrère,

En vous nommant hier notre Associé à l'Académie des Sciences, nous avons voulu affirmer une fois de plus notre admiration pour votre belle vie scientifique.

J'avais demandé que cette distinction coïncidât avec les fêtes inoubliables de Cambridge; les circonstances ne l'ont pas permis à cette époque, et je suis très heureux de voir réalisé aujourd'hui l'hommage qui nous tenions à vous rendre.

J'ai appris à l'Académie, avec une grande tristesse, le deuil cruel qui vous avait frappé cet hiver; je tiens à vous dire combien Madame Becquerel et moi nous prenons part à votre chagrin.

Veuillez agréer, cher et vénéré Confrère, l'expression respectueuse de mes sentiments de haute considération.

HENRI BECQUEREL.

LENSFIELD COTTAGE, CAMBRIDGE,
5 *April*, 1899.

DEAR MONSIEUR BECQUEREL,

I am informed that you have accepted the invitation which you have received to the celebration of the Jubilee of my professorship. I write to ask if you will do me the honour to come to my house, along with Madame Becquerel, if she accompanies you, so that you and madame may be our guests on that occasion.

Lady Stokes is far from strong, but she will do what she can to make the visit of Madame Becquerel agreeable. We neither of us are in the habit of speaking French, but I hope we may be able to speak it sufficiently well to make ourselves understood. I have a pleasing recollection of the kind way in which you received me during the celebration of the centenary of the Institute of France.

I remain with the highest esteem,

Yours very faithfully,

G. G. STOKES.

LENSFIELD COTTAGE, CAMBRIDGE,
22 *July*, 1899.

DEAR M. BECQUEREL,

On my return from London I found the beautifully bound address, composed in such gratifying terms, which was presented to me by the Académie des Sciences de l'Institut de France and signed by you as their delegate.

It is many years now since the Academy did me the honour of electing me a Corresponding Member, since which time I have regularly received the *Comptes Rendus* as they were published. It was quite a surprise to me to receive at the hands of M. Cornu during the celebration of the Jubilee, the Arago Medal which the Academy have done me the great honour of awarding to me.

I requested M. Cornu to convey to his colleagues my most sincere thanks for that mark of their appreciation of my humble services towards promoting the progress of science. I should wish however to express through you my thanks to them in a more formal manner.

Permit me, M. Becquerel, to mention the great interest I have felt in your own remarkable researches in subjects in which we have both been engaged, and to express the high esteem with which I subscribe myself,

Your friend and colleague,

G. G. STOKES.

The letters referring to the earlier occasion are as follows:

PARIS, 57, RUE CUVIER,

le 10 juin 1879.

MON CHER MONSIEUR,

Vous avez été élu hier correspondant de l'Académie des Sciences de l'Institut de France dans la section de physique, et j'ai eu grand plaisir de pouvoir y contribuer et de faire connaître vos importants travaux. Je m'empresse de vous adresser mes félicitations bien sincères et de vous prier de vouloir bien agréer l'assurance de mes sentiments les plus distingués et dévoués.

EDMOND BECQUEREL.

PARIS, le 9 juin 1879.

CHER ET TRÈS HONORÉ MONSIEUR,

Je veux être un des premiers à vous annoncer que l'Académie des Sciences vient de vous élire Correspondant de la Section de Physique par 44 voix (sur 51 votants).

Il y a bien des années que vous étiez désigné : à cette époque vous auriez été honoré par cette élection : aujourd'hui c'est vous qui honorez l'Académie.

Veuillez donc agréer, je vous prie, cher et très honoré Monsieur, avec l'expression de mes félicitations celle de mon respectueux dévouement.

A. CORNU.

LENSFIELD COTTAGE, CAMBRIDGE,

4 July, 1900.

DEAR DR LARMOR,

I write to return you and the Council of the Cambridge Philosophical Society my best thanks for the beautifully bound and got-up volume which the Society did me the honour to put together in commemoration of the jubilee of my professorship. As I look on so many of those junior to me going above my head in mathematics and physics, I cannot help feeling my unworthiness of all the honour that has been done me.

To pass to another subject. I got some time ago a post card asking whether I could attend a meeting at your rooms relative to Section A at the Bradford meeting of the British Association.

I have not decided whether I shall go to the Bradford meeting, but I think it rather probable that I may. If the meeting is only for those who *are* going, I am disqualified. But I intend to be in Cambridge on the 7th; and if my going should be of any use, I am ready to attend on knowing the hour.

Yours very truly,

G. G. STOKES.

Telegrams of congratulation reached Sir George Stokes on the day of the Jubilee festival in great numbers: the following are selected at random as of public interest.

Congratulations and good wishes from friends and admirers in Holland.   VAN DER MEULEN, DIBBITS, FRANCHIMONT, JULIUS, KAMERLINGH ONNES, W. KAPTEYN, KLUYVER, KOSTER, LOBBRY DE BRUYN, LORENTZ, MAC-GILLIVRAY, MULDER, RAUWENHOFF, E. F. VAN DER SANDE BAKHUYSEN, SCHOUTE, SISSINGH, VAN DER WAALS, ZAAIJER, ZEEMAN.

Congratulations and best wishes.

PICKERING, Massachusetts.

Dem altberühmten Physiker gratulirt in herzlicher und denkbarer Ergebenheit die junge Collegin, die Physikalisch Technische Reichsanstalt, mit dem schmerzlichen Bedauern dass plötzliche Erkrankung mich gehindert hat die Glückwünsche der Akademie der Wissenschaften und der Reichsanstalt mündlich rechtzeitig zu überbringen. Hoffe ich, Hochverehrter Herr Jubilar, dass Sie dies gestatten wollen, wenn ich nachträglich darum bitten werde.

KOHLRAUSCH.

Quoique absent, mon ésprit est avec vous et prend une vive part aux manifestations dont votre nom illustre est l'objet.

GENERAL FERRERO, LL.D., Milan.

Die Kaiserliche Akademie der Wissenschaften sendet die wärmsten Glückwünsche ihrem Ehrenmitgliede.

SUESS, Wien.

Dem verehrten Altmeister sendet die wärmsten Glückwünsche das Physikalisch-Chemische Institut zu Leipzig.

OSTWALD.

Ich gestatte mir Ihnen ein Zeichen meiner ganz besonderen Werthschätzung und meiner Theilnahme an Ihrem Jubelfest zu senden.

RÖNTGEN, Würzburg.

Congratulate you most warmly.

Royal Society, New South Wales, Australia.

Best congratulations and wishes to Sir George Stokes.

SIR CHARLES TODD, Adelaide, S. Australia.

To Prof. J. J. Thomson.—Pray forward to Sir George Stokes on fit occasion warmest and most respectful congratulations with the regret of being prevented from personally bringing them.

P. LENARD, Kiel.

Herzlichste Glückwünsche dem grossen Gelehrten.

EBERT, München.

Many letters of regret came from scientific men who could not attend.

Prof. Mascart was unable to leave Paris. "Je serai avec vous de cœur en cette circonstance, et je m'associerai, à distance, aux félicitations et aux vœux dont vous recevrez l'hommage de la part de tous ceux qui apprécient l'importance de votre carrière scientifique. Les relations personnelles que j'ai eu l'honneur d'avoir avec vous, et la bienveillance que vous n'avez cessé de me témoigner, augmentent mes regrets d'absence."

Dr E. Schär, Professor of Pharmacy and Toxicology, sends on behalf of the Faculty of Mathematical and Natural Science of Strassburg, their congratulations. "I have the more pleasure, as Dean of our Faculty, to send you this document, as your researches have thrown so much light on several departments of Pharmaceutical and Physiological Chemistry."

"Prof. Voigt" of Göttingen, who was present, "has the honour of presenting to Sir George Stokes, with the most respectful congratulations for his Jubilee, some views of the place where C. F. Gauss and W. Weber were living and working."

Prof. L. Koenigsberger sent the warmest congratulations of his Faculty at Heidelberg.

Prof. G. Veronese, of Padua, writes from the Chamber of Deputies, Rome: "vi giunga anche il modesto mio plauso per l' opera vostra e il mio sincero augurio che siate conservato alla scienza e alla famiglia ancora lungo tempo."

Prof. F. Lindemann regretted that his duties as Dean of the Philosophical Faculty at Munich compelled his absence from the festival.

Prof. P. G. Tait writes:—"Though I shall not be present at your Jubilee, I am sure you will believe that my admiration of your career is as great, and my best wishes as sincere, as can be felt or expressed by any who may appear in person."

From Sir W. Huggins came a message that though to his very great regret he had to forego the pleasure of being present, "during those days I shall be nevertheless wholly with you in spirit.... No one will join more cordially..., for no one can have a deeper appreciation of your great life-work for the advancement of Science, not only *directly*, in the way of original investigation and discovery and in your professional teaching, but in a thousand indirect ways:—amongst others by the encouragement and assistance you have always been so ready to give to other workers in Science, and by your invaluable secretarial work for so many years at the Royal Society."

Among lifelong friends, Mr W. T. Kingsley* wrote from S. Kilvington Rectory:—"As I cannot be present at your Jubilee and do not want you to think you are forgotten, I send you the two Munich prisms that you did such good work with in old days, when the far end of the spectrum came further and further into view as the days grew longer.

"Do you know what is the material in the Jena glass that gives a nearly rational spectrum; you used to think titanium was the only substance likely to do so. I tried a telescope of Cooke's last summer, and found it nearly perfect; very little outstanding colour even for an edge of the object-glass.

"The object-glass made for Sir R. Ball was finished, but not mounted, so I did not see how it performs.

"I am all but stone deaf and have many other infirmities, but on the whole have much to be thankful for....

"With my best love to Lady Stokes."

* Formerly Fellow of Sidney Sussex College, brother of Charles Kingsley.

Mr C. E. Douglass writes from Brighton:—"I ask for nothing but that you be conscious of the kind and interested feeling with which an old friend of 1841-2-3 regards the incident. I fear that we are now the only two who remain of the little party who used to play at bowls in Pembroke garden, and once—I think you were with us—rowed down together to Ely to see the Cathedral. You will understand my feeling as I think of the many dear friends who have passed away since that time, Woodford, Venables, Power, Cayley, and others."

Mr J. S. Hoare writes from the Rectory, Godstone, Surrey:— "I don't know that I ever had the pleasure of speaking to you : but in 1846 I had the honour of being examined by you...when I was placed as sixth wrangler among a number of friends now, alas, nearly all departed."

From Mr W. M. Spence, formerly Fellow of Pembroke:—"I am very sorry to find myself unable to take part in person.... From the time when in 1870 I heard your lectures, onwards through the lengthy time, nearly twenty years, when I was proud to be a Fellow of Pembroke with you, I could not but feel an increasing admiration for your constant kindness and pleasantness to me, and I am glad to find this opportunity of expressing it. Others will speak more fully of your capacities in other directions, but one who has sat through many college meetings with you can speak better for your kindly consideration for your juniors and their often crude opinions."

Congratulations arrived also from his *confrères* É. Picard and H. Moissan, of the Institute of France; from Prof. A. Brill, of Tübingen; from the Salters Company of London, of which Sir George Stokes was an Honorary Member; and from his native village of Skreen.

Finally, there came the following from Shrewsbury School :—

Illustrissimo et venerando Baroneto GEORGIO STOKES
Praepostores Regiae Scholae Salopiensis S. P. D.

Tibi valde gratulamur omnes quod iam per quinquaginta annos Cantabrigiae studiis mathematicis praefuisti.

Nec nos soli: nam ab omnibus regionibus Angliam sapientes properant qui te maximis afficiant honoribus. Plurima de tua beneficentia et liberalitate eximia audivimus : pro qua te sup-

19—2

plices rogamus ut a magistro nostro, cum sapientissimo tum iucundissimo illo viro, impetres ut nobis ferias agere liceat. Hoc quidem te maiore fiducia rogamus, quod complures causae hanc scholam tibi caram coniuncte reddunt. Quae vero si feceris, nos tibi semper grati erimus. Fac valeas. Dabam Salopiensis princeps Scholae,

DESMONDUS F. T. COKE, pro praepostoribus.

Nonis Juniis.

(The following draft of a letter is dated the week of the Jubilee celebration.)

LENSFIELD COTTAGE, CAMBRIDGE.

6 *June*, 1899.

DEAR SIR,

I meant to have written to you in reply to your letter of May 20, to explain that I meant no disrespect by my comparison of " Egyptian Hieroglyphics." I do not recollect how I addressed the letter, but the enclosure shows that an earlier letter failed to reach its destination.

I quite agree with you in the opinion you express, that in biology we have need of special laws affecting organisms and not represented in the inorganic world. I am not a biologist, and do not know what the general opinion among biologists on that subject may be. I should have supposed that those who imagine that we want nothing more than the laws relating to the [in]organic world were only a rather small minority.

I am not sure that I understand you in what you said about heredity and inertia. I understood you to mean that heredity was to be accounted for by the physical law of inertia, applicable to the inorganic world. In this I could not agree with you. But as nobody has been able thus to account for the law of heredity, and as the assumption that it was thus to be accounted for is illegitimate unless you assume that all the laws of the organic world are merely results of the laws which hold good in the inorganic world—an assumption which you repudiate, and I think rightly repudiate—it seems to me very probable that I misunderstood you, and that all you intended was to draw a sort of analogy between the law of inertia in the inorganic world (though organised matter is subject to it equally with inorganic) and a certain law (that of heredity) belonging to the organic world. This of course would be a different thing altogether, and I should not object to it, though I confess that the analogy seems to me somewhat far-fetched.

Yours faithfully,

G. G. STOKES.

D. Gordon Johnson, Esq.

For the part of the following correspondence which was written by Sir George Stokes the Editor is indebted to the kindness of M. Becquerel.

16 *August*, 1899.

DEAR M. BECQUEREL,

I write to suggest an experiment, in case you should be willing to try it, bearing on the nature of what we call "Becquerel Rays." I think it due to you as the discoverer of these rays, that I should mention it first to you, but if you think it not worth while to try I can have it tried in the Cavendish Laboratory here.

The theoretical ideas which I entertain regarding the nature of these rays lead me to think it very probable that the efficiency of a substance (suppose metallic uranium) which emits them would depend very materially upon its temperature. I would propose to use the time of discharge of a charged gold leaf electroscope while the leaves sink from the first to the second of two chosen inclinations, as the indication of efficiency, and to effect the discharges by a disk of metallic uranium, at one time heated, at another time cooled. It might be heated about as high as it will bear without risk of spoiling the surface by oxidation, and cooled by a freezing mixture, or even, if the result of other experiments should seem to make it desirable, by liquid air. As you have shown that the effect of the uranium depends on the action of the Becquerel rays on the intervening gas, it would probably be necessary to prevent as far as possible the circulation of currents of convection, so as to make the experiments at different temperatures fairly comparable. I am inclined to think it probable that the uranium would be more efficient at a higher than at a lower temperature, but as the effect of a high temperature on air which had thus been rendered active would probably be to make it normal again, it might be desirable not to heat the uranium as high as I mentioned. These however are precautions which can only be learned by experience.

You may perhaps have noticed phenomena which indicate that nothing is to be expected from the experiment I have suggested, or, even if this be not the case, it may not be convenient to you to try the experiment. In that case I should feel obliged if you would kindly inform me; for then on the first supposition I would lay the thing aside, and on the second I might have the experiment tried in Cambridge.

I am not sure whether I gave you a copy of my "Wilde Lecture" on the Röntgen rays when you were in Cambridge. If not, I shall be happy to send one.

Yours very truly,

LE CROISIC,
25 *août* 1899.

CHER ET VÉNÉRÉ MONSIEUR STOKES,

L'aimable lettre que vous m'avez fait l'honneur de m'écrire m'est parvenue au Croisic où je suis depuis un mois, et où je dois rester encore une huitaine de jours avant d'aller passer six semaines à la campagne. Je ne retournerai à mon Laboratoire que vers la fin d'octobre. C'est vous dire que je ne pourrai reprendre avant cette époque les expériences dont vous avez l'obligeance de me parler.

Il y a trois ans, au début de mes recherches sur le rayonnement de l'uranium, guidé par l'idée que l'émission de ce corps pouvait être une phosphorescence invisible, j'ai soumis l'uranium et ses composés à des élévations de température diverses, puis à des refroidissements ne dépassant pas − 20° C., soit en les exposant en même temps à la lumière solaire ou électrique, soit en les maintenant à l'obscurité, et je n'ai constaté aucun changement appréciable dans le rayonnement émis par ces corps revenus à la température ordinaire. Mes autres expériences ainsi que celles de M^r et Me Curie ont montré que pour l'uranium et les autres corps radio-actifs, les diverses manipulations physiques ou chimiques qu'on peut leur faire subir, n'altèrent pas leur pouvoir émissif spécifique. Reste la question que vous me proposez si gracieusement d'étudier, et qui est de reconnaître si l'émission dépend de la température de ces corps.

Il résulte de toutes les expériences que je connais, qu'aux températures de l'hiver et de l'été dans les laboratoires, c'est à

dire, depuis quelques degrés centigrades au dessous de zéro
jusqu'à 20° à 30°, ni les actions photographiques, ni les mesures
de la conductibilité de l'air, n'ont manifesté aucune variation
notable dans l'émission des différents corps radio-actifs.

Il y a deux ans j'avais enterpris la recherche de la mesure
de l'émission de l'uranium métallique à diverses températures,
mais je n'ai pas publié les résultats de mes expériences parce
qu'elles ne m'ont pas semblé concluantes, et que j'attendais le
moment où je pourrais reprendre cette étude en employant des
abaissements de température considérables. Je n'ai pas ici mes
cahiers d'expériences et je ne me souviens pas des mesures que j'ai
obtenues ; je vous les indiquerai quand je serai de retour à Paris.
Voici la disposition que j'avais employée : Une sphère d'uranium

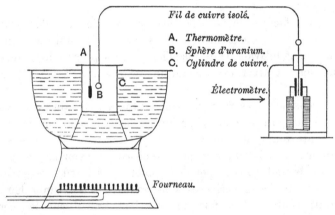

Fil de cuivre isolé.

A. Thermomètre.
B. Sphère d'uranium.
C. Cylindre de cuivre.

Électromètre.

Fourneau.

métallique *B*, était suspendue par un fil fin de cuivre, bien isolé,
et relie à un électromètre de Hankel. La sphère d'uranium était
suspendue à l'intérieur d'un cylindre de cuivre *C*, que l'on pouvait
échauffer ou refroidir.

Si le cylindre est électrisé, la sphère prend le potentiel du
cylindre en une ou deux minutes. Si le cylindre est en communication avec le sol et si l'on charge la sphère à des potentiels
déterminés, celle-ci se décharge et l'on peut suivre la vitesse de
déperdition de la charge à diverses températures. Je n'ai pu
opérer qu'entre les températures de + 100° et − 20°, en entourant
alors le cylindre *C* d'un mélange réfrigérant. Les variations m'ont
paru attribuables aux variations de la conductibilité de l'air par
convection. A cette même époque j'avais montré que la décharge
des corps électrisés se faisait par l'intermédiaire des gaz ambiants,

et que la sphère d'uranium, placée dans le vide, ne se déchargeait pas.

Dans cette étude, j'avais pensé en outre pouvoir substituer l'action photographique à l'observation de la conductibilité de l'air, mais j'ai été arrêté dans cette voie par la difficulté d'échauffer ou de refroidir les corps actifs, sans faire varier notablement la température des plaques photographiques; car une élévation de température modifie leurs propriétés, et aux très basses températures les sels d'argent ne sont plus réduits par la lumière. Si l'on avait constaté que l'action photographique de l'uranium est nulle à la température de l'air liquide, il se pourrait que ce résultat fût dû, non à l'absence de rayonnement de l'uranium, mais à l'insensibilité de la plaque photographique à cette température.

Cette difficulté de faire la part des modifications des sources radiantes, et celle des modifications des agents qui révèlent le rayonnement, m'avait arrêté jusqu'ici, et je me propose le reprendre la question cet hiver, maintenant que je puis avoir à Paris un peu d'air liquide.

Je vous suis très reconnaissant d'avoir eu l'amabilité de vous adresser d'abord à moi, pour rechercher la solution de cette importante question. Je compte m'y remettre dès que je serai de retour à mon Laboratoire, mais je ne saurais avoir la prétention de rester le seul à m'occuper de ce sujet; et s'il vous est agréable de faire faire les expériences sous vous yeux à Cambridge où Mr Rutherford a déjà fait un très bon travail sur les corps radiants, je serai le premier à applaudir au succès.

Dans tous les cas, dès mon retour à Paris, je vous enverrai le résumé des résultats inédits que j'ai obtenus il y a deux ans.

Je vous remercie de l'offre que vous me faites de m'envoyer un exemplaire de votre " Wilde lecture on the Röntgen rays." Je n'en possède pas; je serai très heureux de le recevoir et de méditer sur les idées que vous avez émises au sujet de ce nouvel ordre de phénomènes.

De mon côté, j'ai obtenu avant de quitter Paris, d'assez belles épreuves de mon expérience sur la dispersion anomale de la vapeur de sodium, et je vous en envoie deux dans le cas où elles vous intéresseraient.

Je saisis cette occasion pour vous demander, au nom de Madame Becquerel et au mien, de vouloir bien présenter nos respectueux hommages à Lady Stokes, et je vous prie d'agréer, avec tous mes remerciements, l'expression de mes sentiments de haute considération et de respectueux dévouement.

HENRI BECQUEREL.

LENSFIELD COTTAGE, CAMBRIDGE.

4 Sep. 1899.

DEAR M. BECQUEREL,

I am very much obliged to you for the interesting photographs you kindly sent me of the anomalous dispersion of vapour of sodium. I had seen mention of it in books, but I had not any actual photograph in my possession until I received those you sent.

I thought it very probable that you had already made experiments on the effect (if any) of the temperature of the uranium on its radio-activity. I was not aware however that you had actually done so, nor could I have been, since as you inform me your results have not yet been published. It is hardly worth while now to undertake the experiments I suggested, which hardly differ from those which you mention as having been already tried. I had indeed contemplated trying greater differences of temperatures, but you have already employed a difference of 120° C. which is considerable.

It may be well that I should explain the order of ideas which led me (ignorant as I naturally was of your as yet unpublished results) to propose the experiment. When I first read of radio-activity, my thoughts naturally turned, as did yours, to a sort of phosphorescence resulting from previous exposure. But your subsequent communications to the Academy, which appeared in the *Comptes Rendus*, showed me that the effect was a permanent one, and must be otherwise explained. As to Röntgen rays, my belief is that they consist in a vast succession of independent "pulses" in the ether, each produced when a molecule (or ion, if such it be) reached the anti-cathode. As the properties of the Becquerel rays place them between the Röntgen rays and those of ordinary light, we must suppose the former to consist of a vast succession of series of vibrations in each of which the disturbance, though beginning to be somewhat of the value of a regular periodic disturbance, does not last long.

To fix ideas we may think of such a disturbance as would be represented by a sine or cosine, multiplied by an exponential with a negative index. I imagined the period to be probably short, and regarded the phenomenon as analogous to Tyndall's "Calorescence," only instead of the medium (uranium suppose) which emits the rays being made active by an enormous concentration of ethereal waves of low refrangibility, I suppose the source of activity to be the ethereal vibrations which are always going on at ordinary temperatures. I suppose the molecular structure of the uranium (or other similar body) to be such that a portion of the structure is capable of being thrown into agitation of a, roughly speaking, independent kind by the movement of a larger portion, which latter movement is excited and sustained by the ethereal vibrations which emanate from the various bodies in the environment*.

The experiments I contemplated were qualitative rather than quantitative, and were such as could be made in quite a short time. It does not seem desirable to engage in the more difficult and laborious problem of obtaining quantitative results unless preliminary qualitative experiments have shown that there is really something that it is desirable to measure. It had occurred to me that it would be desirable to guard, as well as conveniently may be done, against the effect of currents of convection. But it might perhaps be well to go a step further, and interpose a screen of thin white paper between the plate which receives the radiation and the uranium ball presented to it, so that the atmosphere

[* When these letters were written it had not yet been recognized that the phenomenon of radio-activity discovered by M. Becquerel consisted largely of the expulsion of cathode and other particles from the material, at enormous speed, in addition to the emission of radiation.

When, after their discovery of radium, M. and Mme Curie showed the very copious emission of heat from it, one mode of explanation that was offered is thrown into sharper relief by these remarks of Sir George Stokes. It was suggested that the heat emitted is in fact collected as energy of radiations incident on the radium from the surrounding space: and the mechanism of the collection might be as described above. A rise of temperature of the radium consequent on this would not violate Carnot's principle, provided the incident radiation belonged to a source at higher temperature,—in the manner exemplified by the trapping of such radiation by the glass of a greenhouse. But the structure by virtue of which the radiant energy in the course of being trapped ejects α and β ions would be on such a view one peculiar to radio-active substances.

But in so far as it has been verified that the emission of heat is permanent, and goes on in the dark, the alternative explanation, supported by Rutherford, has come to be accepted, viz. that the kinetic energy of the α and β particles is intrinsic energy of atomic disaggregation,—being the main part thereof—and that as regards those particles that are stopped in the substance of the radium this energy is transformed into sensible heat exhibited by the radium itself.]

between the screen and the receiving plate may be always the same except in so far as it is affected by the radiation. I am nearly sure (from memory) that I am right in supposing that the Becquerel like the Röntgen rays pass pretty freely through thin paper. I mentioned white paper rather than black with the view of avoiding any significant warming of the paper by heat-radiation from the ball, so as to prevent the air between the screen and the receiving plate connected with the electroscope from being warmed by contact with the paper; a thing which might still further be prevented, only that it would probably be a superfluous precaution, by causing a current of air to play against the outer face of the screen.

In connection with my ideas about the mode of disturbance in the uranium molecule, which results in the emission of Becquerel rays, I will mention a phenomenon which I noticed more than 40 years ago, when I was working with fluorescence. When a little of salt of uranium was introduced into a blow-pipe bead, I noticed that when the bead was thoroughly oxidised in the outer flame, it glowed with a greenish light which recalled the colour of the phosphorescent light of uranium glass, but when the oxidation was slightly incomplete, the glow had not this greenish colour. I conjectured that when the uranium was in state of sesquioxide, there was a molecular group vibrating in periods corresponding to a considerable portion of the spectrum up to green inclusive, and that this mode of vibration was more persistent than usual, the molecular disturbance remaining roughly speaking isolated; but that whenever a small proportion of the uranium was present in the state of proto-salt, the agitation of the sesquioxide group passed away by being communicated to the protoxide group.

The phenomena of phosphorescence of uranium compounds and of radio-activity of uranium itself, seem to point to the existence of a molecular group which is roughly speaking isolated, in the sense that vibrations going on in it are not very quickly communicated to the neighbouring structure*. I have not de-

[* In an addition to *Math. and Phys. Papers*, Vol. III. (1892), Prof. Stokes' views reverted towards the notion of merely general degradation of the incident radiant energy, analogous to what occurs when a body warmed by solar radiation emits the energy as heat-waves: but recent observations by Wood, Morse and others, as to the sharpness and definiteness of the fluorescent spectra, seem to be strongly in favour of his earlier view, which as appears from this letter he never abandoned. Prof. Stokes was there guided by an observation that as the fluorescent light faded

scribed the glow-phenomenon in print, as it seemed too small a matter.

If you should have left Le Croisic before this letter arrives, I suppose it will be forwarded. I am sending my Wilde Lecture by book-post. The style is rather colloquial; it was in fact delivered *ex tempore*, and taken down by a shorthand writer.

Pray present my respects to Madame Becquerel and believe me,

Yours very faithfully,

G. G. STOKES.

Early in 1900 Sir George Stokes published a letter in *Nature* remarking on peculiar appearances shown on photographs of electric lamps that had been sent to him. A suggestion followed from Prof. R. W. Wood that he was familiar with similar appearances due to movement of the camera, but at first Sir George Stokes thought that as the result of his inquiries such a cause was excluded. It turned out however on trial that it was not so. The paper referred to below appeared in *Nature* of date Mar. 23 but *without the postscript*, which may have been withdrawn in correcting the proof.

LENSFIELD COTTAGE, CAMBRIDGE.

26 *March*, 1900.

DEAR SIR NORMAN [LOCKYER],

I have been making some observations on an arc lamp in the middle of Parker's Piece, using a hand mirror which I moved, a deep blue glass, and a pair of *concave* spectacles chosen of such power as to give clearest vision of distant objects with blue light. I require convex for reading, and for distant objects my eyes are almost exactly right; I am merely what I may call infinitesimally *long* sighted.

I was able to see the discontinuity (shown in the photographs by the beading) but what I was not prepared for is plain from a comparison of what I saw with the photographs, namely, that by very far the greater part of the actinic rays from the lamp come, not from the incandescent poles, but from the gas inside the glass globe which protects the poles from wind and from combustion.

I think it would be worth while to add a P.S. to the short

it became redder : which however may be the analogue of his own result in the theory of sound, that the same vibrator radiates short waves much more intensely in proportion than long ones.]

paper I sent in for *Nature*, the object of which was to put on the white sheet. I enclose an addition if you think it worth putting in.

Yours very truly,

G. G. STOKES.

The following letter from Prof. Horace Lamb forms part of a correspondence that went on during the preparation of the enlarged second edition of the *Treatise on Hydrodynamics*.

ESKDALE, *viâ* CARNFORTH.
July 28, 1900.

MY DEAR SIR GEORGE STOKES,

I quite agree that the formula $c^2 = gh\,(\tan mh)/mh$ is 'exact' on the assumptions stated. The only limitation I see is that mh (regarded as an angle) must be well within the first quadrant. As regards the assumption (2) that at the outskirts the amplitude decreases in geometric progression, I think it very probable, but I should be rather puzzled to give a decisive mathematical reason for my faith*.

The remarks on the history of "group-velocity" are very interesting. I do not recall the formula you mention in Cauchy and Poisson, but I have not the memoirs here. I had for some time been thinking that it would be worth while to study these memoirs again by the light of group-velocity†. There is a hint in a paper by Lord Kelvin, quoted on p. 380 of the *Hydrodynamics*. If I remember rightly, the solution for a limited initial disturbance is that (subject to certain ranges of time and distance) the waves found at time t in the neighbourhood of any point P are of such length that the corresponding *group*-velocity would bring them in time t from O (the origin of disturbance) to P. Probably the formula you refer to (which must I think be original) embodies the same result.

I recommended the subject lately to a Smith's Prize Candidate who was hard up for a subject for an Essay; but I fancy he chose something else.

I lately wrote a small note (which I think I sent you) on

---

* See *Hydrodynamics*, 1895, p. 421. The subject is the motion at the outskirts of the solitary wave. The passage is reproduced in *Math. and Phys. Papers*, Vol. v. p. 163, along with an extract from the letter to which this is a reply.

† This intention was carried out in Prof. Lamb's Presidential Address to the London Mathematical Society, published in their *Proceedings* for Nov. 1904.

"group-velocity."* The phenomenon is often referred to as "well-known" by observation, but Scott-Russell is the earliest writer by whom I have seen it mentioned†.

With many thanks for your letter,

I am, Yours very faithfully,

HORACE LAMB.

Prof. Stokes learned with great interest towards the end of his life‡ of the recent development of metallurgy, based on microscopic examination conjoined with the application of Gibbs' Law of Phases, a subject to which he had been introduced through the experiments of Messrs Heycock and Neville at Cambridge, of which some results formed the Bakerian Lecture at the Royal Society for 1903. The following letter is taken from a correspondence of considerable extent.

LENSFIELD, CAMBRIDGE.

16 *April*, 1902.

DEAR MR NEVILLE,

The facts of the Cu-Sn series, so far as I am acquainted with and have digested them, rather lead me to the original view which I entertained when first I saw your joint work in that series, I mean that in your first paper on that series in the *Phil. Trans.* This was that there are definite alloys, H (CuSn), E (Cu3 Sn), and C (Cu5 Sn), and that D is formed by the union of C and E, D being Cu4 Sn or rather Cu8 Sn2.

I think that the halt at C in cooling is due to the crystallisation of C, which takes place for a composition a little richer in copper than Cu5 Sn at the second freezing point. I think that the alloy Cu5 Sn has very strong crystallising (as distinguished from mere solidifying) power, and that it crystallises in very thin laminæ, the direction of which tends to be perpendicular to the gradient of composition, I mean the direction of most rapid variation of composition. The uniformity of composition belonging to the ingot when the whole is molten and well stirred, is disturbed in cooling by the separation of solids in cooling down to C. I think that C crystallises in laminæ, leaving mother liquid in the intervals between them.

---

* *Manchester Memoirs*, XLIV. 1900; also *Proc. Lond. Math. Soc.*, I. 1904, p. 473.

† See also letters from Mr W. Froude, F.R.S., printed *infra*, Vol. II. p. 157.

‡ Cf. however his graphical representation of the relations of ternary alloys contributed in 1891 to a paper by Dr C. R. Alder Wright, F.R.S., *Math. and Phys. Papers*, Vol. v. p. 226.

FINAL TRIBUTES.

The incidents of the final years of his life have been recorded with authority in a previous chapter. His death on Feb. 1, 1902, attracted wide attention, and many formal obituary notices of his career were published. The notice in the London *Standard*, speaking with knowledge, referred to the galaxy of mathematical talent which appeared in Cambridge in the decade when Stokes took his first degree (1841). The writer finally gave his impressions of the man and his character.

"In the previous year the Senior Wrangler was R. L. Ellis, who, if health had permitted, would certainly have fulfilled his friends' expectations, high as those were. The Second Wrangler in the same year was Harvey Goodwin, in his middle age a great moral and intellectual force in Cambridge, who died Bishop of Carlisle. In 1842 Cayley, almost unrivalled for his mastery of Pure Mathematics, was Senior Wrangler, and the next year Adams, the astronomer, who was said to have distanced every competitor in the examination, and discovered the planet Neptune before he became a Master of Arts. These three men, each of whom has left an enduring mark in his own department of Mathematics, were afterwards teaching together in their University as Professors. Again, in 1845, William Thomson, now Lord Kelvin, was Second Wrangler and First Smith's Prizeman.

"In person, Stokes was of middle height, rather strongly built, with clear and sharply cut features, a faint smile often softening his expression. Simplicity characterised his mode of living, as it did the man. His taciturnity was proverbial, but he could express himself well enough when he had to make a speech, and abstained from words because he did not care to waste them. No one who broached a favourite subject, or went to Stokes with a difficulty, could complain that he grudged either pains or time. However small might be the inquirer's knowledge, Stokes made things as plain as was possible. Never was there the slightest assumption of superiority, though he spoke decisively on matters which he understood better than others. As the late Professor Tait once said of him, ' Stokes has attacked many questions of the gravest order of difficulty in pure mathematics, and has carried out delicate and complex experimental researches of the highest originality, alike with splendid success. But several of his greatest triumphs have been won in fields where progress demands that these distinct and rarely associated powers be brought simultaneously into action.'"

On the day after his death, Feb. 2, the following special leading article appeared in the *Standard*, which spoke with obvious knowledge of his work at Cambridge and elsewhere:

"By the death of Sir George Gabriel Stokes, in his eighty-fourth year, Cambridge loses one of its most distinguished sons. He was the last of that little group of Professors—Stokes, Cayley, Adams, and, for a shorter time, Maxwell—who, during most of the Victorian era, made their names and University familiar wherever the higher Mathematics are studied. Others there were under them, like Tait and Kelvin, but their work was not done in Cambridge; or like Lord Rayleigh, Maxwell's successor, whose stay was much shorter than for its own sake the University desired; and the old legend of the Golden Bough is being fulfilled at the present day by younger men. But, like Whewell and Sedgwick in their day, that quaternion of intellectual strong men was specially typical of Cambridge and its chief study. Their University owes them much, yet a chance passer-by, looking into their lecture-rooms, might have pronounced their work a failure. Instead of an entranced throng of students, he might have seen a class of half a dozen, but these would have represented the very ablest young mathematicians, who were often not all Undergraduates. To them the Professor was making clear some of the most difficult subjects in physical mathematics, or feeding them with the concentrated essence of his own researches. Stokes's characteristics as a teacher—and the remark must not be limited to the lecture-room—were lucidity and suggestiveness. He gave those who listened to him the impression of having his subject clear before him, and having seen through and beyond it to an extent impossible to most men. Thus his work, both as a teacher and as a pioneer, in more than one branch of physical mathematics, never appealed to the crowd, but was appreciated at its true value by those who were themselves in the front rank of scientific investigators. His chief laurels were won in the sphere of physical optics; but the remarkable breadth and penetration of his mind carried him, always with success, into subjects not less difficult, which in some cases had but a remote connection with his principal study. Here, again, the value of his services, owing to his unobtrusive character and method of work, was little known to the world at large. But when we turn over the volumes of the Philosophical Transactions and the Proceedings of the Royal Society, which he served for so many years continuously, first as Secretary then as President, we discover how often authors thankfully admit their debt to him for important criticisms and fruitful suggestions. To the really able student, this man of few words and reserved demeanour was at once a tonic and a stimulant. But in his own University, and in a much wider circle, Stokes had hardly less influence in educational than in scientific questions. In these, as in all others, as was natural, his mathematical habit disposed him to caution, but his grasp of truths was too great to allow him to be an unreasonable obstructive. An aptitude for business is often not a characteristic of a master in philosophical studies, but the clearness of thought, calmness of judgment, and tenacity of memory which won Stokes his scientific eminence made him hardly less efficient as a member of a Committee. In his own University it was rare to find one appointed to discuss any matter of great importance of which he was not a member. The same valuable qualities were conspicuous in his work as Secretary to the Royal Society. Though he sat for a few years in the House of Commons as one of the members for the University, for the strife of Party politics he can never have greatly cared, and he spoke only on questions upon which he was an expert. But his influence, both on Cambridge and on many outside its boundaries, will be felt for long years to come. Simple in habits, strenuous in work, free from all taint of self-seeking, Stokes leaves the example of his life as his legacy to those among whom it was spent."

And on the next day, Feb. 3, the *Times* in an equally authoritative leading article summed up his career as follows:

" It is merely true, though it may sound paradoxical, to say that there are men so great that their greatness is not readily demonstrable. Such a man was Sir George Gabriel Stokes, whose death we chronicled yesterday. We may enumerate his scientific papers, we may expatiate upon his work in optics or hydrodynamics, we may dwell upon his masterly treatment of some of the most abstruse problems of pure mathematics, yet only a select body of experts can really understand how great he was in these various directions, while possibly not all the experts understand how much greater was the man than all his works. He presented the rare combination of extraordinary penetration with extraordinary breadth of intellect, producing a lucidity and completeness of treatment which masked the difficulties so triumphantly surmounted. He saw his problems clearly, and he saw them whole; the finish of the solution and the apparent lightness of the effort tending to hide from men's eyes the real magnitude of his achievement. Yet the work given to the world in his name represents but a small part of what he did for science. It is not too much to say that for half a century he was more or less behind the best work done by other men. Acknowledgments of indebtedness to Sir George Stokes are common in the best scientific literature, and private acknowledgments of help, stimulation, or pregnant suggestion are still more numerous. He was absolutely free from the exclusiveness and the secretiveness which are sometimes met with in men of science. He would no more have thought of disputing about priority, or the authorship of an idea, than of writing a report for a company promoter. It was enough for him that science was being advanced, and his knowledge was always at the command of any one who seemed able to carry on the work. There were problems enough for him which no one else could grapple with, and the derivative researches he was glad to assist others to make. It is impossible to calculate the influence of this single-minded and generous aid freely given throughout a long life; but that it must have been enormous will be most fully conceded by the most eminent scientific workers of the age. He was in the largest sense a great teacher, always inspiring and suggestive to those who knew enough to avail themselves of his commanding insight.

"Thus the life work of Sir George Stokes is not to be sought merely or chiefly in his direct contributions to science, valuable and unique as these are. He was a vivifying influence ever operating upon the plastic intellect of his time, and his work is to be looked for in the minds he stimulated, helped, and directed. It need hardly be said that to play this great part in the intellectual life of his time a man must be great by character as well as by intellect. Sir George Stokes was as remarkable for simplicity and singleness of aim, for freedom from all personal ambitions and petty jealousies, as for the breadth and depth of his intellectual equipment. He was a model of what every man should be who aspires to be a high priest in the temple of nature. It is sometimes supposed—and instances in point may doubtless be adduced—that minds conversant with the higher mathematics are unfit to deal with the more ordinary affairs of life. Sir George Stokes was a living proof that if the mathematician be only big enough his intellect will handle practical questions as easily and as well as mathematical formulas. No man was more in request for management of the ordinary business of his University or of the Royal Society, none was a greater authority upon educational questions in general, nor did any bring to the perplexities of everyday business a sounder judgment or a more imperturbable serenity. He had other speculations than those of pure science, and he adhered to convictions which he commended and illustrated by a life of unblemished integrity and nobility. A great man has passed away from among us. Let us all in our several degrees imitate so far as we may a splendid example of singleminded devotion to great ideals."

In the following week (Feb. 12), an enthusiastic tribute to his scientific work by Lord Kelvin occupied the front place in *Nature*. The following extracts will prove of interest, as additions to the appreciations from Lord Kelvin already quoted:

"Stokes ranged over the whole domain of natural philosophy in his work and thought; just one field—electricity—he looked upon from outside, scarcely entering it. Hydrodynamics, elasticity of solids and fluids, wave-motion in elastic solids and fluids, were all exhaustively treated by his powerful and unerring mathematics.

"Even pure mathematics of a highly transcendental kind has been enriched by his penetrating genius; witness his paper 'On the Numerical Calculation of a Class of Definite Integrals and Infinite Series*,' called forth by Airy's admirable paper on the intensity of light in the neighbourhood of a caustic, practically the theory of the rainbow. Prof. Miller had succeeded in observing thirty out of an endless series of dark bands in a series of spurious rainbows, for the determination of which Airy had given a transcendental equation and had calculated, of necessity most labori-

* *Math. and Phys. Papers*, Vol. I. pp. 329—357. From *Camb. Phil. Soc.*, March 11, 1850.

ously by aid of ten-figure logarithms, results giving only two of those black bands. Stokes, by mathematical supersubtlety, transformed Airy's integral into a form by which the light at any point of any of those thirty bands, and any desired greater number of them, could be calculated with but little labour, and with greater and greater ease for the more and more distant places where Airy's direct formula became more and more impracticably laborious. He actually calculated fifty of the roots, giving the positions of twenty black bands beyond the thirty seen by Miller.

" With Stokes, mathematics was the servant and assistant, not the master. His guiding star in science was natural philosophy. Sound, light, radiant heat, chemistry, were his fields of labour, which he cultivated by studying properties of matter, with the aid of experimental and mathematical investigation.

" The greatest and most important of all the optical papers of Stokes was communicated to the Royal Society on May 27, 1852, under the title ' On the Change of the Refrangibility of Light*.' In this paper his now well-known discovery of fluorescence is described, according to which a fluorescent substance emits in all directions from the course through it of a beam of homogeneous light. The periods of analysed constituents of this fluorescent light, in all Stokes's experiments, were found to be longer than the period of the exciting incident light. But I believe fluorescent light of shorter periods than the exciting light has been discovered in later times.

"Stokes found that the fluorescence vanished very quickly after cessation of the incident light. A beautiful supplement to his investigation was made by Edmond Becquerel showing a persistence of the fluorescent light for short times, to be measured in thousandths of a second, after the cessation of the exciting light.

" Stokes's fundamental discovery of fluorescence is manifestly of the deepest significance in respect to the dynamics of waves, and of intermolecular vibrations of ether excited by waves, and causing fresh trains of waves to travel through the fluorescent substance. The prismatic analysis of the fluorescent light for any given period of incident light was investigated by Stokes for a large number of substances in his first great paper on the subject, and was followed up by further investigations by Stokes

* *Phil. Trans.* and *Math. and Phys. Papers*, pp. 259—407.

himself in later years, of which some of the results are given in his paper 'On the Long Spectrum of the Electric Light' (*Phil. Trans.*, June 19, 1862).

"Stokes's scientific work and scientific thought is but partially represented by his published writings. He gave generously and freely of his treasures to all who were fortunate enough to have opportunity of receiving from him. His teaching me the principles of solar and stellar chemistry when we were walking about among the colleges some time prior to 1852 (when I vacated my Peterhouse fellowship to be no more in Cambridge for many years) is but one example. Many authors of communications to the Royal Society during the thirty years of his secretaryship remember, I am sure gratefully, the helpful and inspiring influence of his conversations with them. I wish some of the students who have followed his Lucasian lectures could publish to the world his *Opticae Lectiones*; it would be a fitting sequel to the 'Opticae Lectiones' of his predecessor in the Lucasian chair, Newton.

"The world is poorer through his death, and we who knew him feel the sorrow of bereavement."

This notice called forth a letter in *Nature* from a well-known and distinguished chemist, as follows :

"The eulogy of Stokes by Lord Kelvin, contributed to your columns in terms so appropriately simple, a eulogy so sincere, as we all know, and more authoritative than could be pronounced by anyone else in the world, furnishes an incident that must impress the minds of all true lovers of science. It is not my purpose to intrude where I have no business, but I do feel most keenly the strong call there is to English men of science to see that the hidden work of Stokes does not remain any longer concealed. There is not the least doubt that his greatness and true worth escaped the observation of contemporaries outside the circle of real scientific workers, and there has been one conspicuous occasion quite recently when the order of his merit has been signally ignored.

"About ten years ago the attention of Stokes was attracted to some work in which I was engaged, and this started a correspondence. I had no previous personal acquaintance with him, and I am sure he had no previous scientific acquaintance with me, but notwithstanding this he immediately placed the vast powers of his mind at my disposal, and assisted me with encouragement and advice that from my best friend would have been liberal in amount, whilst in value they could have been equalled from no other source. The abundance, lucidity and punctuality of his correspondence were amazing. I have had as many as three letters from him in one day, and on a particular occasion a telegram in addition, to say that he feared he had expressed himself in one of the letters with too much confidence. I was naturally not a little proud of this connection with a great man, but if my pride had tended to assume the form of vanity, that would have been frustrated by the discovery I was ever afterwards making of the apparently endless number of scientific workers who have received from Stokes the same unstinted help.

"I wish, therefore, to express the hope that in any memoir of Stokes that is published there should be some attempt to gather the unostentatious testimony that would be so cheerfully given by those who are so much beholden to the great and good man who has passed from among us. It seems to me to be at the least a duty to scientific history to help our posterity to see clearly that the order of Stokes's merit as a man and a philosopher was that of Faraday and Newton."

CHEMICUS.

MEMORIAL IN WESTMINSTER ABBEY.

The funeral at Cambridge had been attended by weighty delegations from the learned Societies, who represented all that was best in the scientific life of Great Britain. Directly afterwards the Royal Society resolved that the duty devolved on it, in response to a desire widely expressed, of consulting the University of Cambridge with a view to joint action in providing a national memorial. This resulted in the issue of the following preliminary circular.

"The subject of providing a public Memorial to commemorate in a fitting manner the scientific career of the late Sir George Gabriel Stokes has recently occupied the attention of the University of Cambridge and of the Royal Society, on both of which bodies he conferred distinction for so long a period by his labours; and a joint Committee has been constituted in order to take the matter into consideration and devise measures for carrying it into execution.

"A preliminary meeting of this Committee was convened in the rooms of the Royal Society, Burlington House, London, W., on Thursday, March 12, 1903, at 3 o'clock. There were present The Duke of Devonshire, Chancellor of the University of Cambridge, Rev. Prof. Chase, President of Queens' College, Vice-Chancellor, Prof. Sir Richard Jebb, M.P., Prof. G. H. Darwin, Prof. Sir R. S. Ball, Prof. A. R. Forsyth, and Mr W. Burnside of Pembroke College, as representatives of the University of Cambridge; and Sir W. Huggins, President of the Royal Society, Lord Kelvin, Past President, Principal G. Carey Foster, Vice-President, Mr A. B. Kempe, Treasurer, Sir Michael Foster and Prof. J. Larmor, Secretaries, Dr T. E. Thorpe, Foreign Secretary, Prof. G. D. Liveing, and Prof. A. E. H. Love, as representatives of the Royal Society. Lord Rayleigh and Prof. J. J. Thomson were unavoidably absent.

"On the proposal of the PRESIDENT OF THE ROYAL SOCIETY, the Chair was taken by the DUKE OF DEVONSHIRE.

"The following resolutions were adopted unanimously:—

(I) Proposed by the PRESIDENT OF THE ROYAL SOCIETY, seconded by the VICE-CHANCELLOR OF THE UNIVERSITY OF CAMBRIDGE,

> That it is desirable to place a Memorial of Sir George Gabriel Stokes in Westminster Abbey.

(II) Proposed by LORD KELVIN, seconded by Prof. Sir RICHARD JEBB, M.P.,

> That a joint letter, as now read in draft, be sent from the Chancellor of the University of Cambridge and the President of the Royal Society to the Dean of Westminster, requesting the authority of the Dean and Chapter to place a Memorial in the Abbey in the form of a Medallion relief portrait of Sir George Gabriel Stokes, of the same general character as the memorials of Charles Darwin and other scientific men now in the Abbey.

(III) Proposed by Sir MICHAEL FOSTER, seconded by Principal G. CAREY FOSTER,

> That a Sub-Committee, consisting of the President of the Royal Society, the Vice-Chancellor of the University of Cambridge, the Treasurer of the Royal Society, Prof. Liveing, Prof. G. H. Darwin, Prof. Forsyth, and the Junior Secretary of the Royal Society, be constituted (in case the consent of

the Dean and Chapter of Westminster is obtained) to send out a circular inviting subscriptions for this purpose, to Fellows of the Royal Society, and to Members of the University of Cambridge, and to such other persons as they may think fit: that the Vice-Chancellor of the University of Cambridge and the Treasurer of the Royal Society be requested to act as Treasurers, and that Prof. Forsyth and the Junior Secretary of the Royal Society be requested to act as Secretaries.

That the Sub-Committee be empowered, in consultation with the family of the late Sir George Gabriel Stokes and with the Dean and Chapter of Westminster and such other persons as they think fit, to take the initial steps for the production of such a Memorial, and to report to a meeting of the Subscribers to be summoned for a subsequent date.

"We have now to report, on behalf of this Sub-Committee, that in response to the letter above mentioned a reply was received from the Dean of Westminster expressing his general assent to the proposal and his willingness to take detailed plans into consideration. It has therefore been arranged by the Sub-Committee, after consultation with the Dean of Westminster, that a commission be offered to Mr Hamo Thornycroft, R.A. to execute a Medallion, the material to be bronze and the head to be in high relief; and Mr Thornycroft has undertaken to proceed with the work.

"Accordingly, we invite subscriptions towards placing a Memorial of the late Sir George Gabriel Stokes in Westminster Abbey of the character above described. It is estimated that a sum of £400 will be required for this purpose. Subscriptions should be made payable to Messrs Barclay & Co., Limited, Bankers, and should be sent either to them at their Cambridge Branch or to the Treasurer of the Royal Society. A preliminary list of subscribers is annexed.

A. R. FORSYTH } *Secretaries to the*
J. LARMOR   }   *Sub-Committee.*

May, 1903.

All the arrangements were promptly carried out, and a meeting was held on July 7, 1904, in the Jerusalem Chamber, Westminster Abbey, to formally transfer the memorial to the custody of the authorities of the Abbey. The following report is taken from the official "Statement of Proceedings" issued by the University of Cambridge and the Royal Society jointly.

Meeting at Westminster Abbey.

The Memorial of the late Sir George Gabriel Stokes, which has been erected in the North Aisle of the Choir of Westminster Abbey, was unveiled on Thursday, July 7, 1904, by the Duke of Devonshire, Chancellor of the University of Cambridge, and formally transferred to the Authorities of the Abbey. The ceremony was preceded by a meeting in the Jerusalem Chamber, the invitations to which were sent out in the name of the Duke of Devonshire and the President of the Royal Society, and were confined with a few exceptions to the subscribers to the memorial. This restriction was necessitated by the limited accommodation both in the Jerusalem Chamber and in the part of the Abbey where the memorial has been placed. The Dean of Westminster presided; and among those who accepted the invitation were the Duke of Devonshire, Sir William Huggins (President of the Royal Society), the American Ambassador, Lord Kelvin, Lord Salisbury, Sir R. C. Jebb, M.P., Sir John Gorst, M.P., Mr Bryce, M.P., Sir Arthur and Lady Stokes, Dr and Mrs L. Humphry, the Bishop of Bristol, the Vice-Chancellor of Cambridge University, the Master of Trinity, the Dean of Christchurch, the Master of St John's, the Mayor of Westminster, the Archdeacon of Taunton and Mrs Askwith, Canon Duckworth, Bishop Welldon, Canon and Mrs Beeching, Canon and Mrs Hensley Henson, Sir F. Bridge, Mr and Mrs Adam Sedgwick, Mr Hamo Thornycroft, R.A., Mr Aldis Wright, Mr A. P. Humphry, the Rev. W. F. Stokes, the Rev. Gabriel Stokes, the Rev. Dr H. P. Stokes, Mr Marlborough Pryor, the Rev. J. F. Bethune-Baker, Mr A. Hutchinson, Prof. G. Sims Woodhead, Dr D. MacAlister, Mr S. Skinner, Mr J. Hora,

and the following Fellows of the Royal Society: Prof. Clifford Allbutt, Prof. J. Attfield, Dr Shelford Bidwell, the Rev. Dr Bonney, Principal Carey Foster, Sir W. Crookes, Prof. G. H. Darwin, Sir James Dewar, Prof. H. B. Dixon, Dr E. Divers, Dr Dupré, Dr F. Elgar, Prof. Ewing, Sir Joseph Fayrer, General Festing, Prof. Forsyth, Mr F. Galton, Mr Percy Gilchrist, the Rev. R. Harley, Mr J. B. N. Hennessy, Mr C. T. Heycock, Mr W. H. Hudleston, Prof. J. W. Judd, Mr A. B. Kempe, Prof. H. Lamb, Prof. J. Larmor, Prof. G. D. Liveing, Sir Norman Lockyer, Major MacMahon, Mr H. F. Newall, Sir Andrew Noble, Prof. J. Perry, Dr Isaac

Roberts, Dr E. J. Routh, Sir Arther Rücker, Sir B. Samuelson, Dr R. H. Scott, Prof. J. J. Thomson, Dr T. E. Thorpe, Capt. Tizard, R.N., Prof. F. T. Trouton and Mr W. C. D. Whetham.

Dr Larmor, Secretary of the Royal Society, reported the receipt of letters of apology from persons unable to attend. Mr Balfour had accepted the invitation, but expressed a doubt that engagements elsewhere might ultimately prevent his attendance. Lord Goschen regretted that it would not be possible for him to attend. Lord Lister greatly regretted that he was unable to be present. And among others from whom letters of regret had been received were Sir Joseph Hooker, Sir Michael Foster, the Earl of Belmore, Lord Avebury, Mr Chamberlain, the Bishops of London, Rochester, Worcester, Wakefield, and Stepney, Lord Justice Stirling, Sir E. Thornton, and the Master of Pembroke College, Cambridge.

The DEAN OF WESTMINSTER said :

My Lord Duke, My Lord High Steward, My Lords and Gentlemen :

We meet on one of those rare occasions, which must become rarer still, unless further provision is presently made for the reception of memorials of our most distinguished countrymen in this place. When we shall have filled the vacant space next to the medallion which we uncover to-day, it will be difficult to point to other places on our crowded walls where such a tablet, small as it is, may appropriately be set up. This is not the time to dwell on this subject. I do no more than reiterate a warning already given by my predecessor.

We meet to give and receive the memorial of a great and good man. Last year I received a communication from the Duke of Devonshire and Sir William Huggins, on behalf of a joint Committee representing the University of Cambridge and the Royal Society. This was a preliminary enquiry as to the possibility of erecting a memorial in Westminster Abbey to Sir George Gabriel Stokes. The application was rested upon "the high distinction of Sir George Gabriel Stokes, which received remarkable testimony from the entire scientific world at home and abroad, on the occasion of the Jubilee of his professorship at Cambridge in the year 1899, his enduring services to learning and in public affairs, and his lofty personal character." The issue of that application you will see to-day in the excellent portrait-medallion by Mr Hamo Thornycroft. You will find that it is placed next to the memorial of his life-long friend Professor Adams, and next but one to that of Charles Darwin. The Cambridge calendar shows a unique trio of Senior Wranglers in three successive years: 1841, Stokes; 1842, Cayley; 1843, Adams. It was an age of Mathematical giants. We are honoured to-day by the presence of a great survivor of that generation; for Thomson, now Lord Kelvin, took his degree in 1845. I could have wished that this memorial could have borne a few descriptive words, like the epigrammatic sentence on Adams's tablet, "Neptunum calculo monstravit." But our efforts in this direction were baffled. No single achievement could be named, typical of the whole of Stokes's work, which might give the passer-by a keynote of his distinction. To emphasise a single point would have been to depreciate his greatness, which was the greatness of a master who held undisputed sway over some of the most active intellects in the scientific world. He had the generosity of greatness, and its humility; and his work, which would have added to his fame if he had published it himself, often lies beneath the work of other men as the foundation of their securest results. If I might borrow another metaphor from a sacred source, I believe that I am justified in saying that some of the discoverers and of the interpreters of the most profoundly interesting modern discoveries have again and again at a critical moment heard "a voice behind them saying, This is the way." His is the kind of greatness which the world is slow to learn, but not slow to honour when it is declared on unimpeachable testimony. For such testimony I fearlessly appeal in this historic room to-day. I have said that he was good as well as great. The generosity and humility to which I have alluded belonged to both qualities at their highest. His lofty character was the outcome of a Christian life. Like his great *confrères* Cayley and Adams he was not only a firm believer in the Christian revelation but also a godly and devout man. He was bold in speculation on religious topics, and deeply interested in the underlying harmonies of religion and science. Averse from speaking, he gave a silent and impressive support to very many of the good causes of religion and philanthropy. Perhaps no man has ever sat silent on so many platforms. The cause which secured his presence was visibly strengthened by the countenance of this sober, peaceable, and conscientious son of the Church of England.

Sir WILLIAM HUGGINS said:

Mr Dean, My Lord Duke, My Lords and Gentlemen:

It is my duty to state in a few words the part which the Royal Society has taken in the promotion of this memorial, and also the grounds on which the Royal Society considers Sir George Stokes eminently worthy of the high public recognition of a place within the adjoining Abbey. Sir George Stokes, during the fifty-three years of his Fellowship, rendered the Royal Society exceptional services. His labours were incessant on its behalf; and his noble example and the inspiration of his personal influence were strongly felt within the Society during the thirty-one years that he acted as Secretary, and the five succeeding years during which he presided over the Society. But, grateful as we are to Sir George Stokes, it was not on account of the work within the Society that the Royal Society considered it right to take steps to secure a lasting public memorial to his memory, but as a great man who had rendered signal services to his country by fundamental discoveries in physical science. It seemed to us fitting that we should be associated in this undertaking with the University of Cambridge, which some four years before had conferred upon Sir George Stokes the unusual honour of a festal commemoration of his fifty years' Professorship, and struck a medal as a memorial of the occasion. The University returned a favourable answer to our suggestion; a joint Committee was constituted, and, in response to its application, permission was cordially granted by the Dean of Westminster for the placing of this memorial in the adjoining Abbey.

The studies of Sir George Stokes ranged through the whole domain of physical science, and whatever he touched he illumined and enriched with the penetration and power of his mind. But it was in problems relating to optics and to æther that his mathematical work opened out new vistas in exact science, and laid no small part of the foundation of the physical science of the nineteenth century. Rich as were his discoveries in experimental science, perhaps they were more than equalled by his masterly advances in the application of mathematics to physics, which, in the combination of technical skill and physical intuition, remain models of the mathematical analysis of scientific questions. It may be said that his lucidity, and the perfection of form of his treatment of scientific subjects, often masked the difficulties which he had overcome, and so rendered less immediately obvious the greatness of his achievements. As with his work, so with Sir George Stokes himself, the completeness of his mastery perhaps to some extent rendered less obvious and strong the impression which he made upon the men of his time: in the same way as a great work of art through its perfect symmetry and harmony may impress ordinary beholders less strongly than a work that is of smaller merit. It may be said that the distinguishing characteristic of Sir George Stokes was the even balance of his powers, which left no one quality of his mind unduly prominent. His extraordinary penetration was associated with great breadth of view, and was kept in check by remarkable soundness of judgment. If he erred, it was perhaps in the direction of excess of caution, which, it may be, occasionally prevented his pushing home his investigations to their natural conclusion. The influence of Sir George Stokes upon the mind of his time was what it was because he was as great in character as in intellect. He freely gave generous help to all who sought it. Distinguished by simplicity of living, by pure honesty of purpose, by freedom from all the pettiness of personal ambition, he lived nobly the strenuous life, and his name liveth evermore. Mr Dean, it is on these grounds that the Royal Society considered Sir George Stokes worthy of a memorial in our national shrine by the side of those of Newton, Herschel, Darwin, Adams and Joule.

Lord KELVIN said:

In his scientific work and thought Stokes ranged over the whole domain of natural knowledge. Hydrodynamics, elasticity of solids and fluids, wave-motion in elastic solids and fluids, were all exhaustively treated by his powerful and unerring mathematics. Even in pure mathematics, he was recognised as a fruitful worker by the whole scientific world. But with him mathematics was the servant and assistant, not the master. His guiding star in science was Natural Philosophy. Sound, light, radiant heat, chemistry, were fields of labour which he cultivated by studying properties of matter, with the aid of experimental and mathematical investigation. His earliest results, published in papers communicated to the Cambridge Philosophical Society in 1842 and 1843, were mathematical and experimental investigations in fluid motion, and on the viscosity of fluids. From this he went on to the elasticity of glass, iron, and other natural and artificial solids. He thus laid the foundation of the whole modern science of elasticity, involving as it does much thoroughly demonstrated, and brilliantly illuminating, contradiction of views given to the world by great mathematicians and experimenters, who had preceded him. One of his most beautiful results was the theory of the suspension of clouds, and the change through which they become rain. An astonishing and beautiful application of this theory has been made within the last few years in electrical researches carried out at Cambridge under Professor J. J. Thomson and his brilliant research-corps.

While still occupied with his work on the physical properties of fluids and elastic solids, he gave to the world, in 1850, a magnificent paper on water-waves, containing a thoroughly original and masterly investigation of a most difficult problem, the determination of the motion of *steep* deep-sea waves. The greatest mathematicians of Europe had previously been only able to investigate waves of such gentle slope, and such slightly curved surface, that they could scarcely be perceived by an unaided eye. Stokes boldly attacked the problems of real ocean waves, and found that the steepest possible of regular waves has a crest of 120°, with a slope of 30° down from it before and behind; an admirable triumph of mathematics!

Great as was all that work on fluid motion, more than enough for a lifetime of scientific research, Stokes' greatest province was light and optics. Time forbids my speaking of this in detail at present. I can only say in short that he gave overwhelmingly strong reason for believing that the vibrations of æther, constituting light polarized by reflection from water or glass or other transparent substance, are perpendicular to the plane of incidence and reflection. He laid the foundation for Lord Rayleigh's admirable theory of the blue sky. He discovered fluorescence, afterwards found by Edmond Becquerel to be continuous with phosphorescence, on which his father Antoine and he himself had worked so much, and which his son Henri has now found to be fundamentally connected with the marvellous optical properties of the radio-activity discovered by him.

The scientific work and scientific thought of Sir George Stokes is but partially represented by his published writings. He gave generously and freely of his treasures to all who were fortunate enough to have opportunity of receiving from him. His teaching me the principles of solar and stellar chemistry when we were walking about among the colleges some time prior to 1852, when I vacated my Peterhouse fellowship to be no more in Cambridge for many years, is but one example. Many authors of communications to the Royal Society during the thirty years of his secretary-ship remember, I am sure gratefully, the helpful and inspiring influence of his conversations with them.

For sixty years of my own life, from 1843 to 1903, I looked up to Stokes as my teacher, guide, and friend. His death was for me truly a bereavement.

Lord RAYLEIGH said:

That after what had fallen from the paramount authority, Lord Kelvin, it would not be necessary for him to say very much. Lord Kelvin, although he had modestly disclaimed it, could speak as an equal and almost as a contemporary of Stokes. If he followed, it would be as one who was in a very real sense his pupil. It was just 40 years since he, in common with most of the aspirants for mathematical honours of the year, attended Stokes' lectures on physical optics. It was a delight to him, with his youthful scientific enthusiasm, to learn from one who was so completely a master of his subject, and who was able to introduce into his lectures matter fresh from the scientific anvil. Among the experiments he showed them in that course was a fundamental one in physiology, now very well known, on the spectrum of blood. Blood in its natural condition showed, as was known to most of them, a remarkable absorption spectrum, two dark lines in the region of the green. If that blood was subjected to the chemical influence of such a substance as a ferrous salt, which could deprive it of oxygen, the absorption spectrum was transformed, and instead of the two separate bands in the green they found one single and more diffuse and broad one. The experiment was shown to them, as undergraduates, before it was known, he believed, to the scientific world, or to the Royal Society itself; for it was in that year that those results were first published. The appliances used were always of the simplest; and that was a lesson which many modern enquirers might profit by, for elaborate apparatus was not always required, and was even sometimes in the way. It was the same with other things. Lord Kelvin had alluded to Stokes' great discovery of Fluorescence. Well did he remember one of the leading experiments on that subject, in which a solution of the green colouring matter of leaves was made to glow, under the influence of dark radiation, with intense and blood-red colour. With regard to this influence of a beam of solar light the original detection of the mere phenomenon was due to Brewster, but its profound significance, extraordinary and unexpected as it was, was first made out by the searching analysis of Stokes.

Lord Kelvin had alluded to Stokes' early anticipations on the subject of what was now called Spectrum Analysis. Stokes was always very modest upon this subject, and almost repudiated the credit which Lord Kelvin and others wished to give him. All one could say was that the thing lay between Lord Kelvin and Stokes. The letters which passed between them in 1854, and which had recently been published in the Collected Papers, showed plainly enough that the ideas were there. Lord Kelvin told them he got his inspirations from Stokes. It might be, and was likely enough, that he developed them; but at any rate, between the two correspondents they had the whole theory.

Optics, no doubt, was Stokes' greatest subject; but in other departments of physics they learnt much from him. There was one short paper which was constantly in his own mind, when, as scientific adviser to the Trinity House, he had opportunities of observing the extraordinary vagaries of sound-signals as heard from a distance. It was well known that the sound was less easily heard up wind than down wind, but the explanation had been a mystery. Stokes showed that it depended not so much upon the wind in general, but upon the fact that a wind usually blew more powerfully overhead than it did at the surface of the ground. In consequence of this, a sound wave proceeding outward, which would naturally progress in a vertical plane, was diverted out of that plane; in proceeding up wind, the upper parts of the wave were retarded more than the lower part, and thus the direction of the propagation was diverted upwards and so passed overhead. It was only in about a page of print that Stokes developed his view, but there it was for all to understand. There was another problem connected with acoustics to which his own attention had been a good deal given. Solid bodies vibrated without much change of volume, and in consequence of that, while some parts of the surface were advancing others were receding. Considered as sources of sound, the various parts of the surface might be regarded usually as contributing as much positive as negative. It followed at once that there must be great interference. If all these positive and negative sources were situated close to the neighbourhood of the same point they would neutralise one another; it was only because that condition was not fulfilled that we got any sound whatever. It is known that the sound of a bell vibrating in a jar is largely stifled when most of the air is pumped out of the jar: and it was the remarkable observation made by Sir John Leslie that it is still more stifled when the partial vacuum is replaced by hydrogen that directed attention to this subject. Stokes found the key in the more rapid propagation of disturbance in hydrogen, which allowed the positive and negative sources above mentioned more completely to counteract each other; and of this principle he made far-reaching applications.

Stokes' papers dealt mainly with mathematics and physics. Sometimes one read a paper belonging to that department of science in which one failed to find anything either of mathematical interest or of physical interest. In Stokes' papers he need not say one or other interest was ever present, and very often both. If they reviewed his work they must conclude, he thought, that in hydrodynamics and in optics his achievements were fundamental. Instinct amounting to genius, and accuracy of workmanship, were everywhere apparent; in scarcely a single instance, as the Dean had said, had his advances failed to lead in a right direction. It gave him pleasure to express, even though feebly, the almost unbounded admiration which, for more than forty years, he had felt for Sir George Stokes, especially of course as a physicist, but also, perhaps, not much less as a citizen and as a man.

The Vice-Chancellor of the University of Cambridge (Rev. Dr CHASE*) said:

The President of the Royal Society, Lord Kelvin, and Lord Rayleigh have now with singular felicity and with fullness of knowledge and authority, borne their testimony to the great and permanent worth of the contribution which Sir George Stokes made to science. It is, of course, because it is universally acknowledged that he held a unique position in the very front rank of scientific men, that you, Mr Dean, have felt able so readily to accede to the request that a medallion of him should have a place in the Abbey. Speaking on behalf of the University, if his Grace the Chancellor will allow me to do so in his presence, I desire to thank you, Mr Dean, for your kindness and your sympathy in this matter, which we in Cambridge have had so much at heart. But quite apart from his scientific greatness,

* Now Lord Bishop of Ely.

there are reasons why the University of Cambridge welcomes with peculiar satisfaction the rare and signal honour which is to-day accorded to the name of Sir George Stokes. Continually resident as he was in the University since the time when he came up as a freshman, I think in 1837, his life in Cambridge spanned that notable epoch of change and of development which has transformed the University. As you, sir, have reminded us, he was the senior and last survivor of that great triumvirate of professors, who so long, and with such rare distinction, represented the Mathematical School at Cambridge. For five years he sat in Parliament as the Member for the University; and consistently all through the many years of his Cambridge life, all alike, whether they could or could not duly appreciate his scientific and intellectual greatness, recognised the simplicity and the truthfulness of his character, the dignity and the impressiveness of his presence—the outward symbol of the calm and generous spirit, his patience and his lofty conscientiousness in all details of daily business, his courtesy and his unfailing kindness to all with whom he had to do—not least to the young and the undistinguished, his unswerving loyalty and devotedness to the University, and to his own college, of which for five short months he was Master, his deep reverence for all that is most worthy of reverence in human life. On two occasions, which no one who was present at them can ever forget, the deep respect and regard which men felt for him found definite and emphatic expression. The University rejoiced at the time of his Jubilee as Lucasian Professor that the most modest of her sons should have decisive proof in his lifetime in how high honour he was held by scientific men the whole world over. The scene in the crowded University Church at his funeral, was, I believe, unique. The example of Sir George Stokes, both in what he did and more especially in what he himself was, is the abiding inheritance of the University which he loved, and which he served so long and so faithfully; and that University rejoices with no common joy to-day that his name will now have a permanent memorial which all men can read on the historic walls of this great national Church.

The DEAN:

It is my privilege, my high honour, to lead you into the Church, and ask the Duke of Devonshire to unveil the memorial that is there placed.

The company then proceeded to the Abbey, where, after a brief devotional service, the medallion was unveiled by the DUKE OF DEVONSHIRE, who said:

Speaking on behalf of the subscribers I offer this medallion to be added to the memorials and to be preserved in the Abbey Church.

The DEAN in reply said :

Speaking in the name of the Dean and Chapter of Westminster I accept this medallion to keep and preserve amongst the memorials of the good and great men in this place.

The Medallion is of bronze in high relief and is a striking portrait of the late Sir George Stokes.   It is affixed in the north aisle of the Abbey, adjacent to similar memorials of Darwin and Adams, and not far from the graves of Newton, Herschel, and Darwin.   It bears the inscription

GEORGE GABRIEL STOKES, 1819—1903.

Another memorial, even more durable, has now been completed. The University of Cambridge had arranged that the reprint of his Scientific Memoirs, which had been proceeding with valuable additions for many years during his lifetime, should be carried on with all the care that was possible.   This task has now been finished, by the addition of two more volumes, including the obituary notice written by Lord Rayleigh for the Royal Society, and embellished by selected portraits; the collection consists of five volumes in all.

It can hardly be said that Stokes was the founder of a school, in the ordinary sense of the term, as illustrated by the manner in which Lord Kelvin in early days pushed on the dynamical side of the discoveries of Faraday, or the way in which Maxwell infected his contemporaries with aethereal and molecular science. His interests were too wide, his attitude of mind too cautious, and his restraint on what he called speculations far too severe. In his writings we rarely get a glimpse of anything but the finished product; as in the case of Gauss and other masters of the same calm temperament, the human environment has to be reconstructed as best we can.

The urgency of his contemporaries too, following upon the unique scientific position which he held in Great Britain, compelled him to take all physical science to be his province, not by the fruitful method of passing from time to time from one department to another as they interested him, but by a simultaneous grasp of the whole.   The stretch of mind which this involves is hard to combine with the reflecting repose that eats its way into the heart of a single problem.

Not the least of the interests of the letters, from which a selection is here published, is that they reveal the fascination of human endeavour alongside the formal record of finished achievement. They press home, too, the justice of the claim, long insisted on by Lord Kelvin and others who were on terms of continued intimate association with him, but perhaps not so clearly understood by the younger generation to whom he was largely a historical figure, that he was a main guiding and controlling power in the physical science of his time. If a school has not survived him, it is because his work has been absorbed into the spirit of the age. We can now see more clearly how much fruitful effort in others he started and stimulated, to how many investigators he supplied the confidence that comes from an initial success. But it is also evident how much of his time was wasted, even in his prime, by his scrupulous deference to the demands of correspondents, conscientious and estimable people in the main, from whose importunities there should have been some sort of protection.

In University affairs he was implicitly trusted by all parties. A story was current in the eighties of a party manager who was caustically forecasting in semi-serious vein the probable actions of the persons concerned in a certain election, as dictated by party or prejudice: he had got almost through the list, but still one name remained which seemed to baffle his skill; after a pause he continued "and Stokes—will give an honest vote."

SECTION III_A.

SPECIAL SCIENTIFIC CORRESPONDENCE.

LETTERS TO DR ROMNEY ROBINSON, 1875—1879 AND 1880—1881.

The first group of the following letters have been selected from a series belonging to the period 1875-9, carefully arranged in years evidently by Dr Robinson himself. The remaining part is printed from duplicate typewritten copies preserved among Prof. Stokes' manuscripts.

A long correspondence on anemometers of the year 1877 is omitted: reference may be made to Dr Robinson's memoir, *Phil. Trans.* 1878, also to *Math. and Phys. Papers,* Vol. v. pp. 73—99.

CAMBRIDGE, *4th Jan.* 1875.

...I heard to-day from Mrs Harcourt. The instrument she referred to is a finely graduated circle, with telescopes for measuring refractive indices, which she gave to me for my life time, leaving the ultimate destination yet to be settled. I was not aware this had been sent to you, and I am not sure that it was. When I was at Nuneham shortly after Mr Harcourt's death, she expressed a wish to give us each some instrument of his. She *may* have mentioned the refractometer as for you (I cannot now recollect), and I *may* have suggested that you had an equivalent instrument already, and that it might be very valuable to me as I did not possess one. If she had thought of giving it to you she may have recollected the intention and supposed it had been actually sent. I could not suppose that she meant to offer to me an instrument she had already given to you.

I am not sure whether I mentioned that the Chances have made an experiment on a silico-titanic glass. Hopkinson, Senior Wrangler in 1871, who is their scientific adviser, had the

superintendence of it. He sent me a specimen of the glass. I got
it polished to look into, but it proved so bad I did not go further.
However he got a prism made, and measured as well as he could
the refractive indices. Though the diffused image of a bright
line almost filled the field of view, his results with the two better
of the three angles, taken for a stripe of extra brightness in the
compound image, were unexpectedly concordant. It seemed as
if the glass lay (as to irrationality) between crown and light
flint, but much nearer to the latter than the former. This is
what I hoped and expected. I borrowed the prism, and tried
my method of compensating prisms. I made it much nearer to
crown, $\theta = 10°$, θ being 0° for crown and 90° for flint*, and 60°
for a prism I have of light flint. Bad as the glass was, I hardly
think I can be much wrong, for my results agreed well together.
However Hopkinson is going to remelt the glass.

The glass was of a green colour, just such a tint as could be
produced by iron. He thought it must be due to titanium, as the
chemist who prepared the titanic acid thought he had at last,
after great difficulty, got it quite free from iron. (It was prepared
from titaniferous oxide of iron.) I said I had never seen that
colour from titanium, whereas it was just the colour that iron
would give. I asked if an iron ladle had been used. He told me
not, but a copper one. He got the titanic acid tested at their
works, and it was pronounced free from iron. Meanwhile Liveing
had tested for me the glass itself, and found a *small* quantity of
iron in it. I was still sceptical about the purity of the titanic
acid, and begged for a sample. This H. sent, and I found it did
contain iron. I showed this to H., who happened to come to
Cambridge shortly after, and asked him as to the subsequent
process—whether the TiO_2 had been fused with carbonate of
potash in excess. He said it had, with the addition of nitre to
keep up the oxidation, and that a black mass was obtained.
Whence the blackness? I asked for a sample, which he sent me.
I found it contained alkaline sulphides (as sulphurets are now-
a-days called) and sulphates, or rather *a* sulphide and sulphate.
On boiling with water I got a green solution (while the greater
part remained undissolved) with a peculiar spectrum of absorption.
I supposed this must be due to an inferior sulphide of titanium,

* In Dr Hovestadt's treatise on Jena glass (§ 6) the disappointment of Prof. Stokes'
prevision is ascribed to the countervailing effect of phosphoric acid. See *infra*, p. 333.

first casually mentioned in Gmelin as a black sulphide (protosulphide?) in the course of the description of the known sulphide TiS_2. For according to Gmelin protosulphide of iron is not soluble in excess of solution of a sulphide of a fixed alkali. But I found afterwards that it *was* iron, and I obtained an INTENSELY deep green solution, shewing the peculiar spectrum by dropping very dilute protosulphate of iron into a hot solution of sulphide of potassium. The black colour was therefore clearly due to protosulphide of iron. But whence came the sulphur? I suggested from sulphate of potash as an impurity of the carbonate, which possibly might have got released by substances obtained from the fuel. But H. said the carbonate was specially pure. Sulphur however got in in some way, for the skimmings of the glass proved to be almost entirely sulphate of potash. H. wrote suggesting the titanic acid. In the meantime I had dipped a hot borax bead into the acid, and lo it took fire where it was touched, and on putting a portion into a closed tube I distilled off a great lot of sulphur. Evidently the chemist had employed the HS process, and neglected to burn off the sulphur mixed with the TiO_2. I dare say it did not make much difference in the end result.

H. told me the process for separating iron was so troublesome, that the TiO_2 would cost 15/- a pound. I have I believe effected a material simplification. The HS process is all well enough for the laboratory, but is ill adapted for working with stones weight of matter. I did not like the idea of the oxidation by zinc or iron, as such a lot of hydrogen is given off that the reduction effected can be only a very small fraction of the equivalents of the zinc or iron dissolved. So I tried copper, through the mediation of the sub-chloride, and this promises to succeed perfectly. The mixture of TiO_2 and Fe_2O_3 is digested with hydrochloric acid and metallic copper, with a little oxide or carbonate of copper thrown in to start the reaction. The chloride of copper is reduced by the Cu to Cu_2Cl, and the acid solution of this deoxidizes and dissolves the Fe_2O_3, forming protochloride of iron and chloride of copper, and the chloride in turn is reduced to sub-chloride by the Cu, and so the thing goes on. I have tried it on *rouge*, which dissolved perfectly while I was eating my dinner, even without heat, and heat much quickens the process. Some I tried before with simple HCl remained for days or weeks mostly undissolved, while oil of vitriol did not touch it at all.

5th *January*, 1875.

You have I presume heard from H. Grubb of the proposal to erect a giant refractor of 39 in. aperture in California. He has written to me respecting the consequence of a possible defect of annealing, despairing apparently of getting such large disks perfectly annealed, even if in other respects they could be prepared satisfactorily.

You told me of an object-glass that showed stars with tails (it was not I think focal lines), and on examining the object-glass there were found great defects of annealing, or else strain from the glass being pinched in its cell, I am not certain which.

Would you kindly tell me which it was, and if it were want of annealing, then (if you remember) what was the number of tails, and was the depolarization enough not merely to restore the light but to give one or two orders of colours ?

If tails were produced by want of annealing, that shows I think that the disk cooled differently in different directions round the axis. Probably it may be difficult to prevent the cooling being more rapid on the side next the door of the oven than in other directions; and if so it would seem to be advisable, in order to render the defects of annealing that we must put up with as harmless as may be, to have the bed of the disk capable of being turned round its axis, by means of a rod passing to the outside of the oven, so that the cooling might be symmetrical about the axis of the disk.

The mere addition of iron to HCl and rouge makes the rouge dissolve much better than I should have expected, and yet the Fe_2O_3 does not seem to be reduced as it is by the Cu_2Cl. There is something curious about this action of nascent hydrogen. It seems to be on the borders between ordinary chemical action and electrolysis.

27th *January* 1875.

I will look at Savart's paper if I think of it and if I have time.

A fluorescent eye-piece would not do because the rays are rendered visible indirectly only, by their rendering the fluorescent body self-luminous in their path. Whatever rays get through the fluorescent body without being so spent have at emergence the same properties they had at entrance. Hence it is essential that the fluorescent body should be seen in focus, in order to be used as a screen for exhibiting the ultra-violet spectrum.

Uran glass is not a good thing for the purpose, for the reason that it is very tolerably transparent for a portion of the ultra-violet rays, and as it must be thin, to be nearly enough all in focus for the eye-piece, it would be capable of showing only a small fraction of the effect. It begins to be very opaque a little beyond the end of the invisible point of the solar spectrum ; and for the part of the spectrum beyond that, which we have in electric light, it would answer well.

I shall bear in mind the explosives report, though I don't at present know where I could borrow it.

14th Feb. 1876.

I have thought over the matter of reflection from metals, and I am disposed to use polarized light, and fraction the light by using Malus's law.

First, to compare two metals at any angle of incidence not too small.

Place the metal plates side by side, and adjust their planes to parallelism, say they are vertical. Let the incident light, coming horizontally, be polarized by a Nicol in a graduated circle. Examine the reflected light by an analyzer composed

1. Of a double-image prism of small separation, or better a thick block of Iceland Spar, without any diaphragm, with its principal plane vertical.

2. Of a Savart plate, or something equivalent, with its principal planes inclined at $\pm 45°$ to the vertical. Exactness as to this angle is of no moment.

3. Of a Nicol placed so as to extinguish one of the images of the Spar when the Savart plate is removed.

2 and 3 together constitute a Savart's polariscope.

The Spar will give two images of each plate, and near the junction these will overlap. The overlap is the part of the field to be used.

Let A, B be the coefficients of reflection for the two metals when the light is polarized in the plane of incidence, a, b ... perpendicularly to ..., θ the azimuth of the plane of polarization measured from the plane of incidence.

Let 1 be the intensity of the primitively polarized light, M, N

the two metals, and suppose the O of M overlaps the E of N. The intensity of the first will be $a\sin^2\theta$, and that of the second will be $B\cos^2\theta$. Turn the polarizer till the Savart bands disappear. Then these intensities will be equal, and we shall have

$$a\sin^2\theta = B\cos^2\theta,$$

or $$B/a = \tan^2\theta.$$

Turn the polarizer till the bands disappear again on the other side of the plane of incidence. Then 2θ will be the angle turned through, which will therefore be known.

Now let the metals exchange places, or else rotate the whole analyzer through 180°, as may be more convenient. Then the O of N will overlap the E of M and we shall similarly get A/b.

Lastly, remove one metal and move the other into the middle and observe the overlap of the metal on itself. This will give just as before A/a; so that we know the 3 ratios between the 4 coefficients A, B, a, b.

To get the absolute values I thought of this method.

Mount the two metals M, N (or two plates of the same metal) with their planes parallel but *opposite*, so that after two reflections the light is parallel to what it was. Then the further edge of the nearer plate is the line of junction between the part of the field where the light is direct and that where it is twice reflected. These can be compared just as before, and supposing the E of the twice-reflected to overlap the O of the direct, we shall have $AB = \tan^2\theta$.

Knowing $A:B$ and AB we have A and B, and thence a and b. For telescope purposes however we only want to know the ratios.

If the plates are vertical when opposite, the principal plane of the rhomb in the analyzer must of course be horizontal instead of vertical.

The Grubb instrument you gave me may be used for measuring the angle of incidence. I have a small circle for angles of position.

P.S. If it be preferred to see the lights to be compared side by side, instead of superposing them and trusting to the evanescence of Savart's bands or whatever else may be used, we have only to hide the junction by a black slip of such breadth that the two images of it just touch, or to use an equivalent diaphragm, and proceed as before.

By comparing in either of these ways two portions of the same

light we are rid of the effect of possible fluctuations in the source
of light; and as we polarize the light in the first instance it is
immaterial whether the original source be partially polarized or
wholly unpolarized, and we may use lamplight, or the light of the
sky reflected by a looking glass, or the light of white paper in
sunshine, at pleasure.

I see that the ratios A/a, B/b for the *same* metal may be much
more accurately determined in other ways, as for example by my
elliptic analyzer.   I should polarize the incident light at an
arbitrarily chosen, measured azimuth, and determine the elements
of the elliptically polarized light reflected from the metal.   This
would give A/a and the difference of phase.

30th October 1875.

I had intended to order my curved-faced prism with radii of
8 feet and faces about $1\frac{1}{4}$ in. square.   I wrote to the Darkers, but
their long-radius pans had not been in use for a long time and
were out of order.   I then wrote to Ahrens a prism maker.
He had no pans for a radius larger than 4 feet.   So I thought
I would reduce my scale.   The 4 foot scale is smaller than I
should contemplate for actual use, but on the other hand it would
bring out in stronger relief the difference between a curved-face
and plane-face prism ; and if there should be any hitch in the
method less money would be thrown away.   So I ordered a prism
with radii $+ 4$, $- 4$, and $+ 3$ feet, equilateral or nearly so. $+$ refers
to convex, $-$ to concave. Supposing the prism in its position of
minimum deviation, the proper distance of slit from prism for the
$- 4$ ft., $+ 4$ ft. angle would be either about 8 or about 2 feet.
The larger would naturally be chosen; I had contemplated 16 ft.
with 8 ft. radius.   The faces to be an inch or $\frac{3}{4}$ inch square.
The prism arrived this morning.   I don't know when I may get
sunlight, but I tried it on a soda-flame, comparing it with each
of two flat-faced prisms of flint glass of 60°, and without any
collimating lens, at the same distance of 8 feet from the slit.

The double D was shown admirably, the edges of the components so sharp and free from false light.   It very decidedly
beat the flat prisms thus used (of course they are not meant to
be *thus* used); and with only a single prism of 60° I never saw
the double D better, if so well, even with a collimator or else a
distant slit.

A very small change in azimuth of the prism rendered the image less distinct, the proper azimuth altering with the focusing of the telescope. In this respect .it resembles a very sharp telescope, the focusing of which must be exact if the sharpness is to be fully brought out.

I feel little doubt that a spectroscope furnished with 3 or 4 of such prisms (only larger and with a longer radius) would perform excellently. You know it is much easier to mark a spherical surface true than a plane surface. Each prism should be ground concave and convex, to a radius as large as can conveniently be worked with precision. The slit, if used without collimator, should be placed at the calculated distance (about double the radius). If a collimator be used, a telescope should be focused on an object at the calculated distance, and then be made to look into the collimator, and the slit of the collimator shifted till it was seen in focus. The concave faces are all to be turned towards the incident light.

I shall be curious to examine the solar spectrum with the new prism, but at this time of year I may have to wait some time for it.

1st Nov. 1875.

I thought I would try again the prism, using a point of homogeneous light instead of a slit as that would show better the nature of the image. I found that the fault of the plane prisms, as I had tried them before, lay in want of adjustment to make the refraction in a plane perpendicular to the edge. When that was attended to the plane-faced prisms answered quite as well as the other, perhaps better, at any rate *as* well. The performance of the two kinds is so nearly equal that I cannot say at present that either is better than the other. It seems as if the spherical was less affected by a want of perpendicularity to the edge in the direction of incidence.

I thought I had best inform you at once that I was mistaken in supposing the spherical so superior.

I should say that after I had tried to-day with a point (a fine needle-hole) and so made out the ways of the image, I tried the slit for a comparison of the two kinds of prism.

2nd Nov. 1875.

I had some sun this morning, and I tried the curved prism against a plane one, examining more particularly the components of the double *D*. I cannot say which did best : once or twice there seemed a *shadow* of a suspicion that the plane was best, but really I cannot say there was any difference. I tried yesterday the curved on bright *D*, putting the *convex* side towards the light, which is the wrong way, and *then* I got nothing but a broad confused yellow band.

I noticed both yesterday and to-day both with bright and dark *D* that the two components are not quite alike, the more refrangible being rather sharper and brighter, or blacker, as the case may be ; indicating that the absorption is rather more determinate in the case of the more refrangible. I had noticed, as is I believe well known, that one is a good deal more easily reversed than the other. Having forgotten which, I tried again the reversion by the outer parts of the plane itself as well as by an accessory soda flame placed *in front of* the slit, *i.e.* between it and the eye. In the former method, with a strong soda flame and 3 or 4 prisms of different angles instead of one only, I got a fine hair-like dark line down the middle of the more refrangible components, and none perceptibly in the other. I have seen it down both, but my object to-day was to recall to my recollection which came the more readily.

Both prisms (the curved and the single flat of 60°) showed well the triple *b*, the components being sharp and clear as in Fraunhofer's work, but I could not split the 3rd component (as in Kirchhoff's map, drawn from 3 prisms) with either. I must try for a slit a couple of knitting needles tied together at the two ends, and then separated in the middle by a bit of thin mica or a fly's wing or something of the kind.

The behaviour of the flat and curved prisms when turned on a *point* of homogeneous light (I used a soda flame and needle-*point* prick) when put out of minimum deviation, out of focus &c. was naturally very different; but when out of minimum deviation and focused for the primary focal line, both showed the components of *D* well, and about equally well, separated. The distribution of illumination up and down the line was however different in the curved and flat prism : in the latter it was nearly uniform, in the former it showed like ‖ or ⫼ according to circumstances. The two lines belong each to one component of the D.

The angular distance of the components of b_3 with a single flint glass prism of 60° is only about 7″.

4th Nov. 1875.

I got your letter before I left Cambridge.

As to a curved-faced prism, I did not contemplate using it for absolute measures, but only for viewing, or for the determination of places with reference to known lines, which comes in fact to interpolation. However it seems the defect which the curvature was designed to correct is practically invisible even at so short a distance of the slit as 8 feet, and the plane prism is at least as good as the curved as to performance, though perhaps the curved might be rather more easily made.

I went to-day by invitation of Sir George Airy to the Guildhall to see him admitted as freeman of the City of London.

The award of the Medals is Copley—Hofmann; Royal—Crookes and Oldham.

P.S. I meant to imply, though I see that I did not express, that in the use I contemplated a knowledge of the angle of the curved prism was not required.

8th May 1875.

On returning from my lecture yesterday afternoon I found a prism which had come by post from Hopkinson ready cut and polished, and a letter in which he told me that the glass contained about 7 per cent. of rutile. I immediately went back to the Museum, where my compensating prisms were, and determined its character as to irrationality, and after dinner I measured the refractive indices for A, D, G.

The result appears to be conclusive that the introduction of titanic acid into a silicic glass does *not* confer upon it the purple-refracting quality which it confers upon a phosphatic glass. The dispersive power is raised as it is in a phosphatic glass, but so far from the purple-refracting character being elevated far more than corresponds to the elevation of dispersion, as was the case with the phosphatic glasses of Mr Harcourt, the change lies if anything the other way, so that on every account the introduction of titanium seems to be a positive disadvantage.

The experiment was I think well worth trying, though nature's answer seems now to be decisive against the hopes which were entertained*.

6th Nov. 1875.

There is no difficulty either with a plane or curved prism in making the primary focal line long enough to show the spectrum of a star, by turning the prism a little from the position of minimum deviation. It must not be turned far or the line will be too long, and the light consequently too weak. Sir G. Airy preferred this method for the star spectroscope at Greenwich, instead of using as Huggins did a cylindrical lens. The curved prism does not however I find do well for this, for when the line is short enough the illumination is far from uniform, being for homogeneous light concentrated at the ends or middle according to circumstances, whereas with a plane prism the line even when short is pretty uniform in illumination.

I have not yet tried with the two kinds of prism an *intensely* narrow slit. I see that with a VERY narrow slit a condensing lens should be used to get the greatest illumination. With a moderately narrow slit, for which you may regard light as consisting of rays following the course assigned by geometrical optics, if the slit be far enough off from the prism for the prism to be well filled with the sunlight, you gain nothing by a condensing lens: it merely widens the beam coming through the slit, which is already wide enough and to spare by the time it gets to the prism. But if the slit be so narrow that there is a large divergence due to diffraction, the case is different....

When I think of the primary focal line as formed by the light of a star, I of course suppose the light collected by an objective into a focus tolerably near. The rays from an infinitely distant point would of course form a parallel pencil after refraction at a plane prism however placed, and after collection by a lens would proceed towards a focal point, not focal lines.

8th Nov. 1875.

I tried a very narrow slit to-day—a pair of knitting-needles separated by the wing of a house-fly, of which there were plenty lying about dead. With both prisms at the same distance of 8 feet from the slit (i.e. first one and then the other, not both combined) I saw a distinct difference between b_2 and b_3, the latter

* But see footnote supra, p. 325.

having a resolvable look, and I fancied I occasionally saw it double. As the components subtend an angle of only 7″, and even *when magnified* little more than 2′, the separation approaches nearly to the limit of what the human eye can do, however perfect the image.

If there was any difference between the flat and curved prisms it seemed to me to-day that the curved was if anything the better. The performance of both, as seen with my highest available magnifying power (about 20) is so nearly perfect that I cannot say which is best.

12th Nov. 1875.

Thank you for the suggestions. As to the lenses, by applying direct to a prism-maker whose address I happened to get, I got my prism, postage included, for 10/-; and as the lenses would have to be of flint glass and therefore would have to be made express they would hardly have cost less, and in what proves to be so delicate a matter it is more satisfactory to dispense with the additional surfaces. Your suggestion of the deposited gold or silver is very good, and I have no doubt very nice slits as fine as I should want and finer could be thus made. I tried the smoked glass to-day, ruling with a needle. I got good slits, but might possibly want finer. I suspect I should have added a dash or two of turpentine which should have been left to evaporate, or else a little spirit with a very little shell lac. However I can get slits fine enough; for I can start with a very fine one and make an image of this with a cylindrical lens which will do for anything. However the direction in which I am stopped at present is magnifying power. With my highest power (with my small telescope, for which I have got a lengthening piece for focusing on a near object) which is about 20, the performance of both prisms is practically perfect. I had a good trial to-day, which was one of uninterrupted sunshine. I was just able to split b_3 (the third component of triple b) and could see that the more refrangible component of b_3 was the darker. I have got to go to London to-morrow for a meeting of the Eton Governing Body, and I mean to get an adapting lens made to connect Mary's small microscope with the small telescope so as to use the microscope as an eye-piece. This will give a magnifying power of 60 or so without having recourse to the high-power objectives, and then perhaps a difference may be revealed in the performance of the two prisms.　My impression to-day was rather in favour of the curved one.

I have been twice already to London this week, once for E.G.B., and once for R.S.　The first *evening* meeting of R.S. is next week.

17th Nov. 1875.

This morning's post brought me a letter from Dr Stenhouse enclosing a small quantity of " fluoresceïn," a substance discovered by a chemist of the name of Baeyer, and formed by heating together phthalic anhydride and resorcin.　Its slightly alkaline solution is about the most powerfully fluorescent solution I know. It shows an intense absorption-band near F, and the blue is doubtless the most efficient part of the spectrum in producing the fluorescence.　Accordingly the fluorescence is *very* lively even by candle light, by which quinine and æsculin solutions show little.　The fluorescent light is green.　I have not yet had an opportunity of examining it by the solar spectrum.

20th Nov. 1875.

I think I should like to have the wooden stand.　Supposing even I ultimately got a brick support I might arrange it better from having had some previous experience of the working.

I will copy into a memorandum book the curvatures of the objective, interchanging the first two as the flatter surface of the curve should be towards the light.　I don't know whether it was a mere error of copying, or whether in the process of measurement you reversed the glass and forgot you had done so.　When Faris showed it to me 3 or 4 years ago I found it had an enormous *negative* spherical aberration (i.e. the *opposite* to a simple lens) and wondered that Cauchaix would turn out a glass like that. On thinking it over it occurred to me that the Crown might have been reversed.　If the more curved face were foremost, reversing it would increase *its* aberration which is positive, which was the direction of the required correction.　On mentioning to you at breakfast the large aberration, you at once suggested that one of the lenses had got reversed.　On trying it the spherometer showed that the front face was the more curved; and when we reversed the Crown, putting the flatter face in front, the telescope was all right.

Thomas Romney Robinson.

In the Fraunhofer form and in the Herschel form the 4th surface is convex, and Ross I think made it convex too. Cooke's I believe are concave behind, Grubb's nearly flat.

P.S. I had a good try to-day with the prisms, but was terribly dodged by clouds. I have got my adapting-piece, and put on the small microscope as an eye-piece. Under this power the sharpness of the telescope begins to break down, so that I could discriminate between the two prisms only as well as with the simple eye-piece of highest power. With the curved prism in its best adjustment I got an exquisite sharpness (as seen with the eye-piece) surpassing, though not by much, what the flat prism gave.

I am familiar with *b* triple like ┃┃┃ with single prisms, but

to get the third double like ┃ ┃┃ with a single prism of flint glass

requires great sharpness of definition and a very narrow slit.

THE ROYAL SOCIETY, *6th Jan.* 1876.

I lunched with the Sabines to-day. They are quite as well as usual. He has a niece as you know staying with him now, which relieves Lady S. of a good deal. He asked after you as he always does.

I hope your paper (R.I.A.) will be in print in time to be in the hands of those who conduct the experiments.

I have been in communication with Captain Toynbee of the Meteorological Office, who is going to issue instructions for captains as to the observations of waves, especially long swells and rollers. I should be very glad to know the periodic time of the heavy rollers observed at Ascension and St Helena*.

I have lately perceived a consequence of the theory of waves which I have not seen noticed—that the velocity of propagation of *roughness* on deep water is only half the velocity of propagation of the waves of which the roughness consists, and which I suppose to be rought, regular as to wave-length, like the waves of the sea. The waves of a series gradually die away in front and get larger in the rear, so that we must distinguish between the velocity of propagation of a series of waves as a whole and that of the

---

* See the correspondence in Vol. II. pp. 133–158.

individual waves. Thus a series like

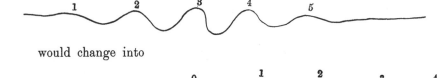

would change into

the same figures marking identical crests[*].

When the depth is not great compared with the length of the waves, the two velocities—that of a series and that of the individual waves—become more nearly equal.

The known expression for the velocity of propagation of the crests, I may as well mention, is

$$\sqrt{\frac{g\lambda}{2\pi}}\sqrt{\frac{e^{\frac{2\pi h}{\lambda}}=e^{-\frac{2\pi h}{\lambda}}}{e^{\frac{2\pi h}{\lambda}}+e^{-\frac{2\pi h}{\lambda}}}}$$

where λ is the wave-length and h the depth.

In the extreme cases (1) of deep water (h/λ large) (2)...shallow ...(λ/h large) the velocity of propagation becomes (1) $\sqrt{\frac{g\lambda}{2\pi}}$ and (2) $\sqrt{gh}$.

ATHENÆUM CLUB, PALL MALL, S.W.
13*th Jan.* 1876.

Tyndall is to read a paper to-night at the R.S. in which I hear he is to put a settler on biogenesis, by showing that when air is simply allowed to stand still in an enclosure for 2 or 3 days till it becomes optically pure by subsidence of the dust, there is no generation of organisms in putrescible mixtures placed previously within it and then opened.

CAMBRIDGE, 19*th Feb.* 1876.

The same post which brought me your letter brought one from Douglas Galton sounding me as to whether it would be prudent to ask Dr Andrews to take the presidency of the British Association at Glasgow this year.

[*] Cf. letter from W. Froude of date Jan. 1873, printed in Vol. II. p. 157: also note by Lord Rayleigh, *Proc. Math. Soc.* 1877, reprinted, *Theory of Sound,* Vol. I. Appendix, where he mentions that the same explanation had occurred to him two years previous.

I did not see him last summer, but I believe he is very well; but I don't know how far the anxiety and work of preparation might be good for him. He is hard at work reducing his last 4 years' work, on which he is to deliver the Bakerian Lecture on April 27.

We should both much like to see Dr Andrews President if it were not for his health, but query would it be prudent ?

You need not return my letter; for though I did not keep notes I do not need them, as I am familiar with the ideas. I don't see my way to getting up to a normal incidence in any case otherwise than by interpolation, but I think I could get nearer to it with two reflections than with one, as the observer's head would be out of the way.

My little elliptic analyzer reads to 6′. I have not yet tried it in the actual experiment. I should think this is as close as I shall be able to observe to.

1st March, 1876.

There is one other point which I have marked in your paper, and which will require on my part somewhat longer explanation, namely the question of aerial tides (p. 430, old 28). But first I may as well explain that Newton's slip in the vortex problem was ...[as in *Math. and Phys. Papers*, vol. i. p. 103].

But the hypothesis from which the result is deduced is not very applicable to the considerable and irregular movements with which we are concerned in eddies of wind.

I come now to the more important matter of aerial tides. As with water, so with air, the disturbing force of the moon would tend to throw the air into the form of a spheroid of revolution with its pole under the moon. The very imperfect equilibrium theory would make the spheroid always prolate. But the dynamical theory shows that it is prolate or oblate according to circumstances. Laplace first showed that supposing the earth covered with a uniformly deep ocean, there would be *either* high water *or* low water under the moon, the which depending on whether the depth lay within such or such limits. If the depth exceed a certain critical value (H say) the tide is sure to be *direct*, i.e. there will be *high* and not low water under the moon; but if the depth be less than H, and not too much less (for then additional waves would come in), the tide at the equatorial regions at least will be inverse. The critical depth H is such that if the water were brought, by a disturbing force, into the form of a spheroid of revolution, and then left to itself by the cessation of the disturbing force, the periodic time of the natural swing (in passing from prolate to oblate and back to prolate) would equal the periodic time of the disturbing force.

Similarly there is a critical height of the homogeneous atmosphere such that if the actual height be greater than this the atmospheric tide is sure to be direct.

The nature of direct and inverse tides may be prettily illustrated by considering the motion of a simple pendulum, acted on by a small disturbing force. If $2\pi/n$ be the period of a natural swing, and $2\pi/m$ the period of the disturbing force, the equation of motion will be

$$\frac{d^2\theta}{dt^2} + n^2\theta = c \sin mt \tag{1}$$

(I suppose the disturbing force expressed by $c' \sin mt + c'' \cos mt$, and the origin of t changed so as to get rid of the cosine). The integral of this equation is

$$\theta = \frac{c \sin mt}{n^2 - m^2} \tag{2}$$

omitting the part $A \sin nt + B \cos nt$ which depends on the initial circumstances, and would disappear by ever so small a friction in the case of a motion going on indefinitely.

We see from (2) that when $n > m$, i.e. when the period of natural swing is greater than the period of the force, the excursion and the force are positive and negative together, and reach a positive maximum together. When the pendulum is to the right the force acts to the right, and when to the left the force acts to the left; and so the force prolongs the swing-time, and compels it to be as great as its own period. But when $n < m$, or the pendulum tends to change more slowly than the force, the pendulum's excursion is a + maximum when the force is a − maximum and *vice versâ*, so that the force acts always *towards* the vertical instead of from it, and quickens the changes that the pendulum goes through.

The first of these is analogous to a direct tide, the second to an inverse. If the height of the homogeneous atmosphere be sufficient to make the natural swing-time of an aerial spheroid, becoming alternately prolate and oblate, less than the period of the disturbing force, the tide will be direct, and there will be high air under the moon.

But whether the tide be direct or inverse—and this is the important point—the westward velocity will be a maximum at high air and a minimum (or the eastward velocity will be a maximum) at low air, and will be null at 3^h, 9^h, 15^h, 21^h, instead of having its maximum (positive or negative) at 3^h, 9^h, 15^h, 21^h and being null at 0^h, 6^h, 12^h, 18^h, that is of course supposing the retardation insensible. It is a general character of wave motion (I refer to a progressive wave motion, not a standing oscillation) that the motion is forwards, or in the direction of propagation, when the fluid is elevated, and backwards when it is depressed. The phase in fact of *horizontal velocity* synchronizes with that of *vertical displacement*.

In deep water for instance, in the case of oscillatory progressive waves like those of the sea, the particles move round and round in circles, and a particle at the surface is moving horizontally forwards (or in the direction of propagation) when it belongs to the top of a crest, and horizontally backwards when it belongs to the middle of a trough, and has no horizontal motion but is moving up or down, when it is at its mean height.

For long waves propagated in shallow water, i.e. water which is shallow compared with the length of the wave, the motion, instead of being confined to the neighbourhood of the surface is sensibly the same from top to bottom, and the vertical velocity is insignificant compared with the horizontal.

Thus in the figure [not reproduced] if the black curve denote the wave at one moment, propagated from left to right, and the red curve at a consecutive moment, volume has passed from left to right of the section AB through the crest, and as the fluid is sensibly at rest beyond the limits of the wave right and left, this can only be by a horizontal transfer of the plane of particles AB.

Just so when the watery or aerial spheroid follows the moon from east to west (the *form*, that is, follows the moon, not of course the *matter*), supposing the tide direct the parts from the meridian of the moon to 90° west are rising and those to 90° east are falling as represented by the arrows in the figure.

The circles may represent sections of the earth and air by the plane of the equator. The vertical velocities are of course invisible compared with the horizontal.

As the reductions have been made the result had best be given, but the introduction to it will require a little alteration. I presume you have the proof to refer to.

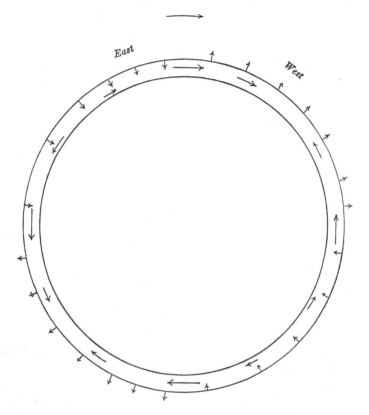

5th *March*, 1876.

Considering that the height of the atmosphere, supposed to be throughout of the same density as at the Earth's surface, would be about 5 miles, and that the friction on the Earth's surface would operate in the first instance only on the lowest strata, and that the tidal period (as to the chief term) is 12 lunar hours, not apparently affording sufficient time for the retardation of the lowest strata to be propagated by means of successions of eddies, to any great heights, I should suppose that the effect of the retardation on the aerial tide *as a whole* would be very small, and therefore that any *barometric* variations referable to aerial tides would be little affected by retardation. It might be different as regards effects on the *anemometer*, for here we have to deal with just the lowest stratum in which the friction takes most effect. The tidal current would however be *gentle*, and the friction therefore small compared with that which we have to deal with as regards ordinary winds. I dare not conjecture how far friction might be sensible in its effect on tidal *winds* observed at the surface of the earth.

As to tidal waves in *water*, it is only in the case of a regularly progressive wave propagated in an ocean, or along a channel of which the depth may be treated as uniform, that the proposition holds good that the phase of maximum height synchronizes with the phase of maximum horizontal velocity. If the wave be propagated up a uniform gulf, uniformly deep, and barred at the end, there will be perfect reflection at the end; the disturbance in the gulf will be what Airy calls a stationary oscillation, and the maximum of horizontal velocity will be at half tide.

Again, take the case of a uniformly broad and deep channel, with short barred (I mean stopped at the end, not with bars at their mouths) side channels opening into it, the channels being short enough to allow us to disregard the difference of height of the water in any one of them at the mouth and end. It is clear that at the mouths and for some distance up the side channels there would be currents up and down them, out of and into the main channel, and these currents would be greatest about half tide, when the level of water at the mouth would be changing most rapidly. About the middle of the main channel the horizontal velocity would be along its axis, and would be greatest forwards at high water and backwards at low water.

A similar difference in the phase of the currents near shore, from what we have in the open sea, would be produced by the shoaling of the water towards the coast. To take the case most unfavourable to variations of phase of tidal currents, I will suppose that at a distance from land it tends to be propagated in a direction on the whole parallel to the coast.......As the tide-wave travels more slowly in the shallower water towards the coast than in the deep water, these curves, loci of similar phase, bend a bit towards the shore. When the tide is rising the water flows towards the land as well as across the channel, and as regards the former component the sea is in the condition of a gulf as already explained. Hence on the whole the phase of the velocity as com-

pared with the phase of the height is not the same as out at
sea. Moreover the currents in the shallow water are much
stronger than in the deep water away from the coast, where too,
even if they had been equally strong, they would escape notice,
as we have no such accurate means of determining a ship's position
as we have close to land. Hence the tidal currents which alone
attract the attention of sea-faring men are just those which don't
follow the law of being a maximum in the positive or negative
direction at high or low water.

But the atmosphere must be assimilated to an ocean covering
the whole earth and of uniform depth approximately, and this is
just a case in which the law I mentioned *would* be obeyed, save
as to the effect of friction.

11th Dec. 1876.

It has probably occurred to you that the increase of friction
÷ pressure obtained when the velocity is small may have been due in
part to the use of cloth. At the same time we know that the statical
f/p is greater than the f/p when the surfaces are relatively
moving, which last is commonly assumed to be independent of
the velocity; and it is not likely that the law connecting f/p
with the velocity, say $f/p = \phi(v)$, should be a discontinuous one,
giving $\phi(v) = c$ so long as v is finite, but $\phi(v) = c'$, greater than c,
when $v = 0$. Rather it would seem probable that $\phi(v)$ would
change rapidly from c' for $v = 0$ to a quantity sensibly equal to c
when v was anything beyond some moderate value....It would
seem probable that the range within which $\phi(v)$ changes from c'
to a quantity sensibly equal to c would be the smaller, the harder
be the softer of the two materials pressed together. If so, it would
require a lower range of velocities to detect the variation of $\phi(v)$
with metal against metal than with cloth against metal, which
would account for the variation not having been, so far as I
am aware, detected in experiment.

I have been in correspondence with Dr Andrews on the
subject of the elastic yielding of the internal volume of capillary
tubes, and the correction to be made for it*.

* See letter in Vol. ii. pp. 126–9.

19 *Dec.* 1876.

I have long been puzzled why Siemens's bathymeter should show no apparent variation of gravity with latitude. I am disposed to think it must be due to the variation, with latitude, of the vertical component of the earth's magnetic force. In going south this would diminish; the vertical steel tube would be less magnetic by induction, and would less attract the flexible steel bottom. The result on the reading would be the same as if gravity increased and might therefore mask the actual decrease.

12*th May*, 1877.

When a series like

$$V = a + bv + cv^2 + \dots$$

is fairly convergent within the limits of v we have to deal with, its leading terms form a natural approximation to the function we are concerned with. But when it is divergent, and we merely know V numerically for a certain number of values of v, it seems to me that its leading terms have no special claim to consideration as furnishing a mode of expressing the function.

I am aware of course that a function $f(x)$ may sometimes be calculated numerically from a divergent series such as

$$a - bx + cx^2 - dx^3 + \dots$$

even though we know neither the form of $f(x)$ nor the law of the coefficients, but only their numerical values. One way is, either from the beginning, or after a certain number of terms, to transform the series by the formula

$$U_0 - U_1 + U_2 - \dots = \frac{1}{2} U_0 - \frac{1}{2^2} \Delta U_0 + \frac{1}{2^3} \Delta^2 U_0 - \dots.$$

But if the coefficients are unknown, and we know nothing but the numerical value of $f(x)$ for a certain number of values of x, and if we assume

$$f(x) = A + Bx + Cx^2 + \dots$$

and, taking a limited number n of the terms, we determine the constants $A, B, C\dots$ by n given values of $f(x)$ corresponding to n values of x [or it may be by N values of $f(x)$, $(N > n)$, for as many of x, the N equations with only n unknown quantities $A, B\dots$ being reduced to n equations by the method of least squares or otherwise] I don't see that the resulting values of A,

B... have any special relation to a, $-b$, ..., nor that the values of A, B... might not be materially changed by including another term; taking for instance $A + Bx + Cx^2 + Dx^3$ instead of $A + Bx + Cx^2$ only. I was beginning to try a numerical experiment, taking a function $f(x)$ such as $\sqrt{(1+x)}$, or $(1+x)^2$, regarding it as unknown, save as to its numerical value for certain values of x mostly greater than 1, taking a limited number of terms of the series

$$A + Bx + Cx^2 ...$$

determining the constants by the given values of $f(x)$, and seeing how they would come compared with the coefficients in the expansion by the Binomial, and whether they would approach a limit as a greater and greater number of terms was taken, but I found it would cost more work than the thing was worth.

I can well imagine that if a, b, c, d, e were got by least squares from

$$V = a + bv + cv^2 + dv^3 + ev^4 \qquad (\alpha)$$

by reducing say 20 observations, the results might be worse for the set of observations *as a whole*, i.e. including those not included in the 20 reduced, than if you had taken the simpler formula

$$V = a + bv \qquad (\beta)$$

and had determined a, b by least squares from the same 20 observations as before. But it seems to me that the sum of the squares of the errors *for the* 20 *reduced alone* must be less with (α) than with (β) from the very nature of the process. When therefore you speak of the result by (β) being better than by (α) you mean I presume for the *whole* series, not the 20 reduced *alone*; though even for these the errors of formula (β) would probably be *more capricious* than those of (α).

I have been engaged in sundry calculations of late relative to the great telescope Grubb * is making for the Austrian Govern-

---

* In a recent letter from Sir Howard Grubb, F.R.S., to the Editor, in reply to an inquiry on this subject, he refers to Prof. Stokes' zeal and activity in these matters in the following terms:

"After such a long lapse of time I am afraid that any letters of Prof. Stokes' I could get together would be of very little use, as many I fear have been lost amongst the mass of correspondence which accumulated in so many years; they are for the most part disconnected and disjointed, he having written in his spare moments, sometimes in railway trains, sometimes at the Royal Society, and sometimes at home, and it would be very difficult indeed to find anything that would be really useful for publication. He was wonderfully painstaking in answering any queries, so much so that I sometimes hesitated to ask him even a simple question,

ment. He had intended at first to make the crown equi-convex, but he began work with the 4th surface, making it flat. When the dispersion of the two glasses was satisfactorily determined, it appeared that the form required for this, if spherical and chromatic aberration were both to be destroyed, using spherical surfaces, gave the 4th surface, or 2nd of the flint, decidedly concave, and if the 4th surface were to be flat, the crown should be decidedly flatter in front than in rear. Probably he could have mastered the over-correction of spherical aberration for crown equi-convex and flint concavo-plane by the form of the surface; but as the theoretical form of no aberration for the flint concavo-plane brought out the second and third surfaces of the combination severally fitting, it involved a far smaller departure from the theoretical form to make the curvatures of the 2nd and 3rd surfaces the same than of the 1st and 2nd, and this would equally well save the construction of an additional pair of tools; and the minute over-correction of spherical aberration in this form is no more than can be easily mastered. So this form is now fixed on.

Athenæum Club, Pall Mall, S.W.
29th March, 1878.

If an anemometer, held by a spring, were moved backwards and forwards in a large room, the general effect it seems to me must be this.

Behind each cup, especially that the concave side of which faces the wind, there will be a tail of eddies, the disturbance corresponding to which will quickly spread over a space much broader than the cups, partly from lateral adhesion of air to air, partly from the interlacing of eddies and quiescent portions of the air. The eddying portions however will be possessed, not merely of a rotation, opposite in direction behind the two edges of the cup, but with a progressive motion. When they spread out and coalesce the rotatory motions will in great measure destroy one another, but the progressive motions will remain, and in the wake there will be a wind in the direction in which the anemometer is moved. To a minor extent there will be a wind

fearing it would encroach upon his time, for he went so deeply and minutely into every aspect of the question that in some cases I had as many as 5 postcards or letters from him in 24 hours, each describing some new view of the particular subject I had enquired about."

also in front, from the push, but the chief wind will be in the rear. In its return course, the anemometer will at first and for some way encounter this wind, so that on the whole the anemometer will experience more resistance than if it were moved continuously forward in still air with the same mean velocity.

That there must be a considerable wind produced is shewn by a simple consideration. Except near the turning points the whole momentum of the force which impels the anemometer horizontally, since none of it is expended in generating motion in the anemometer which by supposition has reached its normal velocity of transport, must be expended in producing momentum in the air or must be balanced by an opposite pressure. It is in fact partly one and partly the other. The pressure produces velocity in the air, and this in turn occasions increase of pressure on the opposite wall and lateral friction from the side walls. But these are produced mainly by the motion of the air, so that the forward motion of the air is anything but insignificant.

CAMBRIDGE, 27 *Nov.* 1878.

It is near 12 now, so I will not promise to finish this to-night.

I will ask Maxwell if he knows of any measures taken of the resistance in the Voltaic Arc.

If we knew the resistance at the break, and also the intensity of the current, we could, as you remark, measure the heat generated there. I am not sure, however, whether it has occurred to you that this would by no means vary as the light produced. I mean that for equal amounts of heat generated there might, under different circumstances, be very various amounts of light. For a given total amount of radiant energy the amount of light corresponding would depend immensely upon the mode of partition of that energy among radiations of different refrangibilities.

Nov. 28. Thus imagine a voltaic circuit with plates so large and wires leading up to terminals so thick, that the resistance in that part of the circuit may be deemed to be null, and let the circuit be completed, in one by a short very thin wire, in another by a thicker and longer wire of the same resistance. In the former case the wire is rendered it may be white hot, in the latter barely red, or not red at all. Yet the total heat which is generated per second in the wire is the same in both cases, and when a permanent state is attained the loss per second by radiation—for

simplicity I will suppose that there is no conduction away—
is the same in the two cases. While, however, the total energy
of radiation is the same, the distribution of this energy among the
different refrangibilities is very different. Passing now to the
case of the voltaic arc, we see that the amount of light for a given
current and given resistance is not *necessarily* the same under all
circumstances. These no doubt are important elements to know,
but we must have photometric as well as electric measurements.
This no doubt you contemplated in your letter. Still I thought
it might not be altogether superfluous to point out how widely
the ratio of the light might possibly differ from the ratio of the
heat in two different cases.

There is another consideration affecting the yield of light for
a given supply of heat. We ought to choose for poles a substance
which is as far as possible black for light and white for heat.
For we want to avoid as far as possible loss—for our immediate
object—of radiant energy in the form of invisible heat, and to
have it in the form of more refrangible radiations. Hence,
according to the doctrine of the reciprocity of radiation and
absorption, a substance which is white for light, but black—i.e.
freely absorbing—for heat would be a bad one to choose for poles,
but one which was black for light but white, that is freely reflecting, for heat, would be a good one. Probably carbon fulfils these
conditions as well as any substance we are likely to get*, for
Melloni found that it is translucent for heat rays of low refrangibility.

I hope you think I was right in recommending Mrs Sanderson
not to sell out her gas shares during the panic, as she was on the
point of doing. I think we are still far from gas being superseded
by the electric light for general purposes, though it is likely
enough it may be used for public buildings and purposes of that
kind. I said I thought the shares would rise again after an
interval sufficient to allow the panic to subside, though they
might not get up to what they had been; that *then* it might be
prudent to sell out a good portion of them in order to divide
her risks.

I am afraid Adams is not so well, for Dr Latham went out
last night for the third time, and then brought Dr Paget with

* The modern use of the rare earths to secure these conditions is well
known.

him. However I hope to get more recent intelligence before I close this.

4.40 p.m. I have just returned from the observatory. The account is that Adams is not any better, but neither is he worse. Both doctors were out this morning, and they are going again at 6.30. They do not expect any important change before Monday or thereabouts. His case is evidently a serious one, but we may hope for a favourable issue.

Yours affectionately,

G. G. STOKES.

P.S. I was near forgetting to answer about the voltaic arc seen against the sun. Foucault made some experiments about it, and arrived at the conclusion that the arc—or was it the poles?—was I think one-third or two-thirds, I forget which, as bright as the sun. I must refer to his paper. It is pretty sure to be No. 19, "Annales de Chimie," Tome 58, year of publication, 1860, pp. 476—478.

15 Feb. 1879.

On Wednesday I went to the South Kensington Museum, and saw Lockyer. I found him in his Laboratory with one ear, and the part of the neck near it, as red as a turkey cock. He told me that on Monday he had been working with the electric arc given by a powerful Siemens' machine. Knowing the injurious effect of the light upon the eyes, he took care not to look at it, so he stood with his back to the light about 2 feet off. Yet it affected the skin of the back of his neck which was exposed to it, producing something like erysipelas. He said that the day before the skin had been black, and that wanting to write something he was obliged to dictate it, as the inflammation had extended to his eyes. His assistants were also affected, but not to such an extent.

I mentioned the thing to Siemens afterwards, and he said yes, that it blistered the skin.

It cannot I should think be the mere heat. In furnace work or in metal smelting one would often be exposed to more severe heat; in fact, the sensation of heat seemed to have attracted no attention. I should think it must have been the action of the rays of high refrangibility in which that light so much abounds. The light seems far to surpass sunlight in its tanning power on the skin.

I forget whether I asked you how many copies of your paper on the Anemometer you would like to have. 50 are sent in any case, and up to 100 will be supplied without cost; beyond that the paper and printing off are charged up to the limit for author's copies, which is an additional 150, making 250 with those supplied without charge. Would you kindly tell me that we may order the corresponding number of copies of the plates? It will take a week or 10 days before the proofs of the plates are ready to go out.

I send you the number of the *Graphic* containing the drawings of the *Thunderer's* gun after the explosion. The section seems to show great want of longitudinal strength. I do not like depending upon the combined pull of two pieces which are separated by a widish interval. The material is so nearly rigid that it is a chance whether the two pieces pull together, or whether a great part of the strain is not thrown upon one, when if that goes the whole will fall upon the other, and so the resistance is overcome in detail. Again, as regards the portion of the second piece—reckoning from the muzzle end—on which the strength depends, we have another instance of indeterminateness, that is, the strain might fall wholly on the one or wholly on the other of the two projections at the end if the material were absolutely rigid. And if the bearing were chiefly on one of these projections, as it appears from the form of the fracture to have been on the inner one, there would be a longitudinal strain at one side which would be added to the general longitudinal strain, and would tend to start a crack in the metal, which once commenced could hardly fail to be continued.

8 *March*, 1879.

With reference to something that you said in a former letter, I must quote a short passage which I have seen quoted in a book that has been sent me entitled *The Supernatural in Nature*. The quotation is from a work of Huxley's entitled *Scientific Education*. He says, "And I for one lament that the bench of Bishops cannot show a man of the calibre of Butler of the *Analogy*, who, if he were alive, would make short work of the current *à priori* infidelity."

I have never been able to make out what Huxley's religious views are.

15 *March*, 1879.

I am afraid we are a good way off from being able to explain the fluorogenic or non-fluorogenic characters of gases as depending on their chemical characters.   I recollect your telling me that you had found nitrogen eminently fluorogenic, and hydrogen the reverse.   As to the chemical classification of the two, there is no doubt that hydrogen ranks as a gaseous metal.   It seems that chemists are disposed to rank nitrogen with phosphorus and arsenic.   Though phosphoretted hydrogen, the analogue of ammonia, does not like it act as a base—or rather as a base *minus* water, the analogue of an actual base being oxide of ammonium— yet when the hydrogen in phosphoretted hydrogen is replaced by ethyl, we get Hofmann's triethyl phosphine, which in its properties is perfectly analogous to ammonia.

The metals are capable under certain circumstances of giving out rays of very high refrangibility.   Thus the long spectrum of electric light, such as it is given by the discharge of an induction coil with a jar in connexion in the secondary circuit, consists mainly of pairs of dots representing the tips of the electrodes, and depending upon the nature of the metals employed for electrodes.

We may perhaps give some sort of an explanation of the greater brightness of the positive electrode when the arc is passed between carbon electrodes.

If we compare the negative and positive discharges in a moderately exhausted tube, the positive discharge seems to take place from a mere point, the negative gives a glow extending over a very sensible portion of the surface.   The total generation of heat is found to be greater at the negative ; but the portion of matter in which this heat is in the first instance concentrated may be greater at the positive, on account of the greater constriction of the path of the discharge.   This greater concentration may more than make up for the smaller quantity, and so the light may be more intense.   I do not think that combustion in the strict sense of the word has anything to do with it.   The result is much the same *in vacuo*, or in an inert gas.

We can give no reason I think in the present state of our knowledge why there should be the observed difference in length of spark between a point and a disk according as the point or the disk is positive.   Indeed I don't think we have yet got the appropriate idea of what a discharge of electricity means.   I said once to Faraday, as I sat beside him at a British Association dinner, that I thought a great step would be made when we should be able to say of electricity that which we say of light in saying that it consists of undulations.   He said to me he thought we were a long way off that yet.

For enlarging our conceptions of the ultimate working of matter, I know of nothing like what Crookes has been doing for some years. I wish you could see some of the work in his laboratory.

Next week I have got to be in London, from Tuesday to Saturday inclusive, for the meetings of the Cambridge University Commission.

P.S.   I saw Lockyer on Friday, and found he had not yet got quite over the tenderness of eyes arising from the action of the radiation from the arc of a Siemens' machine on the skin of his neck.

17 March, 1879.

I am not aware of any large quartz lens that is employed in a telescope, except that at the Dunsink Observatory.   If the telescope is to be employed as an achromatic, it is not easy to see what the second lens could be made of consistently with retaining the transparency of the quartz for the invisible rays of high refrangibility.   The powerful double refraction of Iceland spar puts it out of the question.   I know of no other crystal which could be obtained in sufficient size, unless possibly Beryl.   I do not know whether it is transparent for the rays of highest refrangibility among those that are found in the radiation from the sun.   But after all it is only in the case of observations with the spectrum that such transparency is required, and it is not as a rule for spectral observations that very large apertures are required.

I am glad that Howard Grubb has determined to give a fair trial to the method of testing the objective, using the existing disks, with homogeneous light, correcting the spherical aberration by adjusting the separation between the lenses.   I really think it will give an almost certain prediction as to what *would be* the performance of the telescope, *if* finished.   The result I think would be, either to lead to the non-rejection of a grand disk on account of a fault which it shows, and which is *only* visible under a very critical mode of observation, or else to show that the disk would not answer, and so save the time and expense and trouble of finishing it in the hopes that it might turn out well notwithstanding that defect.

24 *March*, 1879.

I do not recollect that I mentioned to you a very pretty hydraulic brake of Froude's invention which he showed me. It consists of two pieces, one fixed, the other, which is similar to the former, moveable. Each of these resembles a good deal a nautilus shell cut in two by a plane parallel to its plane of symmetry. The moveable one revolves round an axis perpendicular to the plane of section—in the shell-fish illustration—and the hollow interior is filled with water. The partitions, however, instead of being as in the nautilus perpendicular to the plane of section, are strongly inclined to it. Also their traces on the plane of section, instead of being radial, are considerably inclined to the radius. I think, but of this point I do not feel sure, that the inclinations of the traces in the two halves are the same way when viewed successively from the inside, and consequently opposite ways *in situ* as they face each other. I believe the relative motion resembles that of the two blades of a pair of scissors, only you must think of the blades as coming together only at the very edges, their planes being inclined to the common plane of the blades in scissors.

This brake is intended to measure the power of marine steam engines. Former brakes failed for this purpose because the power was so tremendous that the friction it produced when an attempt was made to curb it by friction was such as very quickly to make iron red hot. There was great tendency with such powerful forces to split the thing in two by pushing out one of the half nautilus shells; but this was very ingeniously and simply obviated by making the instrument double; the moveable part consisted of a pair of shells, I mean semi-shells, placed back to back, that is, with the outsides against each other, and the fellows to each were fixtures. Thus the pressures against the two moveable semi-shells simply balanced each other and produced no strain on the machine, and the strain on the fixed semi-shells did not signify, since the fixed parts could without difficulty be made as strong as might be required. In such a brake the resistance would of course vary as the square of the velocity.

It occurred to me that possibly a sort of electromagnetic brake might be used in experiments of the kind you are engaged in. Suppose the revolving shaft carried an armature capable of being made an electro-magnet, and that opposite to it were the poles of a powerful horse-shoe magnet. I suppose the current passing through the armature to be so arranged that each time a pole was passed the current was reversed. There would, however, I fear be difficulty in getting up sufficient brake power this way unless the revolving piece acted by a cog-wheel and pinion on another piece, and this latter carried the armature, which would complicate the arrangement. The magnetizing current might pass round a coarse galvanometer, through a small adjustable resistance, and by means of the resistance the current might be adjusted to a constant strength as measured by the galvanometer.

It is just post-time now.

6 *April*, 1879.

When I wrote last I had not time to mention what struck me as a very curious phenomenon that Crookes mentioned when he read a paper—last Thursday week—before the Royal Society.

He was examining the phosphorescent effects produced by what comes from the negative in a tube so highly exhausted that the dark space ordinarily seen around the negative extends over the whole tube. One tube had about a tea-spoonful of rubies in it. His assistant had the tube in his hand after the experiment was over, and happening to incline it so that the rubies came in contact with an electrode of the tube which he was touching, or else touching the tube-electrode—I am not sure which, but I think the former—he received so smart a shock that he was near letting the tube fall. It struck me as exceedingly curious that so small an extent of matter as the rubies presented should be capable of getting so highly charged as to give a rather severe shock.

Crookes has really opened out quite a new field of research in these recent experiments of his.

31 *May*, 1879.

I have just heard from Grubb, who tells me that he finds that when the less curved face of the crown is placed first, which was the position we contemplated in the actual objective, the spherical and chromatic aberrations are so nearly corrected together when the lenses are separated 4·5 inches that he can dispense with soda light in the testing, which he is glad of as he is a little afraid to

work with those costly disks at night in the dark.   He says, " the
definition with a power of 350 is by no means bad, and rather
surprises me."

I dare say he may have written to you direct, but I thought
I might make sure of your knowing this encouraging result.

I gave my last lecture to-day.   In addition to my lectures
I have had several things claiming my attention which would not
bear being put off, which is the reason why I have been obliged
to postpone the full consideration of your anemometer letter, as
I saw it required thought.

The Vicarage, Petham, Canterbury,

2nd August, 1879.

I came here to-day from Cambridge to see the family of my
late brother.   I am to return to London on Tuesday, staying there
the night for a meeting next morning at the South Kensington
Museum, and mean to return to Cambridge in the afternoon.

I have been very busy of late with one thing or another,
which is the reason why I have not gone more fully into the
anemometer question.

What you said in your last as to the difference of the results
regarding the comparison of the 24 inch and 12 inch, according
as the whirling machine or actual wind was used, leads me to
suggest that it would be very desirable to try the 24 inch against
the Kew standard when the former was loaded in such a manner
as to increase considerably the moment of inertia, placing flat
horizontal masses near the cups; flat, laid horizontally, in order
to avoid sensibly altering the resistance; near the cups in order
to get up moment of inertia with as little increase of weight (and
thereby friction) as may be.   If the wind were perfectly uniform
(and that is a condition nearly satisfied in the whirling experiments, but not in natural wind, except perhaps on rare occasions)
moment of inertia by itself alone would of course make no
difference, but with a variable wind it *would* make a difference.
The conditions of the 12 inch and 24 inch must be very different
as regards moment of inertia; and by comparing two 24 inch ones
differing as nearly as may be *only* as regards moment of inertia,
we should be able to form an estimate how far *this* influences
a comparison made in natural wind, subject as it is to continual
variations; and we should thus be able to judge whether the

natural difference in this respect between the 24 inch and the 12 inch would be likely to influence materially the result of a comparison made in natural wind.

I have been working strenuously the last day or two at the data for Grubb's new crown glass disk. He told me it appeared to be extremely good glass, though the disk was rather thinner than he would like, on which account he thought of slightly lengthening the focal length of the combination. Ten days ago or more he sent me the deviations for this disk and for a prism supposed to represent the flint glass disk which Feil is making. He said, as far as he had examined, the indices appeared to agree very nearly with the former, so that he should be able to proceed with confidence. This statement led me to attach no great idea of urgency to the matter, and I did not at once take it up. But when I came to calculate out the indices I found that, though there was not much difference of refraction in the crown from the last crown, the dispersion was about 15 per cent. greater than in the second crown; about 8 per cent. greater than in the first. This is *most serious*, for *considerably* deeper curves would be required, which the disk is not thick enough to bear. No great increase of focal length could be permitted, as all the mountings have been prepared for a f. l. of 32 feet or thereabouts. The only possible remedy I can think of is one of so bold a character that I hardly think he would venture on it. It is to reconcile easement of the curves of the crown with retention of the intended f. l. of the combination by separating considerably the two lenses, making the telescope in fact somewhat dialytic in character. Deeper curves would in this way be required for the flint; but the flint in course of preparation by Feil is not I suppose yet moulded, and the aperture might be a little less than that of the crown, and the material thus economised could be utilised for increase of thickness to bear the deeper curves.

CAMBRIDGE, 8 *August*, 1879.

I find on calculation that if a telescope is to be made with crown and flint glasses of given quality, and if the power of the crown glass lens is also given, then the telescope is lengthened by being made of the dialytic form, not shortened as I had supposed at the first blush. So that this plan is not feasible.

It must be very vexatious to Grubb to meet with so many

disappointments. I have rather urged him to think of the existing disks, which have performed so well under trial, not showing any perceptible defect when the telescope is used in its natural way, though when we play tricks with it by throwing the pencil out of focus altogether, the glass is seen to be not absolutely perfect. At any rate he might offer to the Austrian authorities to let them have the telescope provisionally until better disks could be obtained. According to the contract the Austrians are to take and pay for the mounting if the failure should be the impossibility of getting suitable disks. But the mounting would not be of much use to them without the objective, and it would be for their interest to come to some arrangement by which they would have an objective to go on with, which, though not all that could be wished for in point of homogeneity, yet does its work very well, and in point of size is quite unique.

14 *Jan.*, 1880.

I am glad to see that you are able to take such an interest in scientific matters. I got your letter about the refraction question to-day.

I have thought over the method of using a diffraction grating, and compared it with methods of refraction. I have made some rough calculations, but I have not gone fully into it, as I have my Smith's Prize Paper impending.

It seems to me that of the two the refraction method gives a rather larger quantity to measure, especially if a new modification which occurred to me in thinking over the subject were made. The diffraction method has one theoretical advantage over this last, namely that it depends on only one medium, air, whereas the other depends on two, suppose air and glass, or water and air; any change in the refractive index of the glass or water might appear to belong to a change in the refractive index of air. But I do not think there would be much difficulty about that. I have not yet mentioned what this method is, but hope to do so before I conclude.

But it seems to me that both the methods depending on refraction and that on the use of a diffraction grating are put out of court by the superior delicacy and convenience of that depending on the measure of the displacement of fringes of interference. I remembered reading the result in Moigno's *Réper-*

toire d'Optique Moderne, and after a little I found the place. It is at pp. 161, 162. Arago and Fresnel applied the method to determining the difference of index between dry air and air saturated with moisture at the same pressure and temperature, which was 27 C. or 80·6 F. With tubes a metre long a displacement of a fringe and a half was observed, and the calculated difference of index corresponding was 0·0000009, nine in the seventh place of decimals. This is slightly under 0·3 per cent. of the whole difference between the refractive index and unity. It amounts to about as much as the one-eleventh of an inch in the height of the barometer. It would come to about 0″·15 in angle for a heavenly body at an altitude of 45°. The one-sixth or one-seventh of a second is a very small quantity to deal with; and though the refraction would of course be much more for a body low down, on the other hand the definition would not be so good.

The refraction method I was led to think of is this. To magnify the effect it is of course well to work with a very considerable angle of emergence. At such high angles there is great loss of light by reflection at the common surface of glass (or whatever other medium you use to confine the air) and air, and if there be many such reflections the loss is very considerable. In the method of a prism with a considerably obtuse angle, the sides formed of parallel plate, there are four such very oblique refractions, and the factor for the intensity of the refracted light being raised to the fourth power becomes very small when the inclination is very great. It appears from the formulæ that there is an immense gain in point of light by using a single refraction only at that very high angle, and getting up the difference of deviation by making the angle of emergence very little short of 90°. The important boundary of the air in the vessel holding it would thus be a surface of glass or other material forming one side of a prism into which the light would be refracted at a moderate angle, while it emerged almost grazing.

The surface of almost grazing emergence would have to be worked very truly plane, as a small change of inclination of the surface, that is, of an element of the surface, would make a comparatively large error in the direction of the refracted ray. This difficulty I thought might be got over by using for the surface of emergence the free surface of water forming one face of a water prism*. The arrangement would be roughly as in the figure [omitted]. What would be observed would be differences of refraction corresponding to the substitution of one gas for another over the water. By using rock oil, or some such fluid in place of water, dry and moist air could be tried alternately. One gas might be made to take the place of another by displacement, so that there should be no risk of displacement due to a slight derangement accompanying the removal and reintroduction of atmospheric pressure by an air pump.

I merely mention this as a matter of curiosity which I thought might interest you. For actual accuracy and convenience I don't think the method of interferences can be equalled.

If a grating were used, suppose a deviation of even 50° were used for a spectrum of some order n. Then supposing the incidence in the grating perpendicular

$$\sin D = \frac{n\lambda}{\rho},$$

where ρ is the interval of the grating. To find the change of deviation consequent on a small change of index of the gas, and consequently of wave length in the medium, take the logarithm and differentiate. We have

$$\cot D \cdot dD = \frac{d\lambda}{\lambda} = -\frac{d\mu}{\mu}.$$

In the case supposed of the difference between dry air and moist, and with a deviation in the grating as great as 50°,

$$\frac{d\mu}{\mu} = 0\cdot0000009, \quad \tan D = 1\cdot1918$$

$$-dD = \tan D \frac{d\mu}{\mu} = 0\cdot000001 \quad \text{almost exactly.}$$

Now one second is $0\cdot0000048481$, say for round numbers $0\cdot000005$. Hence the change in the angular position of the point observed is only the one-fifth of a second, a quantity too small to perceive.

On the other hand nothing hinders us from using, in the method of interferences, a pair of tubes several metres long. With tubes 10 metres long the displacement produced by substituting moist air, i.e. air saturated at 80·6 F., for dry air, would amount to 15 fringes.

---

* But even so, the increased magnification does not affect at all the purity of the spectrum, by Lord Rayleigh's rule. See *Math. and Phys. Papers*, Vol. v. (1871), p. 354.

13 *Feb.*, 1880.

I returned from London to-day about 1 o'clock. I had occasion to go to the observatory in the matter of the sunshine recorder. It is rather late now to begin a letter to you so as to catch the 5 o'clock, but I will see what I can do before post time.

The first point is as to the objection to the formula for resistance which makes it vary as the square of the relative velocity. The ground of the objection is the discontinuous nature of the formula.

Now if the formula professed to be rigorous, that would be a serious objection. But it is certain that the resistance to bodies moving through a fluid with more than excessively small velocities is intimately bound up with the formation of eddies, and we cannot expect to be able to obtain a rigorous mathematical formula taking in the effect of eddies. The formula must be more or less of an empirical nature. The usual approximate formula, making the resistance vary as the square—therefore an even power—of the relative velocity, but at the same time reverse its direction when the sign of the relative velocity is changed, is such that if we take abscissæ to represent the relative velocity and ordinates to represent the resistance, measured positive in a given direction, the curve of resistance is made up of two half parabolas, such as would be formed by imagining a parabola cut in two by its axis, and one half turned round the tangent at the vertex till after a half turn it gets into its own plane again.

This is at any rate a near approximation to the curve of actual experiment, supposing the body that moves through the fluid to be symmetrical with respect to a medial plane perpendicular to the direction of motion. When the motion is excessively small the resistance varies more nearly as the velocity, so that in the neighbourhood of the point corresponding to a relative velocity *nil* the true curve cuts the axis at a small angle instead of touching it. The general forms of the true curve and of the approximate empirical curve would be related somewhat as in the figure below [not reproduced] where the true curve is supposed to be the black one, and the approximate curve the red one made up of two semi-parabolas. In the figure the ordinate is supposed to represent the resistance measured positive in a given direction, and the ordinate is the same, except as to sign, when the relative velocity is the same except as to sign.

Now it does not strike me as an objection to the empirical curve being a near approximation to the true curve that the radius of curvature in the empirical curve changes *per saltum*. This transition takes place in a part of the curve corresponding to such very minute relative velocities that practically we have little to do with it in all ordinary experiments.

But I think fatal objections may be urged to the $V^2 - v^2$ formula. In the first place, as I think I remarked to you, it is not in accordance with the second law of motion. The formula for the resistance would not remain the same, as it ought to do, when a common velocity is superadded to both V and v. Again, if the formula claim to be true irrespective of sign as well as magnitude of V and v, it leads to results which are manifestly incorrect. Thus suppose we have a stream flowing with a velocity of 10 miles an hour; let a sphere move in it with a velocity of 3 miles an hour, first with stream and secondly against stream. Manifestly the resistance will be much greater in the second case than in the first. Yet we have $v = 10$ in both cases, and $v = + 3$ in the first case, and $= - 3$ in the second. Yet the formula $V^2 - v^2$ gives the same result in both cases. And if it be said that the formula supposes V and v to be both positive, then we want a second formula to express the resistance when V and v are in contrary directions; so that in seeking for a formula that shall express the resistance in a given direction in all cases, we encounter a discontinuity of formula, which is precisely the same objection that you urged against the $\pm (V - v)^2$ formula.

As to the ear trumpet, the theory of it is a matter of great difficulty. We must not too closely associate it with the idea of the reflection of light; for the lengths of the sound waves of average pitch would be about 4 feet, and so, large compared with the dimensions of the trumpet. Perhaps the state of things would be better pictured to the mind by thinking of a long Atlantic swell coming into a creek 20 or 30 feet long. Suppose the water deep; then the rise of the water at the end would depend on the tapering form of the creek. To get a good rise, the taper should be fairly gentle, and there should not be projecting shoulders of rock at which a portion of the swell would be reflected and so sent back. I should say that in the ear trumpet the narrow part of the paper cone ought to be led into the india-rubber tube by as flowing an outline as may be. It should not be like this (1), but rather like this (2). Also the whole should be air-tight, especially in the narrow part.

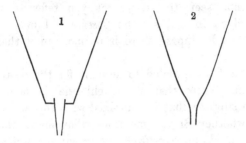

If the edges be merely sewed, I mean the edges of the paper or cardboard which forms the cone, it would be well to introduce a little gum or glue into the chinks to stop the communication between the air inside and outside. Towards the wide mouth it does not much signify, but in the narrower part it should be air-tight.

CAMBRIDGE,
July 11, 1880.

I have just got a collection of phosphorescent substances from Dr Schuchardt of Görlitz. They arrived on Saturday evening. I hope I may have some sun to examine them by in a pure spectrum.

Working at phosphori led me to a new way of examining fluorescent or phosphorescent substances when opaque, by which you can see the spectrum or rather the patch covered by the spectrum in its natural condition, only with the scattered light as it were picked out from it, so that the fluorescent or phosphorescent light may be seen without the scattered light which is mixed with it. In connexion with this I have thought of a way of viewing external objects in their natural forms but by the light of a pure spectrum; so that each point of the object is seen by homogeneous light, the refrangibility changing from one side to the other. I have tried it with some common objects, but hitherto I did not get any result of particular interest. It might possibly be useful for localising any places on the sun which give bright lines*, or for the prominences.

---

* Cf. the spectroheliograph as now employed, first realised about 1892 by Hale, Evershed, Deslandres, and others.

CAMBRIDGE,
16 *October*, 1880.

I had not seen the statement you refer to from Piazzi Smyth. I don't in fact see the *Observer*. I may perhaps look out for it, but I suppose there is no more in it than what you told me.

With you I am puzzled to account for the disappearance of the nitrogen. As to that of the chlorine, I should be rather disposed to suppose that it combined with the electrodes. I do not know whether Smyth mentions what metal the electrodes were made of. It appears from the recent researches of Liveing and Dewar that nitrogen combines under the circumstances of the electric discharge, and even in heated tubes, with metals in cases in which we should not have expected such a result. We knew well before what a stable compound it makes with titanium; but several other of the combinations that they mention, or some other, for I don't know that there were very many in all, are cases which, so far as I know, were not previously known. I don't suppose that in *these* cases the compound was a very stable one.

I should be disposed to look to the electrodes rather than to the constituents of the glass for the disappearance of the chlorine, and also of the nitrogen if it be really true that this latter did disappear, and that the apparent disappearance was not due to the employment of some different kind of discharge, which was fitted to bring out specially the hydrogen lines. My difficulty about attributing the disappearance to the glass is that the metals of the glass are already fully combined.

As to the feeble fluorescence produced by hydrogen in a Geissler's tube, all we can say is that it does not abound with rays of high refrangibility as much as the others, nitrogen for example. Perhaps we might go by conjecture one little step further, and connect it with the comparatively good conducting power of hydrogen. I think I am right in supposing that hydrogen conducts the discharge better than most other gases. If so, the electric tension required to strike across might not be so great as in the case of other gases, and the discharge in consequence not so violent, and not so exciting of vibrations of very short period.

Our poor Master [of Pembroke] is not by any means so well to-day. We did not see him when we called.

CAMBRIDGE,

10 *May*, 1881.

I do not know of any formulæ that have been given for metallic reflection except MacCullagh's and Cauchy's. Cauchy's are substantially the same as MacCullagh's, differing from them only in the details of development. MacCullagh's, as you know, are simply Fresnel's formulæ, calculated on the supposition that the index of refraction is a mixed imaginary, and reduced as Fresnel interpreted his formulæ for the case in which the angle of incidence, internally, exceeds the angle of total internal reflection.

I have never worked at the reduction of experiments on metallic reflection, but I am disposed to think that I should keep in the imaginaries to a later stage, and not get rid of them till it appeared most convenient so to do. I should work on for some way with Fresnel's formulæ.

This seems especially convenient when the data of observation are the principal incidence and the principal azimuth.

Let changes of phase be embodied in the coefficient, so that the factor $a(\cos\alpha + \sqrt{-1}\sin\alpha)$ multiplying the expression for the vibration means that the amplitude is multiplied by a and the phase accelerated by α. Then if p, q denote the factors after reflection for light polarised perpendicularly and parallel to the plane of incidence, unity being the coefficient before incidence, we have by Fresnel's expressions

$$p = \frac{\tan(i - i')}{\tan(i + i')}, \qquad q = \frac{\sin(i - i')}{\sin(i + i')},$$

$$\frac{p}{q} = \frac{\cos(i + i')}{\cos(i - i')} = \frac{\cos i \sqrt{\mu^2 - \sin^2 i} - \sin^2 i}{\cos i \sqrt{\mu^2 - \sin^2 i} + \sin^2 i}.$$

If α be the principal azimuth, at which light polarized perpendicularly to the plane of incidence is retarded a quarter undulation relatively to light polarised in the plane of incidence, and $\tan\alpha = m$,

$$p/q = -m\sqrt{-1},$$

whence

$$\frac{\cos i \sqrt{\mu^2 - \sin^2 i} - \sin^2 i}{\cos i \sqrt{\mu^2 - \sin^2 i} + \sin^2 i} = -m\sqrt{-1},$$

i being here the *principal* incidence.

$$\therefore \quad \frac{\cos i \sqrt{\mu^2 - \sin^2 i}}{\sin^2 i} = \frac{1 - m \sqrt{-1}}{1 + m \sqrt{-1}}$$

$$= \cos 2\alpha - \sqrt{-1} \sin 2\alpha.$$

$$\therefore \quad \mu^2 = \sin^2 i + \frac{\sin^4 i}{\cos^2 i} \left(\cos 4\alpha - \sqrt{-1} \sin 4\alpha\right).$$

From this equation μ can be expressed under the form $a - \sqrt{-1}\, b$, and then for the intensity at a perpendicular incidence we have

$$\text{modulus}^2 \text{ of } \frac{\mu - 1}{\mu + 1}, \text{ or of } \frac{a - \sqrt{-1}\, b - 1}{a - \sqrt{-1}\, b + 1},$$

or
$$\frac{(a - 1)^2 + b^2}{(a + 1)^2 + b^2}, \text{ or } 1 - \frac{4a}{(a + 1)^2 + b^2}.$$

The formulæ of Cauchy, which as I said are substantially equivalent to those of MacCullagh, appear to agree well with observation, but, if I rightly recollect, not quite exactly. This is what we should expect, assuming that they form a correct extension of Fresnel's formulæ. For the latter fail to account for the peculiar phenomena exhibited by diamond and other highly refracting substances in the neighbourhood of the polarising angle, phenomena which have been extended by Jamin so as to embrace nearly all transparent substances. As in transparent substances the failure is chiefly perceived in the case of highly refracting substances, and as metals, apart from their peculiar optical properties as metals, would naturally be classed with highly refracting substances, we might expect Fresnel's formulæ, if rightly extended to metals, to show similar and even more pronounced deviations from exact agreement with observation.

But there is another matter as regards films. Fresnel's formulæ are applicable to a single reflection; and if they are rightly extended, as they were extended by MacCullagh to metals, the extended formulæ ought to apply to metals in mass. We should not expect them to explain the difference between a metallic film and the same metal in mass any more than we should expect Fresnel's formulæ for a single reflection to explain the colours of thin plates.

If the extension of Fresnel's formulæ to metals be rightly made, the same principles ought to lead us to the explanation of the optical phenomena of metallic films, by proceeding as we do in the explanation of the colours of thin plates. We know the principles of the investigation, but the expressions would come rather long, and the numerical calculation would be laborious, not however, I should think, anything very tremendous.

I thought I would work out the numerical value of the intensity from the above formula for one case. So I took the numbers given by Conroy* at p. 495 for silver that had been polished with putty powder, yellow light, namely

Principal incidence 74° 37′; principal azimuth 43° 22′. With these numbers I get from the above formula, for the intensity at a perpendicular incidence ·9349. This is rather higher than the number you quote ·91.

If the principal azimuth were 45°, the formula would give perfect reflection at a perpendicular incidence. I see that Jamin gives a somewhat lower figure for the principal azimuth for yellow light for silver, about 40° instead of 43° 22′ which is given by Conroy. With Jamin's figure the intensity at a perpendicular incidence would come out lower. Conroy's principal azimuths for yellow light for silver that had been polished with putty powder come out about 8 degrees higher than for what has been polished with rouge. As he says that the silver so polished, though very bright, had a reddish tinge in certain lights, I should think his explanation is very likely to be correct, that particles of the rouge too fine to be seen got embedded in the silver.

It is clear that the films, at least the thinner ones, if calculated by the same formula would give a value to the intensity at a perpendicular incidence a good deal lower.

Although Cauchy made no material change in MacCullagh's formulæ, I think he supplied a *rationale* for them, which Mac-Cullagh had not done. I have considerable confidence in them, except that I think that it is probable that the cause which renders Fresnel's formulæ inexact for highly refracting substances in the neighbourhood of the polarising angle operates to a still greater degree in the case of metals†; and as the principal azimuth and principal incidence are measured about what most nearly corresponds to an angle of polarization, those may not be the best data on which to found the calculation.

I heard of the proceedings at the Astronomical Society

---

* *Roy. Soc. Proc.*

† Cf. however Maclaurin's recent papers in *Roy. Soc. Proc.*

à propos to the motion by Sir Edward Beckett, &c. so you need not trouble yourself to send the *Observatory*.

The Adams's are coming home from Brighton to-day. They were to have returned to-morrow, but I heard that Professor Adams had got cold, and wanted to see Dr Latham. I hope it is not anything to signify, but his coming home that way looks as if it were somewhat severe. I heard that he got much stronger at Brighton. We have had very pleasant weather for a person who is strong, but rather a trying kind for persons who are delicate on the chest, fine bright sun and with it a rather keen air.

Balfour Stewart has devised a new form of actinometer, which has now been tried for a year by Blanford at Calcutta. He has sent us the results of the observations, and they are now being printed in extract. The instrument appears to work well, but Calcutta as might be supposed does not seem to be a good place for trying it. With the Bay of Bengal on the south and the Delta of the Ganges on the north, there is doubtless far too much moisture in the air to make it a good place. Thibet would be a capital place, and we are in hopes of getting observations made there. Sir E. Sabine told me long ago that a very serious drawback to Herschel's actinometer was its great fragility*.

With our present knowledge of the constitution of the sun I don't think the supposition that its heat is not supplied from without, but is in fact primitive heat, presents the difficulties it formerly would have presented. We have evidence of the freest interchange between the superficial portions of the mass and those deeper down; so that although the supply of heat from the sun is enormous, the capital to draw upon is so gigantic that ages may pass without any diminution that could be noticed.

I have taken out the volume of the *Annales de Chimie* with Foucault's paper about the comparison of the sun with the electric arc. I have not yet read it, but on dipping into it I see that they—it was really by Foucault and Fizeau jointly— employed photography, fully recognising at the same time that the ratio of the luminous and photographic intensities might be considerably different.

11 *May*, 1881.—I will try another example of the calculation of the intensity. At p. 493, Conroy gives for yellow light, silver polished with rouge,

* Cf. *supra*, p. 182.

Prin. Inc. 75° 53′; Prin. Az. 85° 42′.

I put down all the work

$$\log \sin i = \overline{1}\cdot9867 \qquad\qquad \log \cos i = \overline{1}\cdot3872$$

$$\begin{array}{cc}
2 & 2 \\
\hline
\overline{1}\cdot9734 & \overline{2}\cdot7744 \\
2 & \\
\hline
\overline{1}\cdot9468 & a = 35°\ 42' \\
\text{Subtr.}\ \ \overline{2}\cdot7744 & 4 \\
\hline
1\cdot1724 & 142°\ 48'
\end{array}$$

$$\mu^2 = \lambda^{-1}(\overline{1}\cdot9734) + \lambda^{-1}(1\cdot1724)\{\cos 142°\,48' - \sqrt{-1}\sin 142°\,48'\}*$$
$$= \quad .. \quad + \quad .. \quad \{-\lambda^{-1}\overline{1}\cdot9012 - \sqrt{-1}\,\lambda^{-1}\overline{1}\cdot7815\}$$
$$= \lambda^{-1}(\overline{1}\cdot9734) - \lambda^{-1}(1\cdot0736) - \sqrt{-1}\,\lambda^{-1}\cdot9539$$
$$= \cdot9406 - 11\cdot847 - 8\cdot993\sqrt{-1} = -10\cdot906 - 8\cdot993\sqrt{-1}$$

$$\begin{array}{ll}
\log \sec\theta = \ \cdot1128 & \log 8\cdot993 = \ \cdot9539 \\
\log 10\cdot906 = 1\cdot0374 & \text{,,}\quad 10\cdot906 \quad 1\cdot0374 \\
\hline
1\cdot1502 = \log\text{ modulus} & \overline{1}\cdot9165 = \log\tan\theta \\
& \theta = 39°\ 32'
\end{array}$$

$$\mu^2 = -\lambda^{-1}(1\cdot1502)(\cos 39°\,32' + \sqrt{-1}\sin 39°\,32')$$
$$= \lambda^{-1}(1\cdot1502)(\cos 140°\,28' - \sqrt{-1}\sin 140°\,28')$$
$$\mu = \lambda^{-1}(\cdot5751)(\cos 70°\,14' - \sqrt{-1}\sin 70°\,14')$$

$$\begin{array}{ll}
\log\cos 70°\,14' = \overline{1}\cdot5292 & \log\sin = \overline{1}\cdot9736 \\
5751 & 5751 \\
\hline
\cdot1043 & \cdot5487
\end{array}$$

$$\mu = 1\cdot272 - 3\cdot538\sqrt{-1} = a - \sqrt{-1}\,b$$

$$\text{light lost} = \frac{4a}{(a+1)^2 + b^2} = \frac{5\cdot088}{5\cdot162 + 12\cdot517} = \frac{5\cdot088}{17\cdot679}$$

$$\begin{array}{r}
\text{logs}\ \ \cdot7066 \\
1\cdot2475 \\
\hline
\overline{1}\cdot4591 \\
n° = \cdot2878
\end{array}$$

light reflected at ⊥ incidence $= 1 - \cdot2878 = \cdot7122$.

It appears from this that the rouge-polished silver gives only 71 per cent. at a perpendicular incidence, while the putty-polished gave 93 per cent.

If the difference in Sir John Conroy's determinations was due to the embedding of a foreign substance, it is probable that the formulæ for metallic reflection would not apply. I can imagine that the difference may be real. The rouge would powerfully absorb light, but yet not so powerfully as to be comparable in that respect to silver, and so would not have a high reflecting power like silver.

* λ^{-1} stands for $(\log)^{-1}$.

CAMBRIDGE,
14 *Nov.* 1881.

Thanks for your long and interesting letter about my lecture *.
I was aware that the electromotive forces I quoted from Thomson
were not the highest that could be cited, but I did not know
where to turn for records of higher. It did not, however, much
matter for my purpose, as I had merely to show that the electromotive forces concerned in earth currents were of quite another
order of magnitude from those required to send an electric discharge through such enormous spaces in air as must be done in
auroral streamers.

Balfour Stewart had suggested many years ago in *Nature* that
the magnetic or rather auroral effect of solar disturbances was due
to radiation, but his theory on the subject was quite different
from mine. He suggested that the moving air—moving as wind
—in the higher regions was made a better conductor, or rather
less resistant to electric discharges of the nature of disruptive discharges, which enabled the inductive electromagnetic effect of
the earth's magnetism to produce electric currents, or rather
discharges.

The auroral tube in my experiments was, as you supposed,
only very imperfectly rarefied. I took what I found there, which
happened to be an auroral tube capable of exhaustion by a
common air-pump. The exhaustion was so imperfect that the
discharge, instead of filling the tube, passed through in a stream
no thicker I suppose than one's little finger. However, it sufficed
for my purpose, which was to show the very great difference
between air at ordinary pressures and rarefied air.

We may form some idea of the temperature as you ascend in
the air by the results obtained in balloon voyages; at least up to
the heights that balloons go; beyond that they don't help us.
But the results vary greatly with the circumstances. Quite
recently Captain Templer went up along with one of the persons
employed in the Meteorological Office. At about 6000 feet I think
it was they got into a terribly cold stratum, with the thermometer down to − 7 F. But when they ascended to about 10,000
I think it was they got into comparatively warm air, the thermometer being + 31. It is possible that some formula might express

* At South Kensington: reported in *Nature*; see *Math. and Phys. Papers*,
Vol. v. *Preface.*

pretty well the mean decrease of temperature in ascending. This might be applicable to calculations of astronomical refraction, where you want to combine a great number of observations taken under varying atmospheric conditions. But this would not help much for the aurora, which with us is rather a rare phenomenon. Of course we could get a near approximation to the height of minimum resistance, except it be that the height for which the resistance is a minimum (or I should rather say the density for which it is a minimum) may for aught we know depend very largely upon temperature. Perhaps I may suggest this to Mr De la Rue as a subject worthy of examination.

It does not seem to me likely that in Mr De la Rue's experiments the internal resistance of his battery made much, if any, difference in the striking distance. Prior to the strike, the battery was idle, supplying merely electric tension, not current, except as to the very trifling driblet of a discharge which passed before the spark passed. Of course once the arc was formed, the rate of discharge would depend materially upon the resistance in the battery, unless, as possibly may have been the case, the resistance in the battery was trifling compared with the resistance in the air that the spark had struck across and now the discharge is passing through.

Your suggestion—as I understand you—that the whole earth, atmosphere and all, is charged, like an insulated pith ball, the insulator being the interplanetary vacuum, is new to me, and I must think it well out. I have always felt great difficulty about the production of atmospheric electricity. It appeared from the phenomena to be connected with the condensation of vapour; I mean its more violent manifestations appeared to be so connected; but how?

As to the relation of aurora to cirrus cloud, my notion is that the condensation by which cloud is formed does not give rise to the development of electricity, but that the globules of water or spiculæ of ice, as the case may be, which constitute the cloud, form so many nuclei enabling a previously existing charge of electricity to escape, which it could not do so long as the air was homogeneous, the water mixed with it being all in the elastic state. And possibly the effect of a fog in a kite string is to permit the electricity to pass from the air to the string, which without the fog it could not do, unless the whole length of the string had contained a smouldering match to serve as collector. Even under ordinary circumstances, the difference of potential at two places a moderate height apart is enormous, and perhaps we might get as strong manifestations of electricity as those you mention if we could make each element of the string a good collector of electricity from the neighbouring air. Or rather perhaps if we could make the upper half, or so, of the string a good collector.

There are some other points that I intended to write about, but I am going out with the girls to a concert, and I must go and dress.

14 *Nov.* 1881.

Having now returned from the concert, I will go on with my letter.

I will make some enquiries as to whether auroræ are seen in the arctic regions when the sky is cloudless. I think we see them in this country when there are not more at any rate than very tenuous clouds. I have seen an auroral arch passing nearly through the zenith, and I don't recollect clouds in connexion with it.

I think it quite possible that there may be a development of electricity in connexion with the condensation by which even tenuous clouds are produced, but I should imagine the quantity would not be great enough to feed auroral streamers.

As to the direction of the streamers, Plücker showed that when a strong magnet is presented to a Geissler tube, the negative light proceeds from the electrode along lines of magnetic force. The electricity of a serene sky normally is negative, I believe pretty well without exception, so that if we suppose the auroral discharge to proceed upwards, the cloud or whatever it be that starts it being at the lower extremity, it would be in the condition of the discharge starting from a negative electrode. Further, as to why in both cases a line of magnetic force should be followed, I think Plücker has some speculations, and Reuben Phillips had before him, according to which an incipient discharge starting otherwise than along a line of magnetic force would get torn off as it were by the force of the magnetism on the current. For the force on the element of the current is perpendicular to the element and to the direction of the line on magnetic force passing through the current, and is proportional to the sine of the angle between the direction of the current and that of the line of magnetic force, and therefore could not fail to act laterally on the current unless the current were along the line of magnetic force.

The positive light in a Geissler tube behaves quite differently with respect to the magnetic force. It behaves as a perfectly flexible wire carrying a current would do*.

May the space above into which the auroral streamers extend and then disappear be analogous to the dark space separating the positive from the negative light in an exhausted tube?

As to the action of heat in facilitating the discharge in rarefied air, I had not thought of the change of density it would produce, but only of the specific effect of the heating of the air. At the height of minimum resistance as depending on density, a small change of density would make very little difference.

I found in my reading about the aurora that there is a so-called auroral zone, Nordlicht Zirkel, where it is seen very frequently. This is at a higher latitude, about 70° I think, in the north of Russia, than it is in America where it comes down to, I think, 50° or 60°. I supposed Iceland is comprised in it, for Mrs Magnússon told me that in the winter months there, from about August, the aurora is very frequently going on, I think indeed nearly every night, but I don't recollect exactly what she said of it as to frequency. When you get north of this zone the aurora is much less frequent and conspicuous. Dr Rae told me † that in one of his winter stations, I think it was near Great Bear Lake, the aurora was of the rushing kind, and the magnetic needle was simultaneously affected, while in another of his winter stations, somewhere, I think, N.E. of Hudson's Bay, the auroræ seen were steady, and I think comparatively faint, and the magnetic needle was not affected. These auroræ always appeared to the south, over some open water, or at least there was open water also lying to the south of them. Whether the aurora was just at the distance of the open water, so as to be really just over it, could not, I suppose, be made out.

We had under our consideration the question of using a thermopile. I think the chief reason which deterred us from

* For mathematical theory, cf. *Math. and Phys. Papers*, Vol. v. p. 3; and J. J. Thomson, 'Conduction of Electricity through Gases.'

† See letters from Dr J. Rae, F.R.S., of date 1881, *supra*, p. 237.

using it was the necessity for some skill and training on part of the observer, which it would be difficult to get in an out-of-the-way place like Thibet for example. I have also heard the objection raised of the liability of alteration in consequence of an alteration in the state of the coating of the elements. I have never worked with one myself, and I do not know how far this objection is serious.

I intended to have got you yesterday the *Natures* containing the other lectures at South Kensington, but I was detained till just dinner time.

P. S. About half the above was written this morning—Nov. 15. —I think I mentioned in my former part that my notion of the relation of auroral streamers to light cirrus clouds is that the globules or spiculæ of ice as the case may be form nuclei, permitting a discharge to pass.

As a matter of observation, in clear weather the potential increases positively upwards. At least this is the case in the neighbourhood of the earth, and presumably the same would hold at a good height. There would, therefore, be a constant tendency for positive electricity to pass downwards; but this in general it cannot do in consequence of the resistance to the passage offered by the air. In the high regions this resistance would be less, on account of the lower density; and if the passage be aided by nuclei in the lower portion of this higher region, that may enable the discharge to pass. Besides, there may be difficulty for homogeneous air, free from nuclei, to receive or part with a charge.

CAMBRIDGE, 16 *Nov.* 1881.

I now send you the *Natures* containing the other lectures, except Captain Abney's, which it seems were not published. The first of Lockyer's also was not published. They may not perhaps reach you by this, as they go by book post.

I have written to Captain Abney suggesting that he should try to photograph an aurora. Electric light is so rich in rays of high refrangibility that, notwithstanding the faintness of the light to the eye, I think there is a chance that it could be photographed without so long an exposure as to make the record all confused from the superposition of what took place at different times. If a good photograph could be got, that would seem to offer the best prospect of measuring the height by parallax, of course by means of simultaneous photographs. I am a good deal sceptical about the very great heights assigned. They are derived, I think, mainly from measurements of the altitudes of arches. Now if the arches are foreshortened curtains of streamers, and such seems to have been the opinion of Mr Farquharson—*Phil. Trans.* for 1829—who living far north in Scotland had good opportunities of studying the phenomenon, if, I say, the arches are foreshortened curtains of streamers, the parallax of lines so greatly foreshortened would seem to be very uncertain. I cannot help thinking that numerous measures of the azimuths of streamers, accompanied by an accurate note of the time, and by rough sketches, might lead to more certain results. In a number of observations thus made at two places, it might be possible to pick out pairs in which the object observed could be identified. Of course the lateral motion of the streamers would cause a difficulty, and it would be necessary to select streamers with a view to this difficulty.

CAMBRIDGE, December 13, 1881.

...I will now turn to the different points in your letter.

1. Whatever the negative electricity of the earth may be due to, I do not think that any chemical action between the interior of the solid crust and the fluid inside, if a fluid nucleus there be, would at all account for it. The electromotive force that could be thus accounted for would be utterly insufficient to meet the observed phenomena of atmospheric electricity. When we say that the earth is ordinarily negative, we mean of course relatively to the atmosphere a little way up. According to Thomson, the difference of potential between two points at different heights, in ordinary fair weather, varies from 22 to 44 Daniell's elements per foot of height. No chemical action would account for differences anything like this. I mean of course no chemical action that we can by any reasonable supposition attribute to an action between the solid crust and fluid nucleus, or between the water covering a great part of the earth and the solid parts. A cell or so is apparently the utmost that we could account for in that way.

I had not noticed the statement that the sea is normally positive, not negative like the land, to the atmosphere; that is, that the change of potential in the two cases of land and sea, as we ascend into the atmosphere, is opposite.

It is a very interesting question how the electric potential, or to use the old word, tension, varies as we ascend to great heights away from the surface of the earth altogether, which we still are on even when we are on top of a high mountain. We contemplated, at the Meteorological Office, having electrical experiments made in balloons; but I fear the disaster to the 'Saladin' may knock that on the head. I fear there is small hope now of Mr Powell's safety. If the wind blew the balloon over to France, he probably would come down safely, but we should probably have heard of it before now. I am going up to London to-morrow to a meeting of the Meteorological Council.

I have just been to the Station to get an evening paper, but no news of Mr Powell.

I was aware of the great development of electricity in the condensed vapours emitted by volcanos. It had not occurred to me to enquire about the ashes, nor do I know whether it is a necessary condition of the development of the electricity that the steam should be accompanied by ashes. I should rather suppose not. I cannot help connecting in my own mind the phenomenon with the very rapid condensation of vapour of water in the great cumuli which accompany thunder storms; in fact, which form the thunder clouds. I can hardly help thinking that this condensation has to do with the development of electricity, though at present laboratory experiments have not shown in what way the development takes place, if it does take place under such circumstances.

I have seen it stated that there is a place somewhere on the south-west coast of Ireland where miniature thunder storms are of frequent occurrence. Air laden with vapour comes from over the Atlantic up a valley between two mountains, and gets condensed, and the result is that there are frequent little thunder storms. The vapour is here caught in the act of condensation, and the condensed vapour of giving out electricity.

2. The strength of atmospheric electricity in the colder months seems correlative with the comparative absence of thunder storms; just as thunder storms in the lower latitudes seem to take the place of the aurora in the higher.

In some books on aurora I was lately reading, or rather, about the time when I gave the lectures at South Kensington, it was stated that there was an auroral belt in not very high latitudes where the aurora was more frequent than in higher or lower latitudes. In Asia this belt was in a higher latitude than in America; but whether the belt is found nearly in the same magnetic latitude I do not know. The change of latitude of the auroral belt seems to favour the idea that it may at any rate be at least approximately so.

It seems almost to me that I had written this to you before, but oftentimes I am unable after some time to say whether I have actually written a thing, or merely recollect the intention as if it were a thing that had been actually executed.

Is not the old idea of a second north magnetic pole in Siberia given up as the result of more numerous and more accurate determinations? In Sabine's magnetic maps of the circumpolar regions, there is only one magnetic pole represented, namely, that to the North of America. Of course the actual magnetic system may be decomposed if we please into two magnetic systems, and then each would have its pole; but that is merely like the resolution of forces, and involves an arbitrary procedure.

I cannot lay my hands on a numerical statement as to frequency of aurora in different places, but I came across one lately. It is said that at a given place, at least in moderate latitudes, such as ours, the displays are more frequent about the two equinoxes than at other times of year.

Mrs Magnússon told me lately that in Iceland the displays are very frequent; sometimes for two or three weeks together you will have them every night.

She told me also that on one occasion when she and some other girls were watching a very fine aurora she pulled out a letter written by a lady, a relative or connexion of hers, who writes an extremely small hand, and she was able to read it perfectly by the light of the aurora alone. I asked her whether there was any moon at the time to help her, and she said not.

I asked Mrs Magnússon also about auroral arches stretching from east to west nearly over the zenith. She had not seen such in Iceland.

I hardly think Hansteen's explanation of auroral arches based on the behaviour of an electro-magnet charged with frictional electricity can hold. In such a magnet, of the usual form, that is, with flat poles presenting rectangular edges, there is great concentration, at least I think so, of magnetic force, I mean external magnetic force, near the edges; certainly there is in a permanent magnet, as is shown to the eye by the behaviour of iron filings;

whereas if we were to conceive the earth, retaining its actual distribution of magnetism, to be reduced to a moderate size, and to be strewn with iron filings, no such special concentration of magnetic force about a particular magnetic latitude would be manifested.

I do not think that the experiments you mention as to the different length of the arcs formed in presence of an electro-magnet, according as it is excited or not, enable us to judge of the effect of the magnetism on the conducting power. For the most notable effect consists in the lateral displacement of the arc, due to the magnetism acting as it would on a conducting wire. Thus in the case of a flat platinum plate placed on the flat pole of an electro-magnet, with a platinum point opposed to it, the discharge, when the current round the helix was off, would pass from the point to the nearest part of the plate, and therefore along the perpendicular let fall from the point to the plate. But when the current was turned on it would have to pass to one side of that, in consequence of the lateral action of the magnetic force, and therefore could no longer travel along the shortest line as before, and therefore the distance must be reduced in order to let it pass.

I fear De la Rue is seriously ill. He has been suffering from diabetes, and has been ordered, for the present at least, not to attend to any business. He had an attack once before, but got over it. He was better by the last accounts.

He was working out the density of minimum resistance to discharge when he was taken ill. The pressure for air, at ordinary temperatures, was if I rightly recollect about 0·6 of a millimetre of mercury; no very great exhaustion as exhaustions go nowadays. From our approximate knowledge of the temperature in ascending, we could not be very far wrong in the height at which the density would be reduced to this. The aurora is such a very vague thing, and it is so uncertain whether observers widely separated are seizing on the same phase of the appearance for their measure-ments, that I do not think that great reliance can be placed on the asserted heights.

I have never seen a luminous fog, much less, an appearance similar to that remarkable one that Sabine sailed through. I should suppose it to be a discharge by a vast number of ex-cessively minute brushes; something like what you see in a tube exhausted by a common air-pump when, as the air is slowly let in, the discharge previously a reddish soft-looking thing gives place to a very faint, spider's-web-like luminosity. The current producing this is only a very small fraction of what it was. I imagine that the globules of water or spiculæ of ice in a mist or cirrus cloud facilitate this kind of discharge. Hence I am not surprised at the connexion between lunar halos and auroral displays, though I had not known that the two had been observed to have any connexion.

I should imagine the rushing and the quiet aurora to be analogous respectively to the soft, ordinarily reddish, discharge in a tube when the exhaustion is pretty good, and to the spider's-web-like discharge into which the former passes when enough air has been let in.

I mentioned I think already that Captain Abney had photographed an aurora with some success, and means to try again when he has an opportunity. I think plates have been made much more sensitive since the time he tried.

The arches which I supposed to be foreshortened streamers are not those that appear in the north and form the base of a series of streamers, but the narrow kind extending from east to west which I have seen perhaps twice in my life. In this case a streamer, being parallel to the dipping needle, would be projected on the visual sphere in a great circle passing through the magnetic zenith, or vanishing point of the dipping needle, and would therefore be nearly parallel to the arch.

I said nothing about the parallelism of the streamers to the dipping needle, taking this as having been already explained by Reuben Philips and by Plücker.

I got your letter about the pamphlet for Moigno when I was in London for some days. I sent him a copy after my return.

If my theory of the aurora and its connexion with other phenomena is subject to no more formidable objections than that I take the aurora to be an electric discharge, I have not much fear for it.

Lockyer told me lately that a paper had been written, or rather two papers independently of one another, to prove that the zodiacal light was not, as usually supposed, a cosmical phenomenon, but had its seat in the atmosphere of the earth. He lent me the paper to read, but I have only as yet read a small part.

14 *December*, [1881].

There are one or two more points to mention.

De la Rue has been examining the spectrum of rarefied air under various circumstances, and there is a good deal of change in the lines which come out. I think, but I am not sure, that he told me that he had got the single line of the aurora. However, the circumstances are in some respects so different that a difference in the lines which come out is not to be wondered at. I mean, that out of a set of possible lines, some come out under certain circumstances, and others under others, that is, are brighter; and sometimes the fainter, that is those that under the circumstances are the fainter, become actually invisible. The circumstance that in the aurora the discharge is from air to air, whereas in a tube we have metallic electrodes in communication with the source of electricity, is one difference which might very conceivably affect the character of the spectrum, as to the particular lines which would come out strongly.

I will mention now, for fear I should forget it, that on Thursday I asked Huggins about the spectra of blue components of double stars. He told me that from what he had observed he was under the impression that the blue component had lines in the less refrangible, and the red component in the more refrangible, part of the spectrum, and the late W. A. Miller, who made the observation with him, had the same impression, but the feebleness of the light did not allow him to say more.

I do not think you would gain much by using a large prism for the aurora. The faintness here arises from the small intrinsic brightness of the source, not, as in the case of a star, from the smallness of the source. A large prism implies a high magnifying power, without which the pencil would not enter the pupil of the eye. Neglecting the loss by reflection and absorption, for a given purity and given brightness of the spectrum you may have the spectrum larger, and the apparent breadth of the slit larger, which might enable you to see very faint lines which you otherwise could not see, the object being narrow as well as faint. But the same end can be obtained more economically, at least when once the prism was of fair size, by using two or more prisms. With two prisms instead of one, you can afford to double the breadth of the slit without reducing the purity of the spectrum. You get now the same result as if you had doubled your magnifying power, and doubled the linear dimensions of your prism, which would have involved the use of a single homogeneous piece of glass of eight times the volume.

P.S. I was near forgetting to mention that on Friday my great-nephew Willie, Tom's son, was elected to a Fellowship at Sidney.

Correspondence with PROFESSOR CAYLEY.

The earlier letters of the following series belong to the period when Cayley lived in the Temple, and carried on active legal work as a conveyancer in addition to his mathematical pursuits, when the Modern Higher Algebra, as it was then called, or Theory of Algebraic Forms, was being evolved in concert with Sylvester and Salmon. The letters show Stokes as interested in the general Analytical Theory of Surfaces, and as in his way the equal of Cayley in unravelling their singularities. It is possible that, like his other purely mathematical work, the interest grew out of physical problems, furnished in this case by the theory of the Optical Wave Surface for biaxial crystals, and the phenomena of conical refraction. In turn Cayley gave his attention in a characteristic manner to the theory of Colour Vision. The occasional playful *persiflage* serves to accentuate the perfect understanding which obtained between these great representatives of abstract and physical mathematics. After 1872 Cayley and Stokes were intimate colleagues in Cambridge.

PEMBROKE COLLEGE, CAMBRIDGE,
Oct. 29, 1849.

DEAR CAYLEY,

I had a letter from Thomson to-day, in which he asks me to undeceive you about the Professorship. I beg accordingly to state that it is I and not my brother of Caius who am Lucasian Professor. That the *Times* should have made the mistake is nothing remarkable, but that the same thing should be done in the *Cambridge Chronicle* five days after the election, and with all the particulars of my brother's name, degree, &c. is beyond the beyonds. In fact two Cambridge papers out of the three would have it that the Rev. W. H. Stokes, Senior Fellow of Caius

College, was the Professor. My brother has had lots of joking congratulations.

Thomson and I are at present writing to each other about potentials. I think that potentials may throw light on the interpretation of $f(x + \sqrt{-1}y)$. How horrible you would think it to prove, even in one's own mind, a proposition in pure mathematics by means of physics.

Yours very truly
G. G. STOKES.

2, STONE BUILDINGS,
2 March, 1854.

DEAR STOKES,

The memoir you refer to is I suppose the 'Disquisitiones generales circa superficies curvas,' *Comm. Gott. recent.* t. VI. You will find the memoir reprinted in the fifth edition (by Liouville) of Monge's *Application de l'Analyse à la Géométrie*, and some of Liouville's notes to the work are a good deal connected with the subject. There is a memoir of Bertrand's or Bonnet's, I am not sure whether it is the one 'Démonstrations de quelques theorèmes sur les surfaces orthogonales' Cah. 29 or a memoir in some other recent volume of the *Journal Polytechnique*, which is a very valuable one, and a great many scattered memoirs in Crelle and Liouville. Believe me, yours sincerely,

A. CAYLEY.

17 Oct. 1855.

I am certainly wrong, and I believe you are to a considerable extent right about the surface*—but I am further than ever from seeing my way to the complete discussion of it. I find, even assuming $a^2 > 2b^2$, $b^2 > 2c^2$ it is still necessary to distinguish the two cases $a^2 + c^2 > 3b^2$, $a^2 + c^2 < 3b^2$ and the critical case $a^2 + c^2 = 3b^2$, which probably correspond to very distinct forms of the surface. I am deeply indebted to you for the suggestion of the problem.

---

* The subject of this correspondence was resumed some years later by Prof. Cayley in a memoir "Sur la surface qui est l'enveloppe des plans conduits par les points d'un ellipsoide perpendiculairement aux rayons menés par le centre," *Ann. di Tortolini*, II. (1859), reprinted in *Coll. Math. Papers*, IV. pp. 123—133, in which he remarks that the consideration of the surface was suggested to him some years before by Prof. Stokes. It appears not unlikely that it presented itself as a possible optical wave-surface, and that he was in quest of the singularities of refraction (conical, etc.) associated with it. See Prof. Stokes' letter *infra* of Nov. 10, 1855; also *supra*, p. 122.

69, Albert Street, Regent's Park,
Oct. 19, 1855.

My dear Cayley,

Next Thursday is to be the meeting of the Council, and if you can have what you were going to write about Sir W. Hamilton ready before that time I should feel obliged to you. Do not give yourself much trouble about it, for it will probably come to nothing for the reason I mentioned.

I have not studied the surface since I worked at it just at the end of the E. I.* examination, but I think I have a pretty clear notion of it, except the way in which the degenerate conical points get eaten up when the surface passes towards the form which it assumes when $b = c$ and $a^2 > 2c^2$, which remains to be considered. I got implicitly and in a symmetrical form (symmetrical with regard to a, b, c) the equation of the cuspidal edge, but had not discussed it, and saw how to get the cusps of the cuspidal edge, when I left it. I find lecturing is nervous work, and I cannot settle down regularly to anything while my lectures are going on.

Although the actual cases are numerous, I don't think the transitions which need be considered are very numerous; on the contrary I think you will find they are very few. The transitions at remote parts of the surface may be considered independently. Thus suppose there was a change of form at one place when some quantity A changed from being greater than to less than B, and another change of form elsewhere when C changed from $> D$ to $< D$. We should only have *two* transitions to consider, but there would be altogether four *cases*, namely

$$\left.\begin{matrix} A > B \\ C > D \end{matrix}\right\} \quad \left.\begin{matrix} A < B \\ C > D \end{matrix}\right\} \quad \left.\begin{matrix} A > B \\ C < D \end{matrix}\right\} \quad \left.\begin{matrix} A < B \\ C < D \end{matrix}\right\}.$$

I certainly got the double lines not conic sections, but curves closed like an ellipse or open like a hyperbola according to circumstances; but I do not know whether I looked over my work, and there may be a mistake in the work. I have not got the investigation here, it is at Cambridge. For the plane yz I found (at least I have the equation of what I *think* was the curve I mean)

$$\left(\frac{y^2}{a^2 - b^2} + \frac{z^2}{a^2 - c^2}\right)^2 = \frac{4}{a^2}\left(\frac{b^2 y^2}{a^2 - b^2} + \frac{c^2 z^2}{a^2 - c^2}\right).$$

* East Indian Civil Service.

I think you will find that that part of the cuspidal edge which belongs to the "crest," when it is so young as not to have changed its form from what it had at first, has a form like

(with two cusps) when looked down on from the axis of z. There will be a pair of these, symmetrical with respect to the plane xy. If while a^2 remains $> 2c^2$, b tends to become equal to a, the cusps run round, meet in pairs in the axis of y, and then disappear; the cuspidal edge (this portion of it) changing to a pair of closed cuspless curves running generally more or less nearly parallel to the plane yz, and situated one on each side of it.

The portion of the cuspidal edge which corresponds to the four degenerate conical points is, I think, like four little evolutes of an ellipse, only of double curvature. I have, I think, a clear notion of the degenerate conical points, but I have got to study the form which the surface assumes when, a^2 being greater than $2c^2$, b approaches to c.

By degenerate conical points, I do not mean points, but a finite portion of the surface which, taken altogether, replaces a conical point.

P.S.—Make a plane pass always through the axis of x, and cut the sheet on which the crest lies (the crest being sufficiently young) and pass first outside the crest, then through the cusp of the cuspidal edge, and lastly cut the crest. The sections I imagine will be like

For the section through the cusp the branches meet at an angle of 180°, but the curvature at the point of meeting is infinite.

Oct. 22, 1855.

I said, trusting to memory, that the double line in one of the principal planes might open out like a hyperbola, but I thought afterwards that that would give the curvature in the wrong direction; and I find on re-tracing the curve what I had got before when I was working at it, that in the plane xz it is a figure of 8 curve, like a lemniscate, while in the planes xy and yz it is a closed oval.

If it were not for the strength of your assertion I should say you were not right about the lines being conic sections. I back my curves against conic sections, though I have not my investigation by me to look over.

You get the equation for the plane xz from the equation I gave, namely

$$\left(\frac{y^2}{a^2-b^2}+\frac{z^2}{a^2-c^2}\right)^2=\frac{4}{a^2}\left(\frac{b^2y^2}{a^2-b^2}+\frac{c^2z^2}{a^2-c^2}\right),$$

by advancing one stage in the order $abca...$ and $xyzx....$

These equations and the equations of the plane curves of contact with the tangent plane drop out as matters of interpretation with hardly any work; but my method does not give the double lines which lie out of the principal planes, although it does give the cuspidal edge.

I feel almost sure that when the crest is formed, and from thence on till the next transition, the cuspidal edge has 20 real cusps and probably many more imaginary ones. I think it probable that the total number, real and imaginary, may be 48.

P.S.—I have just thought of the way of getting the rest of the double lines, at least I think so.

Oct. 22, 1855.

DEAR STOKES,

I enclose a note on Quaternions; I have gone on working at the surface, in tracing which I had as I mentioned to you in my last note made some mistakes, but I do not find either any double points or cusps upon the cuspidal line. I have written out the whole investigation and made a carefully drawn figure of the principal sections and nodal and cuspidal lines, and I think I now see pretty well the form of the surface in the case which I have considered $a^2 > 2b^2$, $b^2 > 2c^2$, $a^2 + c^2 > 3b^2$. I shall be very glad to compare results whenever it is convenient to you.

Believe me, Yours very sincerely,

A. CAYLEY.

Mr Cayley's account of SIR W. HAMILTON'S Quaternions.

Sir W. R. Hamilton's quaternion imaginaries are a set of symbols subject to laws of combination different from those of the ordinary algebraical symbols. Definitions in analysis as in any other science are not to be considered as arbitrary; they must satisfy the condition of utility as regards the science to which they belong, i.e. they must be such as to admit of being made the foundation of a system of dependent truths the development of which forms part and extends the limits of the science, and the interest of such resulting theory is a test of the value of the definition. The analytical theory of Quaternions is an eminently interesting and beautiful one, and the beauty is heightened by the singularity of the subject matter, viz. symbols subject to laws of combination different from the laws with which mathematicians have hitherto been concerned. The value of the theory, as a new idea in Analysis, is very great indeed; but the problem which the originator proposed to himself, was, it would appear, to find an analytical theory applicable to lines in space as they occur in geometrical and mechanical questions, and there can be no doubt but that the solution of the problem is contained in the theory of Quaternions, and that this theory does afford the appropriate means for the investigation of such questions—some of the most important and difficult of which are discussed in detail in Sir W. R. Hamilton's work. Both as a part of Analysis, and on account of its applications to Geometry and Mechanics, the theory is one which cannot fail to add fresh lustre to the name of the Mathematician whose researches have already contributed so much to the recent progress of Analysis.

69, Albert St, Regent's Park,

Oct. 24, 1855.

My dear Cayley,

I am greatly obliged to you for your paper on the merits of quaternions. I called on you to-day but you were not at home.

I cannot well compare notes about the surface, for all my notes are at Cambridge. I left them, I believe, partly for fear they might draw me off from my proper work, although it is queer that fiddling with an Atwood's machine should be more proper work for a professor of mathematics than discussing a noble surface.   Yet so it is, for I am an amphibious animal with an ambiguous character, at present, a sort of Jack-of-all-trades and master of none.

The equations of the double lines (I prefer the term nodal edges) which I gave were the result of course of an investigation, but I did not look over it, at least if I did I do not know, and I may have made a mistake; but the existence of cusps in the cuspidal edge was a guess, founded however on a potential analysis. I suppose from your words that you have actually made the investigation.

I should like to finish my investigation, when I have time, in my own way before we regularly compare notes.

Nov. 10*th*, 1855.

I write to remind you that when we formerly talked about the surface you proposed to assume the wave surface of Fresnel as known, and find the envelope of planes drawn at the several points in directions perpendicular to the rad. vectors at these points. You no doubt forgot what you had proposed to do and generalized the plane problem in a different manner.

Dec. 5*th*, 1855.

I have looked again at my own solution.   Whether I contented myself with seeing that the equation I spoke of had equal roots or went a step further I do not now recollect; certain it is that the simplest possible differentiation shows that the hyperbola and principal section *do* touch.   I have taken your own expressions, differentiated them, and arrived at the same result.   The work is not quite so simple as mine, but simple enough.   You must have made some slip.   [*Analysis omitted.*]

So hurrah for the prow; what you said shook my confidence in its importance, but that confidence is now re-established.   It appears to be a point of ordinary occurrence, not a speciality of your or my surface.

For the point of section I find

$$x^2 = \frac{(a^2 + b^2 - c^2)^2 (a^2 - b^2)}{a^2 (a^2 - c^2)}, \qquad z^2 = \frac{(b^2 + c^2 - a^2)^2 (b^2 - c^2)}{c^2 (a^2 - c^2)}.$$

The section of the ellipse with the corresponding principal section is always imaginary, the contact is real when $a^2 > 2c^2 > b^2$.

P.S. The mere inspection of a subordinate parameter-figure shows that the cusp points are, as I supposed, not like an elephant's snout.

PEMBROKE COLLEGE, CAMBRIDGE,
April 26th, 1856.

In practising the methods which I applied first to my and then to your surface, by trying them in other cases, I pitched on a property of the wave surface which may interest you if it should be new to you. It would be an elegant one if the quantities involved were real; perhaps you don't much care about that*. It is this.

Besides the 4 circles of contact with the tangent plane, and the 8 imaginary circles formed from these by symmetry, there exist 4 other plane curves of contact, which are imaginary ellipses, which are defined by the intersection of a real ellipsoid with 4 imaginary planes through the centre. The ellipsoid spoken of has the same centre and principal planes as the wave surface, and the squares of its principal semi-axes are $\left(\dfrac{1}{2}\dfrac{1}{b^2} + \dfrac{1}{c^2}\right)^{-1}$ &c. It passes through the real or imaginary conical points in the principal planes. Call the diameter joining two opposite conical points (real or imaginary) a conical diameter for the sake of a name. Take each of the 4 combinations of one conical diameter from a first and one from a second principal plane. The imaginary planes passing through the two diameters of each combination form the four planes in question, and each intersects the third principal plane in a conical diameter. The intersection of any one of these planes with the wave surface is of course a pair of coincident imaginary ellipses. I should be glad to know if the above property is new to you.

(Undated fragment: probably later.)

...in a general way to those of a quadric surface, though (in the general case) without the symmetry, according as the roots of a certain cubic are 3 real or 1 real, is new to me, but that it may well be without being new to the world. I must either give it or refer to it in the paper you encouraged me to

---

* The curves of contact are all real on Prof. Cayley's tetrahedroid (Liouville, 1846; Coll. Math. Papers, Vol. I.; Salmon, Solid Geometry, §§ 573—4) which is a homographic projection of the wave-surface.

write about the lines of curvature of the wave surface. If old I should be obliged for the reference. I have not found your paper in the *Phil. Mag.*, but Vols. 33, 35, 39, 43 were taken out of the U. Library, and I mean to look in the Library of the C. P. S.

May 1*st*, 1856.

Let $f(x, y)$ be a rational integral function of x and y. Is there any general method of determining whether $f(x, y)$ can or cannot be split into factors so that

$$f(x, y) = \phi(x, y) \, \psi(x, y),$$

where ϕ, ψ are rational as regards x and y?

I have been working again at surfaces notwithstanding my lectures. I have satisfied myself about cuspidal edges, cusps of cuspidal edges, conical points, and curves of plane contact, and also about multiple points in curves, but I have still something to do about nodal edges. Neither multiple points nor nodal edges appear as singularities in my mode of treatment.

I think it very likely that you are aware of the property I mentioned about the wave surface from having discussed the corresponding imaginary conical points at infinity in the polar reciprocal wave surface, though I found it out in working out the lines of contrary flexure, without requiring to know anything about the polar reciprocal.

The caustic surface is of the 12th degree. I see how to get the equation in this and similar cases without direct elimination. [*Sentence deleted here.*]

I see I was going to tell you what was not true, so it is time to stop.

June 21, 1856.

I find for the leading terms of the equation of the caustic surface

$$(a^2 x^2 + b^2 y^2 + c^2 z^2)^3$$
$$\times \{[a^2 (b^2 - c^2)^2 x^2 + b^2 (c^2 - a^2)^2 y^2 + c^2 (a^2 - b^2)^2 z^2]^3$$
$$- 27 a^2 b^2 c^2 (b^2 - c^2)^2 (c^2 - a^2)^2 (a^2 - b^2)^2 x^2 y^2 z^2\},$$

where a, b, c are the half axes of the generating ellipsoid*. This was obtained by a simple elimination applicable only to infinite values of x, y, z.

---

* Perhaps following upon Cayley's "A Memoir upon Caustics," *Phil. Trans.*, May 8, 1856, *Coll. Math. Papers*, II. pp. 336–380, which treats mainly of caustics of circles.

2, STONE BUILDINGS,
26*th Mar.* 1857.

DEAR STOKES,

The following equation is true in some sense or other[*]:

$$\int_b^a e^{x^2} dx = \tfrac{1}{2} e^{a^2} \left(\frac{1}{a} + \frac{1}{2} \frac{1}{a^3} + \frac{1 \cdot 3}{2^2} \frac{1}{a^5} + \cdots \right)$$

$$- \tfrac{1}{2} e^{b^2} \left(\frac{1}{b} + \frac{1}{2} \frac{1}{b^3} + \frac{1 \cdot 3}{2^2} \frac{1}{b^5} + \cdots \right) \quad \ldots\ldots\ldots(1).$$

Write ib for b, $(i = \sqrt{-1})$ and then assume $b = \infty$, we have

$$\int_{ib}^a e^{x^2} dx = \tfrac{1}{2} e^{a^2} \left(\frac{1}{a} + \frac{1}{2} \frac{1}{a^3} + \&\mathrm{c.} \right),$$

the second series vanishing on account of the evanescent factor e^{-b^2}. We have then

$$\left(\int_{ib}^a + \int_0^{ib} + \int_a^0 \right) e^{x^2} dx = \int_a^a e^{x^2} dx,$$

or

$$\left(\int_{ib}^a + \int_0^{ib} - \int_0^a \right) e^{x^2} dx = \int_a^a e^{x^2} dx,$$

where $\int_a^a e^{x^2} dx$ is not of necessity zero; if it were we should have

$$\int_0^a e^{x^2} dx = \int_{ib}^a e^{x^2} dx + \int_0^{ib} e^{x^2} dx = \int_{ib}^a e^{x^2} dx + i \int_0^b e^{-y^2} dy$$

$$= \tfrac{1}{2} e^{a^2} \left(\frac{1}{a} + \frac{1}{2} \frac{1}{a^3} + \&\mathrm{c.} \right) + \tfrac{1}{2} i \sqrt{\pi} ;$$

but retaining the term we have

$$\int_0^a e^{x^2} dx = \tfrac{1}{2} e^{a^2} \left(\frac{1}{a} + \frac{1}{2a^3} + \cdots \right) + \tfrac{1}{2} i \sqrt{\pi} - \int_a^a e^{x^2} dx.$$

The difficulties are to investigate under what conditions and in what sense the assumed equation (1) is true; and to find the meaning in the particular case of the symbol $\int_a^a e^{x^2} dx$.

Believe me yours sincerely,

A. CAYLEY.

---

[*] Cf. Stokes, *Trans. C.P.S.* May 11, 1857, reprinted in *Math. and Phys. Papers*, Vol. IV. (p. 81), for the explanation, on the lines first opened up by Cauchy.

69, Albert St, Regent's Park, N.W.
June 15/57.

My dear Cayley,

I believe I told you one thing wrong. Matches to the colour-blind would not generally be matches to the normal-eyed, but matches to the normal-eyed *ought* also to be matches to the colour-blind.

If x, y, z be the 3 standard colours chosen, and ξ, η, ζ the elementary sensations (supposed), and

$$x = U\xi + U'\eta + U''\zeta,$$
$$y = V\xi + V'\eta + V''\zeta,$$
$$z = W\xi + W'\eta + W''\zeta,$$

and for any colour q

$$q = l\xi + m\eta + n\zeta;$$

then if a normal-eyed person judges that

$$q = Ax + By + Cz,$$

where " $=$ " means " matches," it is necessary that

$$l = AU + BV + CW \quad \dots\dots\dots\dots(1),$$
$$m = AU' + BV' + CW' \quad \dots\dots\dots\dots(2),$$
$$n = AU'' + BV'' + CW'' \quad \dots\dots\dots\dots(3).$$

But if to a colour-blind person the sensation ζ is wanting while ξ and η are perceived it is sufficient that (1) and (2) be satisfied without (3).

Dear Stokes,

The question [*] seems to be, do two colours which match to one colour-blind person match also to another—IF NOT, you may define ξ, η, ζ as the sensations which are respectively wanting to three colour-blind persons P, Q, R, and you may then obtain numerical values for the ratios of U, U' &c., i.e. you may define your sensations of colour in terms of the blindnesses of three other people—which is a very pretty problem for a physicist. In fact if to a colour-blind person (who has not the sensation ξ)

$$Ax + By + Cz = A_1x + B_1y + C_1z,$$

---

[*] Cf. generally Maxwell's letter to Prof. Stokes of 27 Jan. 1857, and the subsequent letters, *infra*, Vol. ii. p. 4 *seq.* See also correspondence with Lord Rayleigh, *infra*, Vol. ii. p. 121 : and letters to Dr W. Pole, *supra*, p. 253.

then
$$(A - A_1)\, U' + (B - B_1)\, V' + (C - C_1)\, W' = 0,$$
$$(A - A_1)\, U'' + (B - B_1)\, V'' + (C - C_1)\, W'' = 0,$$

i.e. $V'W'' - V''W' : W'U'' - W''U' : U'V'' - U''V'$
$$= A - A_1 : B - B_1 : C - C_1 ;$$

and if you take two other people blind to η and ζ you have

$V''W - VW'' : W''U - WU'' : U''V - UV''$
$$= A - A_2 : B - B_2 : C - C_2,$$

$VW' - V'W : WU' - W'U : UV' - U'V$
$$= A - A_3 : B - B_3 : C - C_3,$$

whence
$$\begin{matrix} U, & V, & W \\ U', & V', & W' \\ U'', & V'', & W'' \end{matrix}$$

are proportional to the first minors of the determinant

$$\begin{vmatrix} A - A_1, & B - B_1, & C - C_1 \\ A - A_2, & B - B_2, & C - C_2 \\ A - A_3, & B - B_3, & C - C_3 \end{vmatrix}.$$

Thomson would probably solve the problem by the theory of Potentials.

Believe me, yours sincerely,

June 17, 1857.

Dear Cayley,

I am afraid your suggestion will not do physically, for all the colour-blind appear to agree, at least all whose cases have been accurately investigated. They appear to want the sensation of that tint purer than mortal ever saw which comes nearest to carmine red. To them this tint is equivalent to black.

Were different sensations wanting with different individuals, the physicists would have got hold of it; they would have determined it, but probably without determinants.

A propos to negative colour-sensations. Suppose your thumb placed in a vice, which is screwed tighter and tighter; you have the sensation of a pain greater and greater. Pain may be regarded as a quantity, though its measure by a unit pain may be difficult; but what idea have you of a negative pain?

March 17/58.

Can you recollect whether we came to any decision as to the date to be attached to the author's name: was it the date of reading or the date of the year *for which* the vol. of the *Transactions* was, or did we let the question stand over*? You would much oblige me by a line addressed to 1, Phœnix Place, Blackheath, S.E.

I have left Albert St for good.

LENSFIELD COTTAGE, CAMBRIDGE,
Oct. 23rd, 1858.

If you are fond of having problems suggested to you here is one which I have succeeded in solving†. It is to find the intensity of the light reflected from or transmitted through a polarizing pile of n plates, taking account of the defect of perfect transparency of the glass. As physics are physic to you here is the mathematical problem, or at least one to which it is readily reducible, containing only so much physics as facilitates the enunciation.

Light is incident on n parallel mathematical surfaces, by each of which it is partly reflected and partly transmitted, the quantities incident reflected and transmitted being say $:: 1 : r : t$; it is required to find the intensity of the light reflected from or transmitted through the n surfaces. I need not say that I suppose the light reflected backwards and forwards in every possible way.

The form of the solution is simpler in the case of perfect transparency, for which $r + t = 1$, than in the general case.

Dr Sharpey and I think we had best submit your table to the Council.

March 26th, 1859.

There is to be a meeting of the R. S. Library Committee on Wednesday. Could you make it convenient to meet me at the R. S. on Tuesday at 10 a.m. that we may see about drawing up a report about the progress of the Catalogue? Please address to the Athenæum.

---

* In the Report of the Brit. Assoc. for the Cambridge Meeting 1856, pp. 463–4, there is a Report signed by A. Cayley, R. Grant, and G. G. Stokes "of a Committee to consider the formation of a Catalogue of Philosophical Memoirs." This was the origin of the *Royal Society Catalogue*. The Report was drawn up in part by Prof. Cayley, and is reprinted among his *Collected Papers*, Vol. v. pp. 546–8, with a note added, p. 620, in reprinting.

† *Roy. Soc. Proc.* 1882; *Math. and Phys. Papers*, Vol. IV. 145–156.

Nov. 22, 1861.

Will you write a short account of Sylvester's doings especially within the last 10 years such as would be suitable for reading out when the medal is presented? Of course it must not be too technical. Your account is returned to you to assist you. Please address to Dr Sharpey.

CAMBRIDGE,

21*st March*, ...

DEAR STOKES,

Your differential equation $\dfrac{d^2y}{dx^2} + \dfrac{1}{x}\dfrac{dy}{dx} - \dfrac{n^2}{x^2}y = y$. Supp. to paper on discontinuity &c., *C. P. T.* t. XI. pt 2, p. 414*. On putting therein $x\sqrt{-1}$ for x becomes that of Bessel's I_k^h—so that the series $x^n\left(1 + \dfrac{x^2}{2(2+2n)} + \&c.\right)$, p. 414, is in fact tabulated pretty completely for the values $n = 0$, $n = 1$, and for values of x $0 \cdot 00 \sqrt{-1}$ to $3 \cdot 20 \sqrt{-1}$ in Bessel's paper "Ueber die planetärischen Störungen," *Berl. Abh.* 1824.

I merely mention this in case you should be doing anything with your equation and should happen to require numerical values for the pure imaginary values of x.

I was led to look at your paper by a note from Thomson in reference to the equation $\dfrac{d^2y}{dx^2} + \dfrac{1}{x}\dfrac{dy}{dx} + n^2y = 0$, the solutions of which he wishes to have tabulated.—Believe me, yours sincerely,

A. CAYLEY.

25*th May* [1872 ?].

The theorem is very pretty—it may be stated thus—we may have on equal circles A, B two quadrilaterals $MNM'N'$, $\mu\nu\mu'\nu'$ each the projection of the other in regard to the point O which is the mid-point of the common chord, and such that the corresponding sides meet in points P, P', Q, Q' which lie on the common chord. Believe me, yours very sincerely,

* *Math. and Phys. Papers*, Vol. IV. p. 298.

12*th June*, 1872.

Sylvester asks me to ascertain from you or Maxwell " Whether any or what explanation has ever been given of the loss of half an undulation which is assumed in the undulatory theory of Newton's Rings." I presume every dynamical theory attempts to give an explanation, but that there is not any which can be considered as a received and admitted one. Believe me, yours very sincerely,

27*th Sep.* 1872.

The formula* is a very interesting one—I find however some difficulty about it—in the equation

$$\int_0^x e^{x^2} dx = e^{x^2} \left\{ \frac{1}{2x} + \frac{1}{2^2 x^3} \cdots + \frac{1 \cdot 3 \cdot 5 \ldots 2n-3}{2^n x^{2n-1}} \right\}$$
$$+ \frac{1 \cdot 3 \cdot 5 \ldots 2n-1}{2^n x^{2n-1}} \left\{ \frac{1}{1-2n} + \frac{x^2}{1 \cdot 3 - 2n} + \&c. \ldots \right\},$$

developing on the right hand e^{x^2} in ascending powers and multi-plying out, the negative powers would disappear and the result be identically the same as if on the left hand e^{x^2} had been expanded in ascending powers of x^2, and the series integrated term by term—i.e. the function on the right hand should vanish for $x = 0$—but this is unverifiable, viz. writing $x = 0$ we introduce infinite terms.

In the equation

$$\frac{n}{1} + \frac{n \cdot n - 1}{3x^2} + \ldots = \frac{x^2}{1} + \frac{x^4}{3(n+1)} + \ldots,$$

x^2 nearly $= n$ the convergence is slow on both sides, and the true value of x^2 might, it is conceivable, differ considerably from n: it would be worth while to calculate a few corresponding values of n and x.

Glaisher, when he has done the Bessel functions, is inclined to take up a table of the Jacobian Θ and H functions†. Believe me, dear Stokes, yours very sincerely,

---

* Cf. *supra*, p. 390.

† Of the various reports of the Brit. Assoc. Committee on Mathematical Tables, that of 1875, mainly on tables connected with Theory of Numbers, is by Prof. Cayley (*Coll. Math. Papers*, IX. pp. 461–490). The Committee then consisted of A. Cayley, G. G. Stokes, W. Thomson, H. J. S. Smith, J. W. L. Glaisher.

8*th March* [1873 ?].

You were quite right about the cuspidal lines—besides the ellipses in the principal planes and an imaginary conic at infinity—there are 8 imaginary conics (I think I said by mistake 12) which are the loci of the centres of curvature for the points on the ellipsoid which lie on the imaginary lines through the umbilici; each of the 8 conics touches each of the principal planes, and also the plane infinity—viz. it is a parabola. I have not in the Memoir* used the expression, and did not call the property to mind yesterday evening.

I wrote to Dr Sharpey about the Memoir of General Thompson. I suppose they will hear from him whether there is anything to be paid for the separate copies. Believe me, dear Stokes, yours sincerely,

14*th Nov.*

I thought I had seen something about the number of umbilici —but have just found it by chance—Voss, *Math. Annalen*, t. 9 (1875), p. 241 gives the number for a surface of the nth order as $= n(10n^2 - 28n + 22)$, instead of Salmon's $n(10n^2 - 25n + 16)$, viz. Salmon's number is reduced by $12(n-2)$, and in the case $n = 4$ it is $= 280$ instead of 304. The only use of this to you, will probably be that you will have the correct general expression to refer to. Believe me, yours very sincerely,

26*th Sep.* 1874.

Best thanks. I send the dissertation herewith.

As to the equation $\dfrac{d}{dx}\left(x\dfrac{dy}{dx}\right) + (x-a)y = 0$†, leading to the equation of differences $n^2 u_n = a u_{n-1} - u_{n-2}$; or putting $1/n^2 = l_n$, this is $u_n = l_n(a u_{n-1} - u_{n-2})$, the solution which I gave was in fact a solution of this, l_n meaning any function whatever of n—the solution of the perfectly general equation of differences

$$u_n = a_{n-1}u_{n-1} + b_{n-2}u_{n-2},$$

contains precisely the same number of terms and is of a very similar form—e.g. we have

$$u_6 = 54321 u_1 + 5432 \, . \, b_0 u_0,$$

54321 has a leading term $a_5 a_4 a_3 a_2 a_1$,

---

* Possibly "On the Centro-Surface of an Ellipsoid," *Camb. Phil. Trans.* 1873, *Coll. Math. Papers*, VIII. pp. 316–365.

† A note was promised for *Messenger of Mathematics*.

terms derived from this by changing any pair $a_2 a_1$ into b_1, $a_3 a_2$ into b_2, &c.

terms derived by changing any two pairs $a_4 a_3$, $a_2 a_1$ into b_3, b_2, &c.,

and so on—where the expression a " pair " denotes the product of two consecutive letters,

and so 5432 has the leading term $a_5 a_4 a_3 a_2$, &c.

Believe me, yours very sincerely,

13*th* *Oct.* 1876.

MY DEAR CAYLEY,

I write to thank you for your elementary treatise on elliptic functions which I have been dipping into. It tempts me to make myself master of the subject, which I regret to say I have not hitherto done.

CAMBRIDGE,

30*th* *Sep.* 1880.

DEAR STOKES,

Best thanks for the Mathematical and Physical Papers. I think the *solidus* looks very well indeed and is really a great improvement: it would give you a strong claim to be President of a Society for the prevention of Cruelty to Printers. I ran through the volume to see whether you had adopted exp. x instead of e^x: you do not seem to have any exponents complicated enough to make this necessary or even advantageous. I think you do not approve of cosh and sinh.

Believe me, yours very sincerely,

A. CAYLEY.

CORRESPONDENCE WITH SIR J. NORMAN LOCKYER,
K.C.B., F.R.S.

A collection of letters written by Prof. Stokes has been sent by
Sir Norman Lockyer just in time for a selection to be included
here. They relate mainly to the work of the Solar Physics
Committee, of which Prof. Stokes was an active member.

LENSFIELD COTTAGE, CAMBRIDGE,
22nd Nov. 1872.

MY DEAR LOCKYER,

On Thursday I found your paper at the Royal Society
and read it with much interest.

I have a few criticisms to make.

I think the language in which you describe your method will
require some modification. It reads like hailing as a new discovery
what is already perfectly well known. I allude to the localization
of the metallic lines at the neighbourhood of the electrodes.

I need only refer to the papers by the late Professor Miller
and myself in the *Phil. Trans.* for 1862. I send you a copy of the
latter by book post. I must explain that the figure at p. 606 is so
far diagrammatic that it is intended to represent the *positions* only
of the metallic lines; their confinement to the immediate neighbourhood of the electrodes, especially in the rays of the highest
refrangibility, is described in words. No one who has looked at
the plain spark, without any slit, with a prism and telescope
or even with a prism and the naked eye, no one at least aware of
what Ångström long ago stated, can have failed to notice the
metallic tufts of light at the electrodes, while the air-spectrum
extends right across, and does not vary from metal to metal.

Miller used a slit close to, but of course not touching, the
electrodes. I used the spectral image of the spark itself, without
a slit, as in working by fluorescence there is no intensity to spare.
Consequently I was unable to resolve the band-lines (I mean very
narrow bands) of cadmium, &c., into the fine and very close lines
of which they were made up. Miller's figures represent very
nearly what I saw on the phosphorescent screen, the difference
being that Miller's spectra were a little purer, while mine were

strictly in focus for the spark, whereas Miller's were in focus for the slit and *slightly* out for the spark.

I am nearly sure I lately read or heard of some observations where the observer formed an enlarged image of the spark and examined that piecemeal. I think French was the language, but whether I read it or heard it in conversation with Cornu or Croullebois I can't recollect. I will give you the reference if I come across it. Be that as it may, I am not aware that anyone before you has studied the spark in relation to the variation of length of the metallic lines.

Now as to the more important matter of the substance of the paper. I confess I see no evidence that for a given temperature the spectral appearance depends on *density* of active vapour present rather than *quantity*. What we know of absorption leads me strongly to believe that it depends on *quantity* rather than *density*. Suppose we had two flames of the same temperature, but one ten times as thick, in the direction the light travels, as the other, and suppose the smaller flame had ten times as much sodium vapour in it as the larger; I expect that as to broadening the two *D* lines the poorer flame would make up by its size for its want of richness.

The only evidence I can see you bring forward in support of the opinion that the widening depends on density rather than quantity, is that of the observation occasionally of a narrow bright line in the middle of the widened dark line of a solar spot. But the validity of the inference depends on the assumption that the quantity of heated gas in the prominence is practically sufficient to give out all the light that gas of that density and temperature is capable of giving out, which is by no means clear. If a thickness of 1000 miles of incandescent and highly rarefied hydrogen gave a bright *F* little widened, I can quite conceive that 20,000 miles at the same density and temperature would give a bright *F* very sensibly widened.

I tried last night roughly the experiment I suggested to you— that of seeing if reduction of the light would of itself alone narrow a widened bright line. My dispersion was not sufficient (I used the *D* lines) to allow me to say positively, but certainly as well as I could make out there was no such tendency as I had thought of to narrow the line. In thinking of the cause I was led to rectify (as I believe) my former ideas, and to recede from those of Plücker *towards* yours; but I have not at present gone the whole way, and I have not yet seen evidence that widening depends on density rather than quantity.

Is there a meeting of the Science Commission on Wednesday? The notice mentions Tuesday only, but I thought we were to have discussed the University position on Wednesday. I should rather like to know before I leave home (which will be after the delivery of the morning post) on Tuesday.

23 *Dec.* 1874.

In your conversation with me you spoke of obtaining photographs of the corona, by which I thought you meant of the corona as a whole, so as to get its shape as seen from different places, but I see from your letter you mean the spectrum of the corona.

As to your questions, I had better mention principles, and then you will be able to supply the rest.

Let AZ be the solar spectrum, A, H being the lines so named and Z the furthest limit.

Glass is opaque for the more refrangible rays, and with a glass train you only get about as far as g. The absorption does not come on at all sharply, so that according as you use thin or thick glass, &c., you get a *little* further or less far.

To reap however the advantage of quartz in one part you must have quartz throughout, and quartz is of but low dispersive power.

Speculum metal reflects the invisible as fully as the visible rays. Silver do. *till you get about to s*, where there is a rapid falling off, and from a point about s to z it is hardly more reflective than glass. I know no other metal which has this curious property.

It does not seem *probable* that at a distance from the sun the action will be intense enough to develop a spectrum as long as that of the sun itself. It is possible however that the solar spectrum would extend beyond Z were it not for absorption by the earth's atmosphere. The appearances however are against the supposition that our atmosphere, when the sun is high and the sky clear, exerts any *great* absorbing action on the high rays.

Silvered glass has a great advantage over speculum metal for a siderostat in point of lightness, and it is also more easily got plane. It is very improbable that the small region sz which is sacrificed by employing silver *if we use quartz, or at any rate non-glass, elsewhere* (for otherwise it is gone in any case), will signify :—

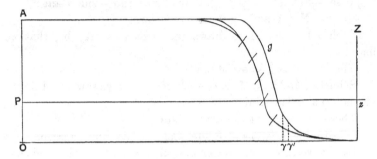

The curve of intensity (as compared with the original intensity) after absorption by a certain thickness of glass, may be like that of the preceding figure*. Suppose OP, the ordinate of Pz, is the least that will impress itself in the photograph. Then we can photograph as far as γ. Doubling the area of the objective, other circumstances being the same, will only carry us from γ to γ'.

If we use a quartz prism we must be content with low dispersion, and if we use a glass prism we may as well use glass elsewhere and silver for reflection.

22 *Jan.* 1875.

I cannot *promise* to come to the meeting of the Eclipse Committee if it be held on Tuesday, as I have a difficult examination paper to get ready which *must* be done and which I have not yet begun. I can only say I will come if I can if Tuesday should be the day fixed on.

25/1/75.

I wrote to you this morning that Main can lend a 10 in. and 10 ft. f.l. Newtonian reflector from the Radcliffe Observatory.

We must take care to marry the observers to their instruments. Bachelors and spinsters are no good, and I would not trust to weddings contracted on the voyage.

---

* The dotted line over γ should meet Pz where the curve crosses it. The lower slanting curve is marked as cancelled.

As you can't come on Thursday could you send a list drawn up thus

man   object   instrument

Suppose :—Schuster—photography of spectrum—siderostat :

No. 1 and Lockyer—reflector.

Such a list should be drawn up as soon as may be, that we may know where we are.

The observers I know of are

Schuster, Meldola, Mr B (the photographer we saw, but I forget his name) and Tacchini.

Those offered that I know of are

Winstanley, a fellow of Exeter Coll. Oxford, and Janssen.

Unless we can reckon on making the expedition fairly efficient I think we are bound to give it up even now. I mean as soon as it is ascertained, should such prove to be the case, that it could not be more than semi-efficient.

2nd Feb. 1875.

You wanted to consult me on some point, but had to go before you could explain what. As time presses I will throw out some considerations on chance of something being useful. With some of them, perhaps all, you are already familiar.

I don't know whether you contemplate using two lenses with a quartz prism, a collimating lens to make the rays parallel as they pass through the prism and then what I may call an image-lens, or one only with the slit and collodion plate in conjugate foci.

If one only, and if the two focal distances are chosen very unequal, it is well to place the prism on that side of the lens which faces the further focus, so that the rays have an approximate parallelism in passing through the prism.

Pour fixer les idées I will suppose that you use two lenses, collimating, of focal length u, image, of f.l. v, but what I say will apply to one if u, v be the distances of the conjugate foci.

If α be the angular aperture of the mirror, or breadth ÷ focal length, the illumination of the aperture of the slit varies as α^2. This supposes that the collimating lens and the prisms are wide enough to take in the whole pencil. The larger the prisms the larger we can afford to make u consistently with taking in the whole pencil.

Let D be the angular dispersion, b the breadth of the slit; then
Length of spectrum $\propto D\,v$.

Breadth of image of slit $\propto b\,v/u$.

Purity of spectrum $\propto \dfrac{\text{length of spectrum}}{\text{breath of image of slit}}$.

$\propto D\,\dfrac{u}{b}$ or $\dfrac{\text{dispersion}}{\text{angular breadth of slit}}$.

Intensity of *continuous* spectrum $\propto \alpha^2 . \dfrac{1}{D} . b\left(\dfrac{u}{v}\right)^2$.

Intensity of *bright-line* spectrum $\propto \alpha^2 \left(\dfrac{u}{v}\right)^2$.

This supposes that the image of the slit is already as broad as it can be made without confusing the spectrum. Beyond this an increase of breadth of the slit would merely widen the lines without intensifying them.

When there is no intensity to spare there is a great advantage in getting up length of spectrum by dispersion rather than long focus of image lens. Thus, if we want to double the length, if we do so by doubling D we get in a bright-line spectrum double length *and double purity* with *equal* intensity (neglecting the loss by reflection in two prisms instead of one), whereas if we get up the length by doubling the focal length v, we get the *same* purity and only *one quarter* the intensity. If we have plenty of intensity to play with this does not much matter, but if it be feeble it matters a great deal*.

2/2/75.

Except as to compactness, there appears to be no disadvantage in a smaller ratio than you contemplate of aperture to focal length of the mirror. The intensity of a bright-line spectrum I gave varying as $\alpha^2 (u/v)^2$. Now u is limited by the size of the prism; we may go on increasing u till the prism is filled. Hence $(\alpha\,u)^2$ may be taken to represent the foreshortened area (A) of the prism, so that intensity $\propto A/v^2$ and a slenderer α is made up for by choosing a longer u. To keep the same purity the breadth b should be altered on the same scale as u. So that ultimately our intensity depends on the size of the prisms (with lens or lenses to match) that we can commend. With a siderostat the prism part being fixed a longer u is of no moment.

* The laws of purity of spectra were established in general terms by Lord Rayleigh, *Phil. Mag.* 1879—80: see also Smith's Prize Question, 1871, in *Math. and Phys. Papers*, Vol. v. p. 354.

8/2/75.

I should anticipate rather the *opposite* result from the shortness of the solar spectrum compared with that of the electric arc. The hotter an electric current makes *a wire* the longer the spectrum *and at the same time* the brighter a part of the spectrum of low refrangibility. The curves of intensity would be something like

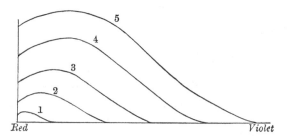

1, 2, 3, 4, 5, curves for increasing temperature.

I would suggest that the *focussing* rehearsals be done *also* with full aperture. For as the focussing is much sharper with full aperture, the observers if they practised only with the 1 inch might learn to be slovenly.

In haste as I have Smith's Prize papers to look over.

SPRINGFIELD, TIVERTON,
9th July, 1875.

My early life was an uneventful one, and I have not much to record. I was born 13 Aug. 1819, at Skreen, a country place about 12 miles south of Sligo in Ireland. My father was rector of the parish of Skreen. I remained at home till I was about 13 years old, when I was sent to Dublin, where I attended the school of the Rev. Dr Wall. In 1835 I was removed to the Bristol College, of which Dr Jerrard was principal. In 1837 I entered Pembroke College, Cambridge, went in for mathematical honours at the normal time, viz. January, 1841, and came out senior wrangler. I was immediately elected Fellow of my College, and held my fellowship till it was vacated by marriage in 1857. A few years ago I was re-elected under a new statute.

You are welcome to make use of any of the above facts which you think of sufficient interest*. I should rather like my two schools to be mentioned, because an account of me has got into print which is utterly false in this respect.

* A notice of Prof. Stokes, by P. G. Tait, illustrated by an engraved portrait by C. H. Jeans, appeared in *Nature*, July 15, 1875.

CAMBRIDGE,
24 *Oct.* 1877.

Perhaps you might like to have the formula giving the deviation by reflection from a grating in terms of ϵ, the interval from line to line of the ruled scratches, and λ the wave-length.

If i be the angle of incidence, and r the angle of reflection for the spectrum of the nth order on the side *from* the normal (*i.e.* so that $r > i$) then

$$\epsilon (\sin r - \sin i) = n\lambda. \qquad (1)$$

The same formula will apply to the other side, *i.e.* from the spectra *towards* the normal, if we make n negative. For example, for the second spectrum on the side *towards* the normal we should have

$$\epsilon (\sin i - \sin r) = 2\lambda.$$

We have from (1) for *small* corresponding changes of r and λ, δr and $\delta\lambda$,

$$\epsilon \cos r \,.\, \delta r = n\delta\lambda. \qquad (2)$$

In consequence of the variation of $\cos r$, the scale is not exactly the same from end to end of the spectrum, nor is the scale in the 2nd, 3rd... spectrum exactly twice, three times... that in the first.

The scale is practically constant for the small portion photographed at once*; and if we wish to preserve the same scale in passing to another point of the spectrum the formula (2) shows that we can do so by letting camera and grating be shifted together, so as to alter the angle of incidence instead of the angle of reflection. Moreover in this way the scale for the 2nd, 3rd.. spectra would be just twice, three times... the scale for the first. I did not notice this result until after I had begun this letter.

* That depends of course on the field of the camera : perhaps it may be too great to allow us to say that. By calculating roughly from formula (1) or by measuring roughly the values of r you work with, and using a table of cosines, you can see at once what variation of scale you are liable to.

20 *Jan.* 1879.

The excess of Na vapour is not absolutely proved by the greater conspicuousness of the D line, nor by the enfeeblement of the K lines. Supposing there is a considerable difference of radiating power between Na and K, in favour of Na, then it is quite conceivable that when the two are mixed together in vapour, the joint heat may go out mainly through the channel of Na-radiation.   If the K were away and the Na at the same tempera-ture to start with, the Na might be less bright because it would cool more rapidly, as containing less heat than the mixture.   If the Na were away, the heat now going out through the channel of K-radiation, the K lines might show much brighter than when the Na was there to cool it.

Too many words illegible to allow me to gather the sense*:

27 Jan. 1879.

I return you Dumas's letter.   I cannot see that it bears directly on the question in dispute.   For the question observe is not, Are the elements compound bodies ?   But, has any satisfactory evidence been now obtained that they are compound bodies ? You would, I imagine, find plenty of chemists, from Prout down-wards, who would regard it as most probable that they were compounded.   I may say that, in common I suppose with multi-tudes of others, I have long supposed for my own part that they were.   I have further speculated on the lines obtained in electric discharges as the most probable source of possible evidence of their compound nature.   At one time I thought the non-com-munity, as I supposed, of the lines with different elements adverse to the supposition that the lines would furnish evidence of the compound nature.   But when I read what A. Mitscherlich had done as to the spectra of compounds, giving evidence that compounds might sustain the heat of the electric discharge without being dissociated, my hopes revived that evidence might some day be obtainable of the compound nature of the elements from the lines in the electric discharge.

But the direction in which these experiments seemed to indicate that the evidence was to be obtained was that of giving some dynamical theory from which the vibration-times of a com-pound could be inferred.

The inference from Mitscherlich's experiments would be that if the elements are compound, but are not dissociated in the arc, then the spectral evidence of their compound nature would involve something at any rate *towards* a theory of the vibration-times of compound bodies.

If they *are* dissociated, then we should expect to find a com-

* Prof. Stokes had himself recently taken to a typewriter.]

munity of certain lines.   But in that case surely we ought to expect to find after continued action of the arc the products of decomposition in sufficient quantity to enable us to detect the presence chemically, at least in the case of elements for which there are delicate means of qualitative detection.

The fact is that spectral analysis so transcends in delicacy our chemical means of detection, that the non-detection of impurities by chemical means does not prove that the impurities are not there in quantity sufficient to show themselves by spectral analysis.   The only way to answer satisfactorily as it seems to me the objection arising from the supposition of impurities, if you depend on chemical analysis, would be to purposely introduce a quantity of the suspected impurity not too small for chemical detection, and then show, supposing, which is not likely, that the fact is so, that the introduction would not account for the phenomena observed.

But if the action of the arc is continued, we might surely expect that we should be able to give some chemical evidence of a minute quantity of the products of decomposition.

Dumas is in favour of the proposition that the elements are or may be compound, but he has not at present, as appears from his letter, scrutinised the evidence that you have brought forward.

I don't think much of his argument as to the atomic weights, because, considering the latitude to be left for errors of determination of the atomic weights, and the large number of elements we have to play upon, the chances are that we could find trial cases of apparent coincidence.   When elements come in groups with allied properties in a chemical sense, as chlorine, bromine, iodine, then if there is also numerical coincidence there may be some evidence in it.

16 June, 1882.

I found your letter here on my return from London.   I had learned the principal points in it at our meeting this morning.

I think a good preliminary trial at any rate might be made with the large prism of 60° of which we spoke to-day.   The faces were, I think, 3 inches square; so that, allowing for the foreshortening, the pencil going through it and falling on the object-glass would be about 1·5 inch broad by 3 inches high. The object-glass would have to be stopped down to about that.

To get an achromatic image, using the spectroscope of seven

prisms of 60 for the spectroscope, and the single prism of 60° for the object-glass, the focal length of the object-glass would have to be seven times that of the collimator; and as the latter is 10 inches the former would have to be 70. You said, I think, it was about 100. This is quite near enough for a trial. The effect of the difference is merely that the magnifications, say of a spot, in and perpendicular to the plane of refraction, instead of being equal, would be in the proportion of 10 to 7. But this amount of distortion would leave the features of the spot perfectly recognisable.

The prism should be mounted in front of the object-glass so that the refracted beam should be centrical or nearly so. The object-glass would have to be stopped down to exclude stray light from the sky. The edge of the prism should be parallel to those of the prisms of the spectroscope, or nearly so, no very great exactness being required. The thin side should be turned to the *same side* as in the spectroscope, that is, as in the first prism of the spectroscope, in order that the prism before the object glass may *oppose* those of the spectroscope.

28 June, 1882.

In case in the use of the method I suggested in studying the arc, you find the necessity of having the prism covering the object-glass exactly in the position of minimum deviation a practical inconvenience, the plan will be to use a collimating lens, or better compound objective (with its crown lens towards the prism), in front of the prism covering the object-glass, so placed that the arc is in its principal focus.

With the sun, as I said, you will have no trouble about minimum deviation, as the want of it does not interfere with the goodness of the image, since the incident rays are parallel.

If the angular width of the arc is most convenient for study when the arc is a good way off, a simple lens will do very well for collimator, as with a considerable focal length for its aperture the spherical aberration of a single lens would be insignificant. If you wish the arc near, it would be better to use the objective of a telescope. When the arc is far you get the whole arc in a portion of the spectrum; when near you get the whole spectrum in a portion of the arc. For photographing you would more especially want the collimator lens or objective, as it would be awkward to have a portion only of the picture good as to outline of the object.

25 *Feb.* 1886.

After breakfast at the Athenaeum I went to South Kensington, so that I did not see your telegram till my return, when I went to the Royal Society.

I am sorry to find that your throat is so much worse. This is very trying weather for a person in that condition. You must keep to the house while the weather is like what it is at present.

I am not sorry that I went to South Kensington, for I have no reasonable doubt that I have found the fault in the instrument.

I tried first the object-glass. Setting at a good distance a pin-hole in a piece of brown paper, illuminated from behind by a jet of gas, and holding the eye in the focus, so as to see the whole of the object-glass filled with light, and then moving slightly the head in different ways, I could see no trace of veins in the glass.

Viewing the image then, with the head a little further back, through a pocket lens, getting it into focus, and then putting it out of focus both ways, I could see no sign of spherical aberration. As however the glass is chemically corrected, and therefore visually shows a good deal of colour, the test was not quite sharp, though very nearly so, this way. So I repeated it after substituting for the gas jet a soda flame, and still I could find no spherical aberration. The correction for s. a. is very good.

I then inclined the glass, to try it for oblique pencils. The form is not *quite* the best for oblique pencils, but very nearly so, and there is no objection to be raised on this score.

I then used a narrowish aperture illuminated by the light of the clouds reflected into a convenient direction, and covered half the object-glass, to try the chromatic compensation. The glass was much under-corrected for colour for the visual rays, as it ought to be; but whether the correction was just what it ought to be to produce the best effect I had no means of judging by this observation, as I have not been in the habit of observing the secondary tint for visual rays that belongs to the best chemical correction.

I have no fault to find with the object-glass, which appears so far as I could judge very good.

I then went into the observatory to examine the secondary magnifier. After some ineffectual trials by another method of illumination, I got your assistant to make the slit narrow, and

illuminated it from behind by a lamp. On examining the image,
I found that there was a *very large* spherical aberration. The
spherical aberration was in fact enormously under-corrected, giving
a very bad image. I took out the second combination, that nearest
to the plate, and the form corresponded to the performance. Here
I doubt not lies the fault. Fortunately it is the smallest compound lens of the lot, so that it would not cost much to get
a proper one. I should think Grubb would do it for 3 or 4 pounds
or less. Why should we go to Paris for a bad article when we
can get a good one in Dublin ? My only doubt is whether Grubb
would be willing to mend the faults of another optician, who
might then get the credit really due to Grubb.

I looked at a sun spot on one of the plates, and certainly the
definition is very bad.

If you are not taking photographs whenever you have a day
when the sun shines, and there is no great reason apparently for
doing so, as it is done at Greenwich, I should rather like to have
the second combination here to examine more leisurely. I need
not keep it long ; but if you are not using it I might perhaps like
to wait for sunshine, though I don't know that I could do much
more with sunshine in the way of examining it than I could by
artificial light. It is quite a small thing, and would easily travel
by either letter or parcels' post.

The combination is a cemented one, but there is no particular
advantage in this, as we have abundance of light ; and by not
obliging ourselves to make it cemented we have an additional
disposable quantity in our hands whereby to fulfil the conditions
that are of the most importance.

I could not get at the first combination without taking the
whole thing to pieces. But it is so close to the focus of the
object-glass of the telescope that its form will tell almost exclusively on the distortion and chromatic aberration of oblique
excentrical pencils, whereas the fault is a want of definition all
over the field. It is evident that the second combination is the
sinner.

8 March, 1886.

I have just heard from Grubb. He will be delighted to
execute the lenses for me, and promises to send me next day the
indices of the glasses he would use.

I fear the edge portions of the field will show much astigmatism in my plan, but the centre, and a good way round it, will, I think, be very good. The plan will be for a general picture to stop down the object glass—the stop may be placed in the cell of the magnifier—till the best general effect is produced; and when specially good definition of an interesting spot is wished for, to take out the stop and bring the spot near the centre of the field. For economy and likewise for convenience of subsequent handling, the camera might be fitted with a wooden dummy holding a plate of say 3 inches square in the middle, and the spot might be photographed on this. Even if we map a series of full pictures good throughout, it would be handy to have also a series of small plates, which could be taken up between the thumb and forefinger, giving the life-history of an interesting spot.

10 March, 1886.

I was thinking of asking you a question or two with reference to the diagnosis of my patient.

The front lens as I told you is not a combination at all, but a simple plano-convex lens. I fancy that the optician thought that it was placed so near the focus of the telescope, it did not much signify whether it were achromatised or not. This is perfectly true as regards the central portions of the field, but perfectly untrue as regards the outer portions. Thus supposing we could unite accurately into a point the rays of one refrangibility coming from a point near the edge—and that is a thing more easily said than done—still the image of a point near the edge would be a linear spectrum about a quarter of an inch long! This surely must give a wide border of blue when the image is focussed on the screen.

I assume that the magnifier was focussed for the spots. If it were focussed for what was deemed, but in consequence of the want of achromatism of the first lens wrongly deemed, the edge of the disk, the spots would be quite out of focus. At least this might make a sensible difference.

For the central portions, assumed to be in focus, I can find no fault but that arising from spherical aberration. This I think would quite account for the bad result; but if this be the cause it shows that the spherical aberration must be very well corrected. Now I don't see my way to doing this without sacrificing some distinctness in the outer portions.

I think it will be worth while to construct my design, even though there might be a better as regards the average goodness of the field. It would not be very costly, for the big brass piece might be retained, and we should merely have to get two identical combinations and cells to hold them, so as to screw into the main piece. I think it would be worth while to have such a thing, even though for general purposes we might prefer to use something else. Its construction would besides be a stepping-stone towards that other something. For the latter we must, I think, have recourse to excentrical pencils and to stops. I assume Janssen's photograph that you mention to have been a direct one, not a magnified image of another photograph. What was the diameter of the picture, and what the focal length of the telescope with which, apart from the secondary magnifier, it was taken?

I made some experiments to-day with a view to try a test I had thought of for exhibiting the general concentration of the rays; but in doing so I came unexpectedly on so curious and pretty a phenomenon that I was led away after that. It exhibits in a very pretty manner the spherical aberration in the existing magnifier. By pretty, I mean in an artistic sense; for there are other, and perhaps better methods of exhibiting and detecting the aberration. It is however a charming thing to look at.

Unless Janssen's picture was decidedly smaller than ours, or else taken with a telescope of decidedly longer focus, I cannot help thinking that a stop was used.

I will sum up with questions.

1. When the sun was focussed on the screen, was there a conspicuous blue border, quite too large—not to mention the different tints—to be accounted for by irrationality of dispersion?

2. Was the instrument focussed on a spot or on the limb?

3. What, about, is the diameter of Janssen's picture, and what, about, the focal length of his telescope?

I am going to London to-morrow morning.

12 *March*, 1886.

To procure a highly magnified image of a solar spot, even to the scale of 5 feet to the whole sun, forms I think far and away an easier problem than to get a good picture of the whole sun of a foot diameter. We only want a microscopic objective of

about an inch and a third focus; but as the angular apertures of
the incident pencils is so moderate, about 8 degrees, I don't think
there would be any difficulty about it, and it might even suffice
to use a single doublet, or achromatic lens. This would require
to be of about an inch and a third focus, which is nothing to the
makers of microscopes. It would not be costly, for the cost rises
rapidly as the focus shortens and the angular aperture widens, and
both are so very moderate as microscopes go that it is I think all
plain sailing.

I assume that you want only individual spots and their
immediate surroundings. To provide for such high magnification
and at the same time a large field would be another matter. It
would be a big spot, would it not, that would be one-tenth the
diameter of the sun, and for such a field as that, or even a fair
quantity larger, I don't think there would be any difficulty.

P.S.—You spoke of the small field you would obtain with the
high magnification as an objection. But surely for studying the
history of an individual spot you don't want anything of a large
field. What field do you aim at? I should think a very few
minutes would be sufficient.

I have got the indices from Grubb, and mean almost immediately to set about the calculation.

For a single spot I would not use the secondary magnifier at
all, but simply a microscopic objective, or even a single achromatic
combination to be applied about an inch and a third from the
focus of the object-glass of the telescope.

14 *March*, 1886.

Last night I made some measurements of angular aperture &c.
of a small microscope by Merz. The objective was composed of
two achromatic combinations, which could be used together, or
one by itself. The latter had a focal length of 1·3 inches, which is
just about what would be wanted in the plan I mentioned to you.
The angle of the pencil it took in was 7 degrees, whereas you
would want only 4—I believe I said 8 in my letter, but that was
a slip—so I began to write you a letter thinking the whole thing
wonderfully easy. But then I felt some misgiving, thinking it
was too easy to be sound, and on reflection I don't think it will do.
In fact, I had been going on the notions of geometrical optics,
and considering merely how the rays would go. But this is just

a case in which it is not lawful to consider light as consisting of rays which follow the paths assigned to them by the laws of reflection and refraction. It is a case in which we must take into account the true nature of light, as consisting in undulations, and not neglect diffraction. I am afraid that diffraction would stand in the way of such very high magnification as you contemplate. So far as I see at present, the best way to get such very high magnification would be to take a photograph as large as can be got thoroughly clear, and then take a photographic copy of that on an enlarged scale. It can only I think be done directly, if done at all consistently with clearness, by the use of excentrical pencils; and the calculation of the conditions of excentrical pencils is a complicated matter. I mean to make some further experiments in this direction with direct sunlight.

A picture of the sun on a moderate scale could easily be obtained. It was in fact thus that Carrington worked, and there could be no difficulty in photographing a similar picture.

P.S. On further consideration I don't think that diffraction would do more harm than in proportion to the increased size of the image, that is, that the imperfections due to diffraction would be more magnified than the image itself. In that case I think the combination would be worth getting. To get the field you mention it might be desirable to combine two pairs of achromatic combinations, as is commonly done in the microscope.

15 *March*, 1886.

On further consideration I have arrived at a neat result, viz. that the diffusion due to diffraction—observe I speak of absolute, not relative diffusion, not that is proportional to the size of the image—varies directly as the breadth of the image and inversely as the breadth of the object glass. This gives the extreme attainable sharpness by any optical arrangement however nearly perfect*.

For example. Suppose you wish to see the extreme attainable sharpness for a 12-inch picture. You can use either the image in

---

* This proposition foreshadowed in the P.S. to the last letter formed the basis of the fundamental paper by Helmholtz, "Die Theoretische Grenze der Leistungs-fähigkeit der Mikroskope," *Pogg. Ann.* 1874, and also of Abbe's earlier paper of the same year. The principle goes back in some measure to Fraunhofer; cf. Lord Rayleigh, *Phil. Mag.*, June, 1886.

the focus of the object glass, or, which will be more convenient if the telescope be furnished with eye-pieces, take a low-power eye-piece, and focus it for such a distance as to give a moderately small image of the sun on a small screen held in the focus. Suppose this to be say 2 inches in diameter, and that you want to study the indistinctness, due to diffraction, in a 12-inch picture. Put a cap on the object-glass with a circular hole of 1 inch diameter. Then the indistinctness in the 2-inch picture will be the limit which you cannot pass in a 12-inch picture, even supposing the optical arrangement is perfect. To get the unavoidable indistinctness in a 60-inch picture, you would have to stop down the object-glass to the one-fifth of an inch, unless you pleased to start with a somewhat larger image.

In short, given the quotient of the breadth of the image of a given object by the diameter of the object glass, the inevitable diffusion due to diffraction will also be given; and by getting the quotient sufficiently high by stopping down the object glass rather than much increasing the breadth of the image, you may make the optical arrangement practically perfect, and observe the indistinctness due to diffraction pure and simple, which cannot be got over, do what you will. One is thus kept from knocking one's head against a brick wall in hunting after an unattainable sharpness.

I mean to try with a small telescope the amount of distinctness which is optically attainable*; but it is only quite recently I deduced the result I have mentioned, and I have not yet had leisure and sunlight to try it.

Observe that when you view the indistinctness in the moderately small image you are not to think of the large image as the small one magnified, indistinctness and all, but are to imagine only the same indistinctness existing in the large picture, supposing the optical arrangements could be made perfect.

By comparing also the experimental indistinctness obtained by stopping down the object glass with the actual indistinctness in the image thrown on the screen prior to photographing it, you can see whether the distinctness of the latter comes near the theoretical limit which you cannot pass, or on the other hand falls greatly short of it. In the latter case the sharpness may be very greatly improved by stopping down.

* Verifications have been made and described by Lord Rayleigh.

If you had plenty of length at command, there would be no difficulty in making the optical arrangement practically perfect. The difficulty arises from the large angular field necessary to give the secondary magnifier in order to bring the focus within reasonable distance.

P.S. The chemical indistinctness due to diffraction will be less than the visual in the ratio of about λ_G to λ_D.

19 March, 1886.

There is one consideration connected with the very high magnification which you proposed that perhaps we have not sufficiently thought of. The scale proposed is 15 times that on which the actual photographs of 1 foot diameter are taken. Consequently the light would be only the $\frac{1}{225}$th part as intense. Therefore it would require 225 times as long an exposure.

Now I don't know what the length of exposure actually given may be. If it were the tenth of a second, the spot with high magnification would require 22·5 seconds. In that time any point of a spot with a magnification such as to make the sun's image 5 yards in diameter would travel on the screen or photographic plate about 30 inches. In the existing arrangement the length of exposure of any point of the sun would be the time the slit takes to travel over its own width. Probably this is much less than the tenth of a second, but if it were even as little as the hundredth of a second, still, with the high magnification and the longer exposure it requires, a point of a spot would travel 3 inches during the 2·25 seconds of exposure. To get a fairly good image, it ought not to travel more than the hundredth of an inch or so.

I am afraid therefore that without a clock movement, which the photoheliograph has not got, it would be useless to attempt such high magnification.

It might be well to try roughly what the length of actual exposure is; but there seems little doubt that some sort of clock movement would be imperative. The movement ought to be very smooth, but the rate need not be extremely exact, as the length of exposure would only be a question of seconds.

19 *March*, 1886.

I wrote rather in a hurry to you to-night, as we were going to a concert.

The length of exposure would have to vary as the [square of the] magnifying power; and as the velocity with which the image moves on the plate varies as the magnifying power, the length of path on the plate varies as the cube of the magnifying power. But if you enlarge a smaller photograph, the increased length of path varies only as the magnifying power.

Thus suppose you had a picture of a moderate size in which the path was barely sensible, and that you wished for a picture on a scale 15 times as great, linear. The path on the picture would be 15 times as long as in the small photograph. But if you got up the desired size by photographing at once on the large scale without clockwork, the length of path would be 3375 times as long as in the small photograph.

I wrote to Mr Grubb not to trouble himself about answering a question about a microscope, as I am sure the plan would not succeed without clockwork; and the requirements of the clockwork would be rather peculiar.

30 *March*, 1886.

I went to S. K. chiefly to see the photographs. I found however that very few had been kept, I mean of those done with the present instrument. I had been thinking of proposing an experiment; but the photographs contained, which I had forgotten, the very thing I had been thinking of trying, and the result was just what I had expected, but showed besides one of two things that I had not thought of, but which completely fall in with theory.

I had been thinking of proposing to try the image of a hair or wire stretched excentrically across the field. I forgot that there are actually wires so placed. I anticipated that though the central wires show clean, these would show confused. This is a simple consequence of the want of achromatism in the front lens. Now the parallel excentric wires *are* confused, but one thing that I had not thought of was that the side of the image of an excentric wire towards the parallel central wire was sharply defined, while the image on the other side trailed off. This is easily explained by the fact that the strong chemical action in the spectrum begins pretty suddenly, as you travel in the direction from red to blue, and trails off as you go on. Another thing of which I had anticipated the existence, but not that there would have been anything to make it visible, was the radial direction of the imperfect image. Now this is shown by the white tufts along the images of the excentric wires. These are doubtless due to dust on the wires, and they all point radially, not perpendicular to the wire.

The images of the central wires on the other hand are very sharp almost to the limb of the sun, and even there are not much amiss. Now spherical aberration would tell alike on the centrical and excentric wires. This shows that the main fault of the secondary magnifier is the non-achromatism of the front lens.

This however does not account for all. It would prevent the definition being good except near the centre of the field, but it would not prevent it from being good there. In the photographs I saw there were no spots near the centre, but you told me the definition was not good even there.

Now there are I think only three things that would account for the combination of goodness of the images of the central wires and badness of the images of the spots. These are

1.   badness of the object-glass;
2.   heterogeneous state of the air;
3.   bad adjustment of the wires to the focus of the telescope.

As to (1) the object-glass as well as I could examine it seemed to me extremely good.

As to (2) it would not account for the difference of performance of the two instruments, unless indeed there were something different in the housing; suppose for instance that the air in the present house got greatly heated, and the house was only opened just before an observation, so that the hot air was streaming out and mixing with the colder air in front of the object-glass.

As to (3) I suspect after all there is much truth in what the sapper said. He certainly said it required different focussing to show the spots best and the wires best.

But this might quite well have been, even supposing the wires had been carefully adjusted, in consequence of the expansion or contraction of the tube due to change of temperature. Suppose the adjustment had been made perfect at a certain temperature, and the temperature then changed 20° F., that would put the wires about the eightieth of an inch out of the focus of the telescope; and as a result, when the wires were brought into perfect focus on the screen or plate, the focus of the spots would be about two inches and a half from the plate, and if the wires had been adjusted when the tube had a temperature of 80° F., and the instrument were used when it had a temperature of 30° F., when the wires were in focus on the plate the spots would be six inches and a quarter out of focus.

I think for accurate work there should be an easy focussing adjustment to the wires, and that just before a picture was taken; provided at least there had been much change of temperature since the last, the wires should be adjusted to be seen sharply in focus at the same time with the spots. This would best be done by means of an eye-piece, the eye of course being defended; best I think by a deep green glass, further reduced in intensity if necessary by a glass of the usual kind, for the advantage of focussing with approximately homogeneous light of not too low refrangibility, as the object-glass is chemically not visually corrected.

If I am right in my suppositions, it will not be at all difficult to get the secondary magnifier to perform well by substituting an achromatic combination for the front lens.

I showed the magnifier to Grubb. I have brought it back with me, and mean when sunlight permits to examine it critically by pencils of the same slenderness as those that actually fall upon it in its place in the telescope.

17 *April*, 1886.

One cannot but regret the destruction of the first difference-engine ever made. I only heard of it a few days ago, being led to look into the matter by an enquiry made by an Italian *savant* about difference-engines, which was sent to Mr Scott at the Meteorological Office.

Adams of Cambridge was astounded when I told him yesterday about 1.30 p.m. that it no longer existed. Before our meeting he had seen his brother, who told him that when it left King's College it went to South Kensington, to the exhibition I think; and I think still later it was in the South Kensington Museum.

Can you find out for me to whose custody it was consigned when it left the South Kensington Museum*? I think the history of its fate should be put on record in some accessible place, for the information of those who may be interested in it some 40 years hence.

3 *May*, 1886.

I don't know whether the telescope of the photoheliograph has yet been restored to its usual condition. The easiest and most certain way of making out the faults of the secondary magnifier would be to observe a star or planet directly with the eye, the head being nearly in the place of the screen, and the image being regarded through an eye-piece or a watchmaker's lens, with the addition of a green or red glass in case it should be found that the chromatic dispersion arising from the circumstance that the telescope is achromatised not visually but photographically should be an obstacle to making out the faults of the image for homogeneous light....

The want of achromatism in the first lens is of course one obvious fault. But as I mentioned, the over-correction of the astigmatism in the second combination due to the first single lens (the front lens of the magnifier) having too short a focal length is I believe even more serious. But whether it is so is not an easy matter to say in the detached condition of the magnifier, unless I were to get some apparatus specially made. I refer to the experimental determination; and the mathematical calculations to which those excentric pencils lead are very laborious, so that having a great many other things on hand I have not ventured to engage in *them*.

26 *Oct.* 1886.

I am going to London to-morrow, to stay till Thursday evening. I intend to bring with me the secondary magnifier, and I propose on Thursday morning to go with it to South Kensington if it would be convenient to you to meet me there, and we could try it. I should like a morsel of red glass, and a watchmaker's lens or in default of that an eye-piece. I need not, I presume, bring these with me.

* Babbage's unfinished calculating engine is still at the South Kensington Museum.

Of course I don't mean to try it by actual photography; but one can tell very fairly what it will be able to do by seeing what kind of images of the wires it gives; I mean especially the excentric wires.

8 *Feb.* 1887.

On returning from London this evening at 6.30, I found your telegram. I think the result is extremely satisfactory; for you say the definition of the wires is very good up to 6 inches from the centre. Now that would carry us to the edge of the sun, assuming the diameter to be 12 inches, which must be near the mark. Any defect of definition due to the secondary magnifier must necessarily show itself in the image of the wires, assuming that you have a transversal as well as a radial wire. I assume that your telegram refers more particularly to the transverse, or excentric wire, the one on which we saw the wing at the R.S. Of course a defect of such a nature that the image of a point out of the centre should be a radial line, would not show itself on a wire passing through the centre, but would necessarily come out in the image of an excentric wire. With the old front there was a strong wing on the excentric wires even at their intersection with the central one, that is, at a distance of I suppose 4 inches or less from the centre.

So far as I could judge by rather rough testing, the object-glass appeared very good, which agrees with the description of it given you by the optician who made it. With a good object-glass and a secondary magnifier giving good definition up to a distance from the centre as great as the solar radius, we ought to get good solar pictures, except in so far as the definition might be interfered with by a heterogeneous condition of the atmosphere outside, or eddies of heated air in the house itself.

I will write to Grubb, and tell him of the good promise of the secondary magnifier.

P.S. If some slight defect should show itself just at the edge of the solar picture, and an interesting spot lay near the edge, of which we wished to get a good image in the general picture, that is, without taking a special photograph for that spot, we might place the sun's image slightly excentric, so as to favour the spot we cared specially for. It would be only rarely that there would be important spots close to the edge simultaneously on opposite sides of the disk of the sun.

14 April, 1887.

Thanks for the telegram. You told me that in focussing on the ground glass you could get at one focus a clear edge to the disk, and at a different focus, about three inches different, an imperfect image of a small spot, which was not seen at all at the first focus. I said this was just what spherical aberration would produce, though I felt rather puzzled to account for the effect, supposed due to spherical aberration, could be so great, for with the magnifier stopped to a quarter of an inch there was hardly any spherical aberration from it, and I had examined the O. G. for spherical aberration, and it appeared to be very well corrected. I have little doubt now that the result was due to chromatic aberration. Aberration of any kind would account for such a result, and as the O. G. is chemically corrected there is much chromatic aberration, visually, remaining.

In the final focussing I already expressed a preference for a spot, or the mottling if it will show it, to getting the edge of the disk clean, but the latter may be used as a near approximation. The effect of the residual chromatic aberration will be far smaller on the photographic plate, or on the fluorescent solution of æsculin, or quinine, than it would be visually on ground glass, and therefore the best focus for a clean edge would differ much less from the best focus for a spot or mottling when the two were examined by photography or suitable fluorescence than when they were examined visually. The condition of getting the edge clean will quite suffice to show whether my supposition as to the cause of previous failures is right. But if we have a spot, or else if the air is homogeneous enough to show mottling, we had best go by that.

15 April, 1887.

I heard this morning from my friend [W.] Kingsley, who was one of the first, if not the very first, to photograph microscopic objects, though he did not publish his results. He found the properties of horse-chestnut bark solution of great use. He writes:

"In the cases of object-glasses in which the blue was left outstanding very far, and where a secondary magnifier was used, I found it would have been next to impossible to have got the focus correct without using some phosphorescent screen."

He got a holder made in which the plate could be put somewhat oblique to the axis; and then when the focus had previously been found nearly, so as to get the best focus somewhere on the inclined plate, the final determination of the place of the focus was made by seeing what part of the image on the inclined plate was most distinct.

I did not know how you worked to get the focus. You wrote of gradually approaching it, and of its being a week's work to get it, from which I supposed that you took a lot of photographs so as to feel your way to it. On the other hand you spoke once of being bound to get the edge of the sun clean on the ground glass—at least I understood that that was what you meant— which looks as if you focussed by eye.

I thought I would let you have Kingsley's testimony as to the great convenience of using a phosphorescent screen. When the relation of the best chemical focus to the focus obtained on the fluorescent screen has been ascertained, I think it likely that it may not be difficult to choose a fluorescent screen the focus on which shall agree with the focus required for best photographic work. But I have little doubt that the focus on the quinine or æsculin stratum will at any rate come very near what we want; perhaps so near that we cannot tell the error.

P.S. Kingsley says: "In the microscope the emerging pencils are so very small that with a secondary magnifier the difference between the visible (ordinary) and chemical focus is very great, but puzzling as this was I found the false light the worst enemy, ghosts of all kinds appearing."

He thinks it would be well to line any tube used in solar photography with black velvet. However, we have no tube now, and the complex combinations in a microscope would be more likely to be haunted by ghosts than the simpler arrangement of a photoheliograph.

17 April, 1887.

When I said your pencil was about twice as slender as mine, I forgot about the stop, which when in makes it about 4 times as slender, and increases the vagueness of focussing proportionally. Hence the estimate, with stop in, would make your mean error of focussing by means of a fluorescent screen about an inch and a quarter.

But in getting a near approximation in the first instance to the chemical focus, it might be well perhaps to take the stop out; replacing it when the focus has been nearly found, so that it remains to put the finishing touch; as the instrument in this case should be in the condition in which it will be used.

The wind is still easterly, so I suppose, notwithstanding our own glorious weather, you cannot do much there on account of the smoke of London. We are nearing the time when the east winds take their departure.

13 *July*, 1887.

Sunday morning's post brought me the photographs which you were so good as to send. I looked at them then and since. As far as I can perceive, the photographs with the trial aperture indicate nothing in the object-glass that would account for the observed imperfection of definition in the secondary image. The various trials seem pretty well to exhaust other sources of error, and leave it to waviness in the air. I should think the surroundings of a telescope have much to say to this when it is a question of taking the image of the sun. I imagine that a grass field might be the best locality for a telescope intended for sun observations, and that flags, pavement or gravel about it would be bad, as tending to produce ascending currents of heated air. I don't know what may be the nature of the gound immediately about Janssen's observatory. You will know, as you have been there.

The appearance of the trail leads me to think that it is likely that the point I had in view in suggesting the aperture might be better tested by an aperture of a different form, see enclosed figure, like a very broad excentric slit rounded off at the ends. The length of the slit, if such I may call it when the diameter is say 1·5 inch, is placed equatoreally. I should propose instead of taking images with the clock going to take trails with the clock stopped. The focus would be altered just as before, the object-glass being capped during the alteration of focus, which would make a little break in the trail, and would show that an alteration had taken place, and the light of the star being allowed to act long enough to give a trail long enough for comfortable examination. It need not of course be long for that. The point would be to see whether the distribution of light at the two sides of the focus would be similar.

As to the spectra, the impression I would get from a general examination of them is that the bulk of the lines were due to some very common element or elements. The true platinum lines are distinguished by showing themselves only close to the electrodes, a consequence of the very sparing volatility of platinum.

But if this be so, where did the numerous lines come from that are shown even in the spectrum from platinum electrodes? May it not have been from very fine dust? Clifton told me that in working with—I forget now whom—they found that they could not work with two elements on the same day. The air of the laboratory got so impregnated with dust from the first that was used that it would persistently show itself in working with the second.

Now would it not be well for controul to test the thing in this way? Let the electrodes be arranged for sparking, and then enclosed in an envelope smeared inside with glycerine, except over a small space left for a window, where we want the glass to be clear. Passing a few sparks to knock off any dust that might be sticking to the electrodes, let the whole be left at rest for say 24 hours at least, and then spark and take a photograph. Or else if it can conveniently be arranged to make the envelope air-tight, exhaust the air with an air pump, and readmit it very slowly, passing it through a tube well stuffed with cotton wool, so as to filter the air that gets in, and let this process be repeated three or four times. I need not say that this would have no tendency to get rid of impurities, if such there be, contained in the platinum itself. If the platinum should have been fused in a vessel of lime or plumbago, it is likely enough to contain as an impurity a little calcium or iron.

I had best I suppose leave the photographs for you at the Royal Society when I am going there, or at any rate the spectra, for I don't suppose you want the other back.

P.S. Perhaps you know the elements to which the non-platinum lines in the photographs are respectively due.

I do not think that the solar corona can be referred to an atmosphere resting gravitationally upon the sun; still less, that its equatoreal extension can be referred to centrifugal force.

First, if we had a mass of gas extending as far as the corona, and resting gravitationally upon the sun, even if it were hydrogen its pressure at the surface of the sun would be something gigantic. Its density would correspond, only of course its temperature would be excessively high, and therefore the density not so great as might at first sight appear. Now a comet at its perihelion passage got so near the sun that it must almost if not quite have got within the photosphere, and yet it was not arrested, nor if I rightly recollect notably retarded.

Secondly, as to the equatoreal extension of the corona, this cannot be referred to the effect of centrifugal force in a gravitational atmosphere. For at the earth's surface the ratio of the centrifugal force to gravity at the equator is only about 1/289, and at the sun's equator the corresponding ratio is about 150 times smaller. Hence the disk is sensibly circular, and even at a distance from the surface of the sun as great as his radius the oblateness of a level surface in a fluid mass resting gravitationally upon and revolving with the sun would be about 20 times as small as the oblateness of the earth; so that the effect of the centrifugal force may be neglected altogether.

Thirdly. The photographs of the corona indicate a structure quite inconsistent with the idea of an atmosphere resting gravitationally upon the sun.

Doubtless the corona indicates the presence of matter. But that matter I take it does not form an atmosphere resting hydrostatically, though it may be greatly disturbed, upon the sun, but rather is in a dynamical condition, so that any small portion of this matter is to be regarded as, so to speak, a projectile.

If different elements show their bright lines extending to different heights from the sun's surface, and as a rule those of higher atomic weight extend less high, that is not, I take it, due to any effect of gravitation effecting a separation by difference of specific gravity. If you mix carbonic acid and hydrogen, and leave the mixture to itself, the carbonic acid will not settle to the bottom, and the hydrogen collect at the top. I think the most probable explanation is that the different elements require different temperatures to keep them in the gaseous form, those of higher molecular weight as a rule requiring a higher temperature; and when they are condensed into liquid or solid, forming a mist, they can no longer give out their characteristic bright lines. Then again some when they get far enough from the sun to be cool enough may enter into chemical combinations, when they would cease to give lines, or would give a different system. And even of those that remain in the gaseous state, the temperature required to make them show their characteristic lines might be different in different cases, so that on this account again there might be a difference in height at which different elements ceased to show their lines.

9 *Dec.* 1887.

I am not sure quite what your ideas are with reference to the tails of comets. I am disposed to accept, as the best theory I know, your supposition that the incandescence of that part of the whole comet—the very nucleus, with perhaps a portion of the root of the tail—which as the spectrum shows is self-luminous, is due to collisions. I do not think that the collisions take place all over the space where the self-luminosity appears, but mainly at the centre of the nucleus. Nor do I think that there is any need to invoke electrical currents to keep up the luminosity. Nor do I think that the tail consists of meteorites, unless you call dust or mist finer than the finest Krakatoa dust which is supposed to have produced the glows, and to have remained in suspension for a couple of years, meteorites. The strong polarisation of the light of the tail of a comet proves that the luminosity of the tail is in great measure if not wholly due to the reflected light of the sun.

I think the telescopic appearance shows that we have to deal with jets of gas from the nucleus, which, whether it be in its original state of gas, or in the state of excessively fine dust or mist into which it is afterwards collected, is acted on by repulsion from the sun, probably, as first so far as I know suggested by Sir John Herschel in his *Cape Observations*, an electrical repulsion, and so driven away from the sun into space with a velocity which soon becomes enormous. A portion of the gas which is near enough to the nucleus to give the spectrum of glowing gas, does not follow the head as it appears to do, but consists of gas which is moving away with very great velocity, but is continually being renewed, so that it appears to be stationary relatively to the head, though the actual matter of which it is composed is receding from the head with a very great velocity. Though it may be a good distance from the centre of the nucleus, the interval of time which has elapsed since it quitted it is but short, so that it has not yet cooled down sufficiently to lose the temperature, or rather the internal molecular agitation, which enables it to show a distinctive spectrum, but on getting somewhat further from the head it becomes too cool for that.

The difference between your view, if I rightly understand it, and that I have endeavoured to give an idea of, is that you think that the luminosity must have been produced where it is seen, whereas I think the luminous gas is moving away from the head so fast that it can retain its temperature, or a temperature sufficient to make it luminous, to a good distance, a distance which however would be got over in a comparatively short time.

P.S. In case you should wish to make anything clearer, I may mention that your paper has not yet been sent to press, and if you write at once to Lord Rayleigh, to Terling, you will be in time.

26 *May*, 1890.

At the Royal Society soirée I intended to have called your attention to one of the exhibits, but you passed on into the crowd, and I was pinned to the door. It reminded me of what you told me a good while ago of having once seen the sun green through steam that was issuing out of the steam pipe of a steamer on Loch Katrine. You probably saw Mr Shelford Bidwell's experiment on the effect of electrifying an issuing jet of steam, but I do not know whether you noticed the effect to which I wanted to call your attention. You must have seen the somewhat orange colour of the shadow of the cloud, but I do not know whether you noticed that sometimes, just at the first moment of turning on the electricity, a faint *greenish* shadow replaced what had previously been no shadow at all. It is to be shown again at the ladies' soirée, so that if you did not see it at the gentlemen's you will have an opportunity of looking out for it.

3 *Oct.* 1892.

You have I dare say had a paper from Prof. Johnson, of McGill University, who quotes you as having in print ignored the fact that Newton himself fully described the method of forming a pure spectrum by a slit and lens combined with a prism.

For my own part I was well aware of this, but I was not aware that Newton had been robbed of the credit that is his due by several writers.

16 *Feb.* 1894.

Some years ago you showed me at S. K. an apparatus with which you made a remarkable experiment. As well as I recollect it was this:—A glass vessel, or as I may call it bulb, connected with some kind of air pump (Sprengel?, Mercurial?) was partly filled with sodium, and repeatedly evacuated after passing an electric discharge, the sodium, as I was forgetting to say, having been previously warmed (I think so) each time. At last the discharge ceased to show the D line. I don't think I have it quite right, for you said I think that what was now in the bulb over the sodium (I forget how examined) could be shrunk up into nothing.

Would you kindly tell me where this is published, if published it be? I think it ought to have been published.

16 *April*, 1898[4].

Thanks for the photographs taken in the late eclipse. You have been very successful with the chromosphere; the sun half could of course be obtained at any time. It shows however how good the former must be; at least if I may assume that there was the same width of slit for the two. I notice that a few of the bright chromosphere narrow bands have what appears to be a very narrow line of reversion in the middle, which appears, in some cases at least, to be a prolongation of a dark line in the solar spectrum. I may perhaps be going to London on Monday; at any rate I am to go on the 29th, to stay till next day.

I am afraid my memory is very imperfect as to the experiment I mentioned. The chief thing I seem to remember about it is an apparently paradoxical result of mercury running up, when it was allowed to do so, so as to fill the bulb where it was to be expected that there was gas, hydrogen I think.

24 *Feb.* 1894.

I think your papers in the Proceedings (vol. XXIX., pp. 140, 266) very important. I have read the part containing the results several times over, having in my head the question of their interpretation, and nearly every time fresh light seems to break in on me.

The supposition that D is due, not to sodium NaNa, but to NaH, removes I think some serious chemical difficulties in relation to flames.

I have had some correspondence with Professor Smithells about flames, and it was out of this that my question to you grew. I wrote to him last night suggesting what seems to be a crucial experiment.

9 *April*, 1894.

When I wrote this morning I forgot to answer your question about sodium. I am in correspondence with Smithells about this and some other questions. One experiment has been tried which is favourable to the supposition that what gives the D line is sodium, and not a hydride, but one or two more are projected which are likely to be decisive. I may be able perhaps to tell you on Saturday. When I say decisive, I mean as to the question, metal or hydride ? Whether the so-called element sodium is or is not a compound, and if a compound whether hydrogen is one of the constituents, are speculative questions which remain as they were.

Troost and de Hautefeuille (I think that is the name of the second chemist) seem to have proved the existence of a definite alloy of sodium and hydrogen, answering to the formula Na_2H_4. (See Watts's Dictionary of Chemistry.) The sodium absorbs about 237 times its volume of hydrogen to form its compound. All the hydrogen can be got out by heating in a Sprengel Vacuum.

30 *May*, 1895.

Most certainly I join in the entertainment to Cornu.

21 *Dec.* 1895.

As you have seen several total solar eclipses, I should be glad of your opinion on one point.

Is the amount of scattered light (which may be judged of by the amount of illumination of the dark moon) sufficient to make it a matter worth caring for whether in taking a picture of the corona (without seeking to get the very faint remote portions) the corona has to contend with twice, instead of only once, the intensity of the scattered light?

I ask with reference to an experimental arrangement which I suggested to Turner as an alternative to the one he was thinking of. The arrangements would be much simplified if we might disregard the difference between twice and only once the scattered light as tending to obscure the corona.

30 Dec. 1895.

Have you seen a paper by Eilhard Wiedemann and G. C. Schmidt entitled "Fluorescenz des Natrium- und Kaliumdampfes und Bedeutung dieser Thatsache für die Astrophysik"? It is of immense importance in relation to solar physics. It opens up the possibility of quite a new explanation of some of the phenomena.

7th Sept. 1896.

I will read the chapter of your Chemistry of the Sun when I get access to it. But I have been going about and out of the way of books.

My notion of the nature of the formation of the prominences has long been that it is much like what (according to my notion) cyclonic disturbances are in the earth's atmosphere (at least as to the *origination* of the latter). I imagine that the cooling by radiation in the upper strata of the sun gradually brings about a condition of unstable equilibrium on a large scale, and masses of gas from the intensely heated interior break through. They ascend like a bubble in water by gravitation, but on account of the gigantic scale on which all takes place the velocities of ascent are enormous.

Have you seen a paper by Mr David E. Parker in the "Photogram" of July? Unless the thing be a mare's nest (and from the statements made I don't think it can be) it is enormously important.

I am on my way from Ireland and am going to Lord Kelvin's in the first instance. I am going to the Liverpool meeting of the British Association.

P.S. The paper of D.E.P. relates to photographing the corona without an eclipse.

29 August, 1896.

Thanks for your paper *re* total eclipse of 1893. As to No. 15, my idea of the nature of the prominences is not that which I should naturally express by the words "are fed from the outer parts of the solar atmosphere," but is perfectly compatible with what you tell us of the lines. From their forms, and all I know about them, I believe them (the prominences) to be masses of incandescent gases which are violently ejected from the sun, rapidly shot up to a great height, where there is an approach to a perfect vacuum, and then fall down again towards the sun by gravitation. The cooling by expansion and radiation, though absolutely speaking the gases would still remain intensely hot, would render possible the formation of compounds which could not exist in the chromosphere, as they would there be dissociated, and *that*, even in the few minutes, as it may be, which have elapsed since the first expulsion of the gas.

I think the solar atmosphere, if by that be meant the gas which rests by gravitation on the sun as our atmosphere does on the earth, extends, in any sensible degree, only a relatively small distance from the surface of the sun, but that masses of gas are forcibly projected far above this, and then return by gravitation, meanwhile expanding more or less, though in such enormously extensive portions of gas the expansion requires a time we should not at first have been prepared for.

APPENDIX.

ADDRESSES OF CONGRATULATION, 1899.

UNIVERSITÉ DE PARIS.

L'Université de Paris est heureuse de vous apporter aujourd'hui ses félicitations et de saluer en vous un Maître éminent de la Physique mathématique et de la Philosophie expérimentale.

Dans votre longue et féconde carrière, si dignement parcourue, vous avez hardiment abordé les questions les plus difficiles de l'Optique, de l'Hydrodynamique, de la Physique solaire et terrestre, et vous les avez résolues avec un égal succès, donnant ainsi l'exemple si rare de la puissance mathématique alliée à l'habileté expérimentale.

L'Optique vous a très souvent attiré. Vous avez suivi dans leur marche les rayons lumineux réfractés dans les corps cristallisés ou réfléchis par les métaux.

Vous avez traité le problème fondamental de la Spectroscopie. Avant vous, les effets mystérieux de la fluorescence semblaient échapper à toute loi ; vous nous avez appris la relation qui existe entre les vitesses des vibrations de la lumière émise et de la lumière génératrice. Vous avez appliqué votre analyse pénétrante au calcul de l'aberration de la lumière. Vous avez étudié les causes de la rotation du radiomètre. Vous avez heureusement contribué à l'exploration du Soleil par la photographie de la couronne solaire et par la mesure du rayonnement solaire.

Les phénomènes compliqués de l'Hydrodynamique ont attiré votre attention : vous avez su définir l'effet retardateur du gaz ambiant sur le mouvement du pendule, et calculer le coefficient de correction dû à la résistance de l'air.

Récemment encore vous avez contribué à élucider les singulières propriétés des rayons de Rœntgen, en les assimilant à des séries d'impulsions isolées, bien différentes des vibrations périodiques qui constituent la lumière ordinaire.

En vous remerciant, illustre Maître, de vos beaux et nombreux travaux, nous émettons l'espérance que vous ne cesserez pas d'en accroître le nombre, et que vous continuerez à enrichir la Science de nouvelles découvertes.

GIRARD, *Le Vice-Recteur, Président du Conseil de l'Université.*
E. LAVISSE, *Le Secrétaire du Conseil.*

UNIVERSITATI CANTABRIGIENSI CANCELLARIUS MAGISTRI ET SCHOLARES UNIV. OXON. S.P.D.

Libentissimo animo delegatos nostros ad Universitatem vestram hodie mittendos curavimus qui vobiscum venerabilem atque ornatum virum Georgium Gabriel Stokes ea qua par est observantia salvere jubentibus partem habeant. Necdum nobis animo excidit quibus laudibus nuperrime simus prosecuti grandaevum virum Baronem de Kelvin, et ipsum a Cantabrigiensium Academia profectum, scientiae studiis ac philosophiae naturali etiam nunc felicissime incumbentem, ita ut credendum sit Naturam ipsam, matrem benignam atque praeceptricem, antistitum suorum et interpretum quasi immortale ingenium servare neque senio obnoxium neque laboribus defatigatum. Quo fit ut nos quidem utriusque studia in re diversa consimilia reputantes par nobile fratrum dissociare nequeamus. Quippe quinquagesimus jam nunc agitur annus ex quo vir insignis Georgius Gabriel Stokes, summos honores atque amplissima praemia in Academia vestra tempestive consecutus, scientiae mathematicae Professor Lucasianus creatus est. Quam quidem annorum seriem (longam illam quidem et multis fetam miraculis) quibus investigationibus, quantis illustravit inventis! adeo ut nulla non in orbis terrarum parte inter se certasse videantur doctissimorum virorum societates, tam illustrem alumnum ordinibus suis adscribere cupientes. Quem si cum antiquioribus philosophis comparare licuerit, nunc quidem Archimedis discipulum habetis, rationes mathematicas ad rem physicam adhibentis, nunc Lucreti interpretem, qui magistri sui commenta de luce solis et vapore planius ac dilucidius expediverit, dum momenta illa describit

> Quae quasi cuduntur perque aëris intervallum
> Non dubitant transire, sequenti concita plaga;
> Suppeditatur enim confestim lumine lumen,
> Et quasi protelo stimulatur fulgure fulgur.

Illustri viro ac de universa fere scientia optime merito salutem nuntiamus amicissimam.

Datum in domo nostra Convocationis die tricesimo mensis Maii A. S. MDCCCXCIX.

UNIVERSITY OF HEIDELBERG.

Q. B. F. F. Q. S.—Viro illustrissimo nobilissimo Baroni Georgio Stokes, qui ante quinquaginta annos Cathedram Lucasianam Matheseos in illustri Universitate Cantabrigiensi adeptus ac per unum lustrum Regalis Societatis Londinensis pro scientia naturali promovenda praesidis munere functus doctrinam mechanicam hydrodynamicam opticam mathematicam per totam vitam excolvit, qui bis terve dignissimus illustrissimi Newtonii successor nova lucis invisibilis vestigia visibilia in corporibus fluorescentibus experiendo comprobavit, qui spectro Newtoniano per lucem electricam clausum campum lato limite patefecit, seni suavi simul ac venerabili, quo die continuam tot annorum felicitatem cum Universitate urbeque Cantabrigiensi concelebrant quicunque mathesi philosophiaeque naturali favent, nos decanus senior ceterique Professores ordinis physicorum in litterarum Universitate Ruperto-Carola ex sententia congratulamur ac bona fausta felicia omnia comprecamur, cuius gratulationis testimonium hanc tabulam esse voluimus.

ERNESTUS PFITZER, *h. t. decanus.* H. ROSENBUSCH.
LEO KOENIGSBERGER. OTTO BÜTSCHLI.
GEORGIUS QÜINCKE. W. VALENTINER.
ADOLPH STENGEL. THEODOR CURTIUS.

P. P. HEIDELBERGAE, die I mensis *Iunii*, a. MDCCCXCIX.

UNIVERSITY OF ST ANDREWS.

The University of St Andrews gladly avails itself of the opportunity afforded by your jubilee as Lucasian Professor of Mathematics in the University of Cambridge, to tender to you its sincere congratulations and to assure you of the grateful respect in which you are held in Scotland by all who are acquainted in any degree with your career. The chair rendered illustrious by the immortal Newton has during your long occupation of it acquired new glories, and signal proof has been furnished by you during the last fifty years of the value to the community of a life devoted to the study of pure science.

The oldest University in Scotland is naturally interested in the honours with which your life is crowned, since it numbers with pride among its former students and Professors such men as James Gregory, John Leslie, David Brewster, Balfour Stewart, James David Forbes who worked with success in some of the fields in which you have laboured and reaped much fruit.

It is the earnest wish of the University that you may continue to have health and strength for many years to come in which to remain the ornament and glory of your ancient University.

In name and on behalf of the University of St Andrews,

JOHN BIRRELL, M.A., D.D., *Clerk to the Senatus.*

THE UNIVERSITY, ST ANDREWS, 1st *June*, 1899.

UNIVERSITY OF GLASGOW.

SIR,—We, the Senate of the University of Glasgow, desire to offer you our hearty congratulations on the auspicious occasion of celebrating the Jubilee of your Professorship in the University of Cambridge. We would express our admiration for that rare combination of mathematical power and experimental skill which has enabled you during the last fifty years to achieve such brilliant success in unfolding to this generation the mysteries of the physical world. Nor are we unmindful of that insight by which you have penetrated the veil of outward things so as to behold with the eye of faith the spiritual basis of the material universe. We recall with gratification and pride that more than half a century ago our predecessors, influenced by a testimonial from your pen, appointed to the Chair of Natural Philosophy here one whose name has been a tower of strength to this University, and whom we have much pleasure in sending along with the Principal to convey to you personally our felicitations.

Signed and Sealed by authority of the Senate,

WILLIAM STEWART, D.D., *Clerk of Senate.*

26th May, 1899.

UNIVERSITY OF ABERDEEN.

Pergratum revera fuit litteras accipere ex quibus rescisse gaudemus festum celebratum iri in honorem Viri Illustrissimi Georgii Gabrielis Stokes, Baronetti, Praelectoris quondam paullisper in hac nostra Academia Aberdonensi, quem ad fastigium Scientiarum in Newtoni Cathedra Professoria pervenisse constat. Proinde, ut fas est, ex animo gratulamur Universitati Cantabrigiensi Collegioque ipsius Pembrochiano propter saeculi dimidium sub tantis auspiciis feliciter exactum. Aberdoniae, Kal. Jun. A.D. MDCCCXCIX.

UNIVERSITY OF STRASSBURG.

Viro nobili Georgio Gabrieli Stokes Libero Baroni Professori Academiae Cantabrigiensis Socio et Praesidi Collegii Pembrokiani artium liberalium magistro iuris utriusque honoris causa doctori scientiae naturalis honoris causa doctori, qui indefesso labore fertili ingenio felici eventu per plus L annos rerum naturae cognoscendae deditus earum rerum quae ad lucem pertinent causas occultas ante ceteros dilucide perspexit, ea ratione atque ea experientia quo modo lux in variis corporibus in varios colores varie distribuatur investigavit ut per ea studia hoc ipso tempore nova et mira de lucis radiis invenire licuerit, quae in sonitu corporumque liquidorum motibus incognita latebant nullo deterritus obstaculo rationibus subiecit, ex mathesi maius firmiusque auxilium rerum naturalium cognitioni petiit, Kal. Iuniis anni MDCCCXCIX festum diem celebranti quo die ante hos L annos Cantabrigiae in Matheseos Cathedram Lukasianam evectus munus Academicum auspicatus est, fausta omnia precatus congratulatur et hoc signum reverentiae esse voluit Ordo Physicorum Universitatis Wilhelmae Argentinensis.

PROF. EDUARDUS SCHAER, M.D.,

h. t. Decanus.

DABAMUS *id. Maiis* MDCCCXCIX.

SENATUS ACADEMICUS, UNIVERSITY OF EDINBURGH.

SIR,—We, George Chrystal, and George Frederick Armstrong, Professors respectively of Mathematics and of Civil Engineering in the University of Edinburgh, have the honour to approach you as the Delegates appointed by the Senatus Academicus of the said University to attend the celebration of your Jubilee, for the purpose of tendering their cordial congratulations on this auspicious occasion and of expressing their genuine appreciation and admiration of your signal services to the advancement of Science. The brilliant success which has attended your attacks upon many of the most abstruse problems of Pure Mathematics finds its counterpart in the remarkable results produced by your strikingly original Physical investigations; and the Senatus Academicus recognise that this twofold series of scientific triumphs was possible only for one who unites the highest mathematical genius to unique skill as an experimentalist. Permit us, Sir, to make special reference to your papers on the "Critical Values of the Sums of Periodic Series" and the "Numerical Calculation of Definite Integrals and Infinite Series," as embodying some of the most important improvements which mathematical methods have received with a view to their adaptation to Physical questions; and in the domain of Physics to your epoch marking treatment of Hydrokinetics and the Dynamics of Elastic Solids, and to your magnificent researches on the Undulatory Theory of Light, on Fluorescence, and on the Foundation of Spectrum Analysis. In virtue of these and many other grand contributions to Higher Intellectual Progress you have not only won a pre-eminent position amongst living men of Science, but have made good your claim to be hailed as the Newton of the Nineteenth Century.

The Senatus Academicus are furthermore sensible that the influence which you have exercised during your long tenure of the Lucasian Professorship of Mathematics has been one of the most

efficient agents in the creation of that splendid school of Natural
Philosophers of which the University of Cambridge is so justly proud.

Nor can they forget that as Gifford Lecturer in the University
of Edinburgh you have published two important volumes on
Natural Theology which strikingly exhibit its relations and
analogies to experimental studies, treat of Evolution and Teleo-
logy and their moral and spiritual bearings in a profound and
suggestive way, show how the findings of Science throw light
on many religious difficulties, and make manifest the unreason-
ableness of a facile acceptance of Materialism or a summary
rejection of the Supernatural.

In the name of the Senatus Academicus we congratulate you
on the approaching completion of your fiftieth year of office, in
a spirit of devout gratitude that to one so endowed so long a
period of active service should have been vouchsafed by the
blessing of Providence; and it is our earnest prayer that a
lengthy term of years may yet await you in which you may
continue to extend the frontiers of Science for the benefit of
mankind, and to your own and your country's lasting honour.

G. CHRYSTAL, *Prof. Math.*

G. F. ARMSTRONG, *Reg. Prof. Engineering.*

1st and 2nd June, 1899.

TRINITY COLLEGE, DUBLIN.

(Illuminated with Celtic ornament.)

Nos Praepositus et Socii Seniores Collegii SS. Trinitatis iuxta
Dublin gaudentes tibi gratulamur quod quinquaginta annos in
cathedra Lucasiana feliciter complevisti. Quanta cum laude hoc
praeclaro tuo officio Academico functus sis, hic dies festus, hic
concursus virorum illustrium satis demonstrat. Neque in tua
Universitate tantum, sed etiam in Societate quae dicitur Regia,
primum socius, deinde ab epistulis, postremo praeses, scientias
fovisti, et multorum qui naturae secreta indagabant ingenia
stimulasti, direxisti, adiuvisti. Omnibus adeo ubique qui disci-
plinis mathematico-physicis operam dedere notum est quanto cum
fructu multa in rerum natura investigaveris, quantum praesertim
lumen, ut ita dicamus, ipsius luminis origini et propagationi
attuleris. Praecipue autem nobis Hibernis gratum est memorare
te in hac insula esse natum, a stirpe propter ingenium et doctrinam
iamdiu inter nos insigni, in hac urbe elementis litterarum imbutum,
ab hac Universitate corona honoraria ornatum, et in hanc nostram
societatem Academicam susceptum. Itaque senectutem tranquil-
lam, faustam, felicem tibi, Vir clarissime, ex animo optamus et
precamur.

Dabamus in Coll. SS. et individ. Trin. iuxta Dubl.
pridie Kal. Iun. post Chr. nat. anno MDCCCXCIX.

LA R. ACCADEMIA DE' LINCEI DI ROMA.

La R. Accademia de' Lincei è ben lieta di associarsi alle onoranze, che Vi son rese da ogni parte del mondo civile nelle feste giubilari del Vostro ingresso quale professore in codesta alma Università cantabrigiense. E se da un lato essa così obbedisce ad una consuetudine di cortesia, bella ed anche doverosa verso uno scienziato eminente che è decoro dell' albo accademico, dall' altro sente che le sue gratulazioni in occasione tanto solenne Vi spettano di pien diritto, perchè assai pochi possono vantare come Voi titoli cospicui per essere annoverati fra i seguaci più schietti de' fini e de' metodi ai quali si inspirarono i fondatori dell' Accademia. Come Galileo, il più illustre degli institutori dell' Accademia de' Lincei; come Newton, del quale ereditaste degnamente la cattedra, avete saputo conciliare la indagine geometrica la più squisita coll' abilità sperimentale la più rara : onde i Vostri lavori sono riconosciuti quali modelli insigni dall' universale e resisteranno al tempo. Senza entrare in una recensione minuta delle Vostre ricerche sull' idrodinamica, sulla variazione della gravità alla superficie terrestre, sull' equilibrio e sul moto de' corpi elastici, sulle radiazioni luminose basterà che qui si ricordi la classica Memoria sulla diffrazione, dove per la prima volta l' Ottica fisica ha ricevuto per opera Vostra un assetto meccanico semplice e rigoroso; Memoria che sarebbe sufficiente da sola a rendere immortale il nome del suo Autore. Pertanto niuna meraviglia se tutti i cultori de' buoni studî ed in particolar modo i cultori della filosofia naturale volgono in questi giorni con grato animo il loro pensiero a Voi. Il saluto che Vi manda l' Accademia de' Lincei, prova che non distanza di spazio o diversità di favella possono attenuare l' ammirazione per la Vostra nobile vita tutta consacrata all' incremento della Scienza.

Il Presidente dell' Accademia,
EUGENIO BELTRAMI.

THE ROYAL SOCIETY OF LONDON.

DEAR COLLEAGUE,—We the President, Council and Fellows of the Royal Society of London rejoice in being able to offer to you our heartiest congratulations, on this happy occasion of the celebration by the University of Cambridge of the fiftieth year of your holding the great historic chair of Lucasian Professor of Mathematics.

We are bound to you by a double tie. Had you been simply a follower of Science, not a Fellow of our Society, we should have still looked upon you as belonging to us. We exist for the

Advancement of Natural Knowledge, and to this your life has been devoted with signal success. For more than half a century, you have again and again shed light upon the secrets of Nature; you have by your genius illumined the dark places of light itself. We are proud of, and grateful for, your work. But you are drawn yet closer to us by another bond. You have been so long a Fellow of the Society that if you live two more years, as we trust you will, we shall be able to celebrate another jubilee, that of your admission among us.

For no less than one and thirty years you held the office of Secretary to the Society, and so great during all that time was your zeal on our behalf, that with our gratitude for your long service is mingled the reproachful thought that, if thereby the researches of others were made more fruitful, the world may have been robbed of something which you otherwise might have done yourself.

For five years you filled to our great content the office, we may without vain-glory say the high office, of President of our Society. As President you sat in the chair in which once sat the greatest of Cambridge men; and this is not the only way in which you have followed him. We call ourselves the Royal Society of London; but at our birth we in part belonged to Oxford, and from the first we have again and again joined hands with Cambridge. We recall with pride that Isaac Newton belonged to Cambridge and to us, and we feel a special joy in doing honour to-day to one who has trod in so many of the footsteps of the author of the "Principia."

Wishing you all your heart's desire, We remain your loving Colleagues, The Fellows of the Royal Society,

LISTER, *President.*

L'INSTITUT DE FRANCE: ACADÉMIE DES SCIENCES.
(ACCOMPANIED BY THE ARAGO MEDAL.)

MONSIEUR STOKES,—L'Académie des Sciences de l'Institut de France, qui s'honore de vous compter parmi ses correspondants les plus illustres, est heureuse de s'associer aux hommages que le monde savant vient vous rendre à l'occasion du cinquantième anniversaire de votre professorat dans la chaire Lucasienne de Cambridge.

L'Académie, en vous priant d'accepter une médaille d'or à l'effigie d'Arago, désire remettre en vos mains un témoignage tout particulier de la haute estime en laquelle elle tient vos travaux, études magistrales sur la mécanique des fluides et sur la lumière. Elle ne saurait oublier vos belles recherches sur le changement de réfrangibilité de la lumière, les vues profondes que vous avez émises sur l'absorption de la lumière, et sur les relations de ce phénomène avec la réflexion métallique ou avec l'émission de radiations lumineuses nouvelles, et enfin l'énoncé de cette loi, qui porte votre nom, et qui, dans les phénomènes de fluorescence, reconnaît des mouvements dont la fréquence ne dépasse jamais celle des mouvements excitateurs.

L'Académie, admirant la vigueur toute juvenile d'une intelligence qui, après une carrière aussi longue et aussi belle, continue à rester à la tête du mouvement scientifique, fait les vœux les plus ardents pour que, longtemps encore, la science conserve un guide aussi précieux.

Le Délégué de l'Académie des Sciences,
HENRI BECQUEREL.

THE IMPERIAL ACADEMY OF SCIENCES OF BERLIN
sent a diploma of full-fellowship of the Academy.

Auspiciis Serenissimi ac Potentissimi Guilelmi Imperatoris Germanorum Borussiae Regis Academiae Scientiarum Borussicae Protectoris Clementissimi Virum Illustrem suisque titulis condecorandum Georgium Gabrielem Stokes Regiae nostrae Academiae socium declaramus eumque honore, privilegiis, beneficiis Academicorum ordini concessis rite ornamus. Cuius rei ut plena fides existat, ex decreto Academiae in acta relato hasce litteras sigillo nostro publico et subscriptione consueta munitas expediri iussimus.

> HERMANNUS DIELS, *classis phil.-historicae secretarius perpetuus.*
> ARTH. AUWERS, *class. phys.-math. secr. perps.*
> JOHANNES VAHLEN, *class. phil.-historicae secretarius perpetuus.*
> GUILELMUS WALDEYER, *classis phys.-mathem. secretarius perpetuus.*

BEROLINI die 22 Maii anno 1899.

YALE UNIVERSITY.

The Corporation and the Faculties of Yale University respectfully acknowledge the receipt of an invitation from the Senate of the University of Cambridge, to take part in the celebration, in June next, of the Jubilee of Professor Sir George Gabriel Stokes, Baronet. They desire to express their thanks for the opportunity of offering their congratulations on occasion of the commemoration of so rare and notable an academic event; and they cordially unite with scholars throughout the world in an appreciative recognition of that brilliant series of researches, extending over a period of fifty years and upwards, and touching nearly every department of Mathematical Physics, by means of which the name and the fame of Professor Stokes have been securely established and new renown has been added to the ancient University in the service of which his life has been spent.

TIMOTHY DWIGHT, President.

FRANKLIN BOWDITCH DEXTER, Secretary.

NEW HAVEN, CONNECTICUT, March 7, 1899.

THE AMERICAN PHILOSOPHICAL SOCIETY.

SIR,—The American Philosophical Society, founded by Benjamin Franklin, has commissioned me as one of its Vice-Presidents, to convey to you on this occasion of your Jubilee year, its warmest congratulations, and at the same time to express to you its admiration for the great extent and the profound scientific character of your labors in original investigation, especially in mathematical physics. From the date of Foucault's experiment in 1842, in which you were the first to discern evidence of the connection between the sun's composition and that of the earth, down to 1898, when your theory of pulses afforded the first rational explanation of the Röntgen rays, your investigations have profoundly influenced our views in æther-physics and have laid the foundation of much of our present knowledge in this direction. May the hand of time deal most kindly with you in the years to come, to the end that science may be enriched still more by the results of your marvellous insight into the mysteries of Nature. With profound respect I have the honor to remain very truly yours,

GEORGE F. BARKER.

PHILADELPHIA, May 15th, 1899.

Prof. Barker also presented an address from the University of Pennsylvania.

ROYAL ACADEMY OF GÖTTINGEN.

Hochgeehrter Herr Jubilar!

Wo immer physikalisch-mathematische Forschung und Lehre eine Heimstätte gefunden haben, wird der Tag, an dem Sie auf eine fünfzigjährige Lehrthätigkeit zurückblicken, mit Gefühlen der Bewunderung und der Dankbarkeit begrüsst werden. Zu den Huldigungen, die Ihnen heute von fern und nah dargebracht werden, drängt es auch die Königliche Gesellschaft der Wissenschaften zu Göttingen, ihre herzlichsten Glückwünsche zu fügen; dürfen wir Sie doch seit nunmehr fünfunddreissig Jahren zu den unsrigen zählen.

Der Name Stokes ist uns vertraut, seit uns in den Anfängen unserer physikalischen Studien zum ersten Mal jene reizenden Erscheinungen entgegen traten, denen Sie den Namen gaben. Die Erforschung der Fluorescenz bildet eine Ihrer schönsten und fruchtbarsten Leistungen, ein klassisches Vorbild einer Experimentaluntersuchung, ausgezeichnet durch die sinnreiche Verwendung der einfachsten Hülfsmittel, durch die vollständige Berücksichtigung aller Einflüsse und Beziehungen, durch die Klarheit der Anschauungen, die Einfachheit der Resultate. In der That, wenn wir die weitere Entwickelung der Lehre von der Fluorescenz überblicken, so müssen wir sagen, dass kaum etwas hinzugefügt ist, was über das von Ihnen gelegte Fundament hinausgeht.

Zur Zeit aber, als jene preisgekrönte Arbeit veröffentlicht wurde, lag schon eine mehr als zehnjährige Periode eifrigster und erfolgreichster wissenschaftlicher Thätigkeit hinter Ihnen. In allen Ihren Arbeiten bewundern wir die Freiheit, mit der Sie über das Rüstzeug des Mathematikers verfügen; zuletzt aber sind es doch die physikalischen Fragen, auf welche das Interesse sich concentrirt. So schon in den ersten Arbeiten, welche Sie veröffentlicht haben; Sie behandeln darin fundamentale Fragen der Hydrodynamik, eines Gebietes, dem Sie Ihre Neigung während Ihrer ganzen wissenschaftlichen Laufbahn bewahrten. Sie begnügen sich aber nicht mit allgemeinen theoretischen Resultaten; vielmehr wird die Rechnung immer so weit geführt, dass die Natur Antwort geben kann auf die Frage, ob die Grundannahmen der Theorie zutreffend seien oder nicht. So kommen Sie, ohne die entsprechenden Arbeiten Poisson's zu kennen und auf einem einfacheren und allgemeineren Wege, zu der Einführung der Reibungsglieder in die Differentialgleichungen der Hydrodynamik. Auf den speciellen Aufgaben, die Sie im Anschluss an die Grundgleichungen behandelt haben, beruhen Methoden der experimentellen Forschung, die noch heute von Bedeutung sind.

Aber nicht bloss auf die Grundgleichungen der ganzen Theorie war Ihre Aufmerksamkeit gerichtet, auch die specielleren Theile der Hydrodynamik haben Sie durch zahlreiche Arbeiten erweitert und vertieft, so vor Allem die Theorie der oscillatorischen Welle und der Einzelwelle, die Lehre von den gemeinsamen Bewegungen der festen Körper und der Flüssigkeiten.

Die Gleichungen der Hydrodynamik führten Sie hinüber zu den verwandten Gleichungen der Elasticitätslehre. Dass Sie Ihr Wissen und Können bereitwillig in den Dienst allgemeinerer Interessen stellten, davon zeugt die Arbeit über den Bruch von Eisenbahnbrücken, in der Sie die Wirkung des fahrenden Zuges auf die Biegung der Brücke so anschaulich dargestellt haben.

An die Grundgleichungen der Elasticität auf der einen, der

Hydrodynamik auf der anderen Seite knüpfen sich Ihre Untersuchungen über die Natur des Aethers, über die Aberration des
Lichts. Sie betreten damit das Gebiet der Optik, welche Sie, von
der schon erwähnten epochemachenden Leistung ganz abgesehen,
durch eine solche Fülle von Arbeiten bereichert haben, dass wir
nur Weniges hervorheben können. Wir erinnern an Ihre Arbeit
über Diffraction, in der Sie jene wichtigen Differenzen von Diffe-
rentialquotienten eingeführt haben, die wir jetzt als den Quirl
eines tonischen Vectors bezeichnen; wir erinnern an Ihre Arbeit
über die Zusammensetzung des natürlichen und des polarisirten
Lichtes, an die schöne Untersuchung über die Brennebenen
gerader Linien, die durch eine doppelbrechende Krystallplatte
gesehen werden. Was die Optik an sicherer Erkenntniss gewonnen,
das haben Sie in den Burnett-Vorlesungen ebenso klar wie ein-
dringend einem grösseren Kreise zugänglich gemacht. Über das
Gebiet der Physik hinaus zu den fundamentalen Fragen der
menschlichen Erkenntniss vordringend, haben Sie Zeugniss dafür
abgelegt, dass auch für die wissenschaftliche Arbeit die schönste
Frucht auf dem ethischen Gebiete liegt.

In der Lehre von der Leitung der Wärme und der Elektricität
verdanken wir Ihnen die Zerlegung des allgemeinen Ansatzes in
die beiden Theile mit symmetrischer und mit antisymmetrischer
Determinante, eine Zerlegung, deren Bedeutung immer mehr
hervorgetreten ist. Die Strömungsvorgänge in krystallinischen
Mitteln haben Sie durch die Einführung des Hülfskörpers auf die
Vorgänge in isotropen Körpern zurückzuführen gelehrt.

Über so Manches, was Sie zu der Lehre von der Gravitation,
zur Kenntniss des Radiometers beigetragen haben, gehen wir
hinweg, um noch mit wenig Worten Ihrer Leistungen auf mathe-
matischem Gebiet zu gedenken. Sie schenkten uns das bekannte
Theorem über die Gleichheit eines Randintegrales mit einem
Flächenintegral, das mit Ihrem Namen bezeichnet worden ist.
Sie klärten uns darüber auf, wie eine Superposition stetiger
Wellenlinien eine sprungweise sich ändernde Linie darzustellen
vermag, indem Sie den so wichtigen Begriff der ungleichmässigen
Convergenz eingeführt haben.

Hochgeehrter Herr Jubilar! Wenn Sie an dem heutigen Tage
auf Ihre wissenschaftliche Laufbahn zurückschauen, so wird dies
mit einem tiefen Gefühle der Dankbarkeit geschehen; Sie waren
Zeuge der wunderbar schöpferischen Kraft Faraday's; Sie hatten
Maxwell, Lord Kelvin zu Genossen Ihrer Arbeit. Sie haben die
unmittelbare Realität erlebt, zu der Hertz die Maxwell'sche
Theorie erhoben hat. Sie blicken auf diesen Siegeslauf der Physik
zurück mit dem Bewusstsein, dass es Ihnen beschieden war, an
dieser Entwickelung als der Ersten einer Theil zu nehmen, dass
Sie zu der Schaar der Geister gehören, von denen das Wort Plato's
gilt:

$$\Lambda\alpha\mu\pi\acute{\alpha}\delta\iota\alpha \ \ \check{\epsilon}\chi o\nu\tau\epsilon\varsigma \ \ \delta\iota\alpha\delta\acute{\omega}\sigma o\nu\sigma\iota\nu \ \ \grave{\alpha}\lambda\lambda\acute{\eta}\lambda o\iota\varsigma.$$

Möge die Freude an dem Errungenen, die Lust zum Schaffen Ihnen noch lange bleiben, möge der freundliche Strahl der Sonne noch lange Ihrem Lebensabend leuchten!

Die Königliche Gesellschaft der Wissenschaften zu Göttingen.

ERNST EHLERS, *d. Z. Vorsitzender Sekretär der k. A. d. W.*

THE AMERICAN ACADEMY OF ARTS AND SCIENCES, BOSTON, MASS.

The American Academy of Arts and Sciences, Boston, Massachusetts, acknowledges the great honour of the invitation of the University of Cambridge to participate in the Jubilee of Sir George Gabriel Stokes, Bart.

The Members of the Academy desire to express their deep sense of his great services to Mathematical and Physical Science; for it can be truly said that he has illumined these subjects by his genius: and his opinions in these branches of learning have been universally considered final. Such distinction is supreme: and demands the homage of men of Science in America. The Academy regrets that it is unable to send a representative to Cambridge: and trusts that this acknowledgment of the life service of Sir George Gabriel Stokes to science may sufficiently express its great interest in the gathering of men of science to honour one who has contributed in such high degree to make Cambridge University a remarkable centre of Mathematical and Physical Science.

ALEXANDER AGASSIZ, *President.*
SAMUEL H. SCUDDER, *Corresponding Secretary.*

BOSTON, 10 *May* 1899.

COLUMBIA UNIVERSITY.

The Trustees and Academic Staff of Columbia University in the City of New York, join me in extending to you most hearty congratulations on the occasion of the fiftieth anniversary of your installation as Professor in the University of Cambridge.

The whole scholastic world rejoices with your renowned Institution in celebrating an event so rare as the completion of a half century of continuous and eminent service in the advancement of learning. And particularly worthy of such commemoration does this event appear to the smaller body of workers who are devoted to the mathematico-physical sciences, and who are more intimately acquainted with the numerous contributions to knowledge which have distinguished your career.

A name so conspicuously identified with all the important advances made since the epoch of the illustrious Lagrange, Laplace, Poisson, Green, Fresnel, and Cauchy, in the theories of hydromechanics, physical geodesy, wave motion, and light, commands the admiration of men of all nationalities; but especially are Americans, who inherit with you the language and the traditions of a common ancestry, delighted to render their tribute of profound esteem on the occasion of your academic jubilee.

With the cordial greetings which Columbia extends to you on this happy anniversary, I beg to express the hope that science may continue to be enriched by your activity for many years to come.

SETH LOW, LL.D., *President.*

May 1, 1899.

UNIVERSITY OF PRINCETON.

We, the Faculty of Princeton University, send our most cordial congratulations to the University of Cambridge on the auspicious occasion which marks the completion of fifty years of distinguished and honoured service by Sir George Gabriel Stokes, Bart., as Lucasian Professor of Mathematics in the University of Cambridge. We recognize the value of his finished and masterly contributions to science, which have so honourably maintained the reputation of his chair and of your ancient University. We recognize also in him those fine qualities of character which have made him eminent not only in the world of science but also in social and public life.

We hope that a representative of our body, who will be subsequently named and commissioned, will convey to you in person our felicitations and will unite with you in the jubilee to which we have received your invitation. We trust that his presence will promote those relations of sympathy and affection which were manifested by you through your distinguished representative at our recent Sesquicentennial Celebration.

In behalf of the Faculty of Princeton University

FRANCIS L. PATTON, *President.*

NASSAU HALL, PRINCETON, NEW JERSEY.
May 1*st*, 1899.

IMPERIAL MILITARY ACADEMY OF MEDICINE, ST PETERSBURG, ACCOMPANIED BY DIPLOMA OF ELECTION AS HONORARY MEMBER OF THE ACADEMY.

On this solemn day, when the University of Cambridge celebrates the anniversary of your fifty years' glorious activity as a scientist and teacher, the Imperial Military Medical Academy of St Petersburg beg you, its highly esteemed honorary member, to accept their most cordial congratulations, and express the hope, that your numerous disciples and friends may have the benefit of your profound knowledge and great experience for many years to come.

President, PASHUTIN. Secretary, DIANINE.

BATAAFSCH GENOOTSCHAP, ROTTERDAM.

SIR,—The Bataafsch Genootschap der proefondervindelijke Wijsbegeerte in Rotterdam delights to gratulate Sir George Gabriel Stokes, for the celebration of the most memorable day in which he remembers the nomination before fifty years to the Lucasian Professorship of Mathematics in the University of Cambridge.

It is not in our way to remember the long list of the most remarkable mathematical and physical researches which you completed in this period. We may be permitted to declare that your labor advanced greatly the science and does honor to your country.

The Bataafsch Genootschap feels very obliged to you for the great honor by counting you so many years among its members correspondents, and hopes that it will be given to you to perform still many years in the interest of science.

F. B. s'JACOB, Praeses magnificus.

G. J. W. BREMER, Director and first Secretary.

ROYAL ACADEMY OF BELGIUM.

The Royal Academy of Belgium sent a diploma of nomination as Associate Member, signed by

Le Directeur, W. SPRING.

Le Secrétaire Perpetuel, LE CHEV. E. MARSHAL.

MANCHESTER LITERARY AND PHILOSOPHICAL SOCIETY.

The members of the Manchester Literary and Philosophical Society, in offering their heartiest congratulations on the occasion of the Jubilee of your tenure of the Lucasian Professorship, are at one with the whole scientific world in expressing their admiration for the signal services which, during these fifty years, you have rendered to the cause of Science.

Although the occasion is rare we do not celebrate to-day merely length of years, since to you it has been given, by opening new fields of research, to add distinction to a Chair already illustrious beyond others.

We admire the force and originality with which you have attacked so wide a range of subjects with such conspicuous success. Your directing influence has been felt and acknowledged by the most notable among the workers in science.

We remark with gratitude that you were among the first to recognise the value of the work of our late fellow-member Joule.

We have counted you as one of our Honorary Members since 1851, and we look back with pleasure to the occasion, in 1897, when you delivered before the Society the first Wilde Lecture on " The Nature of the Röntgen Rays."

It is our earnest wish that you may yet be spared for many years to exhibit those qualities which have made your name famous.

J. Cosmo Melvill, *President.*

Osborne Reynolds, Charles Bailey   } *Vice-Presidents.*
Arthur Schuster,   William H. Johnson }

R. F. Gwyther,   Francis Jones, *Hon. Secretaries.*

25 *April,* 1899.

ROYAL IRISH ACADEMY.

Cum Tu, uir doctissime, per quinquaginta annos, grande mortalis aeui spatium, Scientiae Mathematicae apud Cantabrigienses summa laude summaque dignitate Professor Lucasianus iam praefueris, Academia Regia Hibernica hoc tam fausto die gratulatione, honore, obseruantia Te prosequitur laeta lubens meritum.

Memor enim quam longe lateque Tu Scientiae fines promoueris, dum quibus legibus rotentur liquores liquido, quibus rationibus emittantur radii lucis dilucide docueris, haec Academia, ipsa pro sua parte Scientiae cultrix et fautrix, folium suum coronae illi uiridissimae lubenter addit quae hoc die festo Tibi per orbem terrarum ut dignissimo nectitur: uotumque facit ut Tibi per multos annos superstiti ita in tempore uenturo ut in decem lustris tam honeste peractis omnia sint bona, fausta, felicia.

Scribendo adfuerunt

Rosse, *Praeses.*
Iohannes H. Bernard, *ab actis.*

D. Dublini a. d. iii Kal. Iun. mdcccxcix.

ROYAL SOCIETY OF EDINBURGH.

Greetings and best wishes from the Royal Society of Edinburgh. Address will follow. KELVIN, *June* 2, 1899.

On behalf of the Council of the Royal Society of Edinburgh, we congratulate you heartily on the approaching completion of the fiftieth year of your tenure of the Lucasian Professorship.

We desire to express our conviction that much of the great advance in mathematical and experimental development of Natural Philosophy which has been made in the nineteenth century is directly or indirectly due to you. Your published writings on mathematical and experimental physics form an imperishable monument to your persevering devotion of labour and genius to the increase of knowledge during fifty-seven years.

We rejoice to know that you enjoy good health and undiminished activity in scientific work. We hope that these may be continued to you for many years to come.

KELVIN, *President.*
P. G. TAIT, *Secretary.*

May 19*th*, 1899.

ST EDMUND'S COLLEGE, WARE.

The President and Members of St Edmund's College, Old Hall, Ware, beg leave to assure the Chancellor, Masters, and Scholars of the University of Cambridge of their deep sense of the honour paid them by the invitation to take part in the ceremonies which commemorate the fifty years of eminent service rendered by Sir George Gabriel Stokes, Baronet, LL.D., D.Sc., F.R.S., Lucasian Professor of Mathematics. All who are concerned with the promotion of learning, and more especially those who are privileged to be in some measure associated with the University on this unique occasion, must feel satisfaction that labours at once so honourable and so useful have been thus prolonged.

The highest Academic distinction has been preceded by scientific esearch of the utmost value, and justified by work fruitful in public benefit. In the tributes of honour now about to be rendered to a great and worthy life, this College is glad and happy to be allowed to share.

BERNARD WARD, *President.*
JAMES L. PATTERSON,
Bp of Emmaus, M.A., Oxon., *Delegate.*

June 2, 1899.

L'ÉCOLE POLYTECHNIQUE DE PARIS.

In presenting congratulations in company with Prof. Becquerel, Prof. Cornu spoke as follows:—L'École Polytechnique, qui a produit Malus, Arago, Fresnel, Cauchy est heureuse et fière de présenter à Sir G. G. Stokes l'hommage de sa respectueuse admiration. Elle voulait vivement que pendant de longues années encore Sir G. G. Stokes puisse continuer les beaux travaux si profitables à la Science.

L'ÉCOLE NORMALE SUPÉRIEURE DE PARIS.

L'École Normale Supérieure de Paris est heureuse de s'associer aux fêtes du Cinquantenaire de Monsieur le Professeur Stokes.

Ses admirables travaux sur la fluorescence, sur l'hydrodynamique, sur l'optique et la diffraction, ont répandu sa gloire dans le monde entier.

Ceux d'entre nous qui connaissent le mieux ses œuvres imprimées ne regrettent que l'excessive modestie de l'Auteur : ils savent quel prix les plus illustres de ses compatriotes ont attaché à ses conversations, à ses lettres, et combien d'efforts il a fallu à lord Kelvin pour obtenir de lui la publication, malheureusement trop courte, de ses vues théoriques sur la fluorescence, sur la double réfraction, sur la dispersion anomale.

Mais si vive et si profonde est leur admiration qu'à peine conçoivent-ils comment elle aurait pu s'accroître, si même ils avaient pénétré d'une façon plus directe et plus intime le génie scientifique de celui que fête aujourd'hui l'Université de Cambridge.

Qu'il soit donné à l'Université de Cambridge de conserver longtemps son glorieux doyen ! En l'honorant comme elle fait, elle contribue à la réalisation de la Justice.

Le Directeur de l'École Normale Supérieure,
GEORGES PERROT, membre de l'Institut.

Le Sous-Directeur des Études scientifiques,
JULES TANNERY.

ROYAL INSTITUTION OF GREAT BRITAIN.

The Managers of the Royal Institution of Great Britain on behalf of the Members beg to congratulate Sir George Gabriel Stokes, Bart., F.R.S., D.C.L., LL.D., and the University of Cambridge, on the completion of half a century of their auspicious union.

The Managers share in the general and cordial recognition of Sir George Gabriel Stokes' signal services to Science and Education during his occupancy of the Lucasian Chair of Mathematics, which the fiftieth anniversary of his appointment to that Chair has evoked; but they desire more particularly to recall the agreeable relations which have for a long period subsisted between him and the Royal Institution. On the 18th of February, 1853, he lectured to the Members on "Change of Refrangibility of Light"; on the 4th of March, 1864, he lectured on "Discrimination of Organic Bodies by their Optical Properties"; during May and June, 1886, he gave a course of lectures on "Light"; and in 1886 the Actonian Prize of the Royal Institution was awarded to him by the unanimous vote of the Managers, in recognition of the manner in which his published papers proved, in accordance with the terms of the Trust, "illustrative of the Wisdom and Beneficence of the Almighty."

The Managers earnestly hope that Sir George Gabriel Stokes will be long spared to maintain the splendid traditions of the Lucasian Chair, to intensify the light of Scientific truth, and to enrich his generation with sterling knowledge.

(*Signed*) NORTHUMBERLAND.

UNIVERSITY OF BONN.

Quo die ex omni orbe terrarum eruditi peritique naturalium rerum homines concurrunt ad Te celebraturi peractum rarissimo exemplo quinquagesimum annum, quo profiteri litteras professorque excolere ac propagare coeperas, eo die ne nos quidem, universitatis Bonnensis philosophi, Tibi deesse voluimus, quin gratulatione optimisque votis nostram erga Te benevolentiam significaremus.

Ad eruendas et cognoscendas leges physicas, imprimis opticas, Tu quid rerum gesseris et quantum contuleris, omnibus quorum scire interest adeo notum est, ut loquendo iterare supersedeamus. Tuum enim nomen in perpetuum duraturo vinculo adligatum inhaeret toti illi lucis luminumque generi, quod ad fluorescentiam et phosphorescentiam refertur, quam per omnia apparere divolgatam proximis demum his lustris perspectum est. Et hae tamen commentationes, quibus Tu antecessorum in eodem spatio cursus superasti frustraque conatos vicisti, non faciunt nisi minimam operae tuae ac meritorum partem, quoniam experimentis istis experiundique arti adiunxisti theoreticas disquisitiones pari laude dignissimas.

Hodie igitur Tibi optamus ac precamur ut porro bene valeas vegetaque ac viridi senectute fruaris, idque nos cum Tua gratia tum communium litterarum causa vehementer desideramus, ut quas Te non desiturum esse sciamus egregio labore et palmario successu quamdiu vives adiutare. Itaque in annos plurimos vive valeque.

F. Küstner, *h. a. decanus.*

Bender.   W. Foerster.

Dr von Bezold.   H. Kayser.

D. Bonnae x Kal. Junias a. 99.

CAMBRIDGE PHILOSOPHICAL SOCIETY.

We, the President, Council and Fellows of the Cambridge Philosophical Society, beg leave to present to you our warm congratulations on the completion of fifty years service as Lucasian Professor of Mathematics in the University of Cambridge.

We claim the privilege of taking a very special interest in this event.

For more than half a century your connection with our Society has been of the most intimate kind, and we rejoice in recalling how much we owe to you for the unceasing interest you have shown in its welfare and progress.

To you moreover we are indebted for what we hold to be a chief glory of our Transactions, namely the long series of memoirs contributed by you, which have taken rank among the permanent classical writings of physical science both as records of advances accomplished and as sources of inspiration to other investigators.

At this time, when the University of Cambridge is specially assembled to do you honour, we join very heartily in rendering our homage and in desiring for you a continuance of strength, activity, and happiness in the future.

Signed and sealed on behalf of the Council and Fellows of the Cambridge Philosophical Society on the 15th day of May, 1899.

Joseph Larmor, *President.*

ROYAL ASTRONOMICAL SOCIETY.

We, the President, Council and Fellows of the Royal Astronomical Society offer you our congratulations in this the fiftieth year of your tenure of the Lucasian Professorship of Mathematics in the University of Cambridge.

We rejoice to pay our homage to one who by his rare intellectual powers has done much to advance human knowledge. By your researches in physical optics you have elucidated the laws which govern the propagation of light, and have developed the principles which enable us to study the constitution and movements of the remotest stars. Your investigation concerning the variation of gravity at the earth's surface will always be remembered as an important contribution to the theory of the figure of the planet on which we live.

Astronomers have thus their special share in the appreciation of your work, and in common with many other seekers after truth they owe you a debt of gratitude for the far-reaching influence of the help and sympathy which you have extended to those who have sought your advice and assistance.

Signed and Sealed on behalf of the Council and Fellows of the Royal Astronomical Society this 12th day of May, 1899.

G. H. DARWIN, *President.* H. F. NEWALL, *Secretary.*

ST DAVID'S COLLEGE, LAMPETER.

Quinquaginta annis iam feliciter exactis post labores Artium Mathematicarum Professoris summa cum laude susceptos et tibi et Academiae Cantabrigiensi libentissime gratulamur. Quanta diligentia res naturae occultiores per tot annos indagaveris, quanta subtilitate ingenii rationes explicaveris, quis est qui non audierit? Haec tanta merita vehementer admirati, in laetitiae partem hodie vocati, diem et tibi faustum et omnibus philosophiae studiosis memorabilem haud inviti celebramus. Te Newtoni, illustrissimi viri, successorem nec tali viro indignum, te cum laborum eius aemulum tum famae heredem, salutare gaudemus. Datum apud Lampeter a. d. iv Non. Jun. A.D. MDCCCXCIX.

INSTITUTION OF CIVIL ENGINEERS.

The Members of the Institution of Civil Engineers tender their warm congratulations to their Honorary Member, Sir George Gabriel Stokes, Baronet, F.R.S., on the occasion of the celebration of his fifty years' tenure of the Lucasian Professorship of Mathematics in the University of Cambridge; and desire to express their warm appreciation of the distinguished services rendered by him in the advancement of those Sciences which form the basis of modern Civil Engineering.

W. H. PREECE, *President.* J. H. T. TUDSBERY, *Secretary.*

1 *June*, 1899.

KING'S COLLEGE, LONDON.

Nobilissime, et Collegiorum Praesides et Scholares, hunc diem in honorem Viri doctissimi, Georgii Gabriel Stokes celebrantibus consociari liceat nobis, Collegii Regalis Londinensis Professoribus et Praelectoribus. Quamquam nos, quod ad aetatis annos attinet non modo prae vestra memoria prorsus pueri infantes sed Huic etiam Viro aliquanto sumus impares. Septuagesimum enim nunc maxime annum sibi nostrum adnumerat Collegium ; Hic vero puer incunabula nostra vidit. Idem, ex quo quinquagenarii muneris Academici cursum incepit, haud paucos hujus Collegii alumnos doctrina sua ad docendum instruxit. Itaque ut Academiae Canta-brigiensi, sic Huic ipsi ejus patri patronoque talem nos debemus reverentiam qualem Mentori Telemachus per ignota navigans vel Nestori juventus vix adulta. Atque ita quidem vix sumus juven-tatem egressi ut nondum Collegio nostro status contigerit Acade-micus. Majores idcirco Huic amico nostro tribuimus gratias, qui, ut prius Academiae vestrae novis instituendae legibus contulit operam, sic fautor erat gravis ac disertus Academiae Londinensis ex simulacro in veram Universitatem promovendae. Quod opus arduum sane et impeditum inchoantibus ipse facem praetulit. Universitatem enim studiorum diversis suis studiis tum in mundi physici rationibus, tum in Theologiae Naturalis nova via exqui-renda, haud mediocriter auxit.

DABAMUS LONDINI, Kal. Iun. MDCCCXCIX.

BRITISH ASSOCIATION FOR THE ADVANCEMENT OF SCIENCE.

The Council of the British Association for the Advancement of Science desire to offer you their cordial congratulations on the completion of fifty years of your tenure of the Lucasian Professor-ship in the University of Cambridge.

You have been a Member of the Association for more than half a century, and have served it in many capacities during that period. You were appointed Secretary of the Section of Mathe-matical and Physical Science in 1845, and continued in this laborious office until 1851. In the two following years you were a Vice-President of the Section, and became President in 1854 and again in 1862. Many times Vice-President, you were President of the Association in 1869, at the meeting in Exeter, and have been a permanent Member of the Council for the last thirty years.

Your services to the Association and to the cause for which it exists are far from being fully told by a mere enumeration of the offices you have held. In 1852 you gave an Evening Lecture to the Members at the Belfast Meeting on a branch of Optics which has been chiefly elucidated by your own researches; and from 1845, the first year of your membership, till the meeting last year at Bristol, the Reports of the Association have been enriched year by year by your contributions. Your celebrated reports·on "Researches in Hydrodynamics" published in 1846, and on

"Double Refraction" in 1862, are constantly referred to as classical writings by the cultivators of those branches of Physics, and have conferred abiding lustre on the publications of the Association.

Of your other conspicuous services to the cause of Science it is almost needless to speak, but your association with the Royal Society as Secretary for thirty-one years, and subsequently as President, has given you a place which is without a parallel among those who, during the last half-century, have fostered the progress of Science.

That you may long continue among our leaders in the advance of knowledge is the earnest desire of the Association.

Signed on behalf of the Council,

WILLIAM CROOKES, *President.*

UNIVERSITY OF DURHAM.

The University of Durham desires to present its heartiest congratulations on the completion of the fiftieth year of his Professorship.

While those faculties which have been most cultivated in the University of Durham are, perhaps, remote from the Sciences which Sir George Stokes has so brilliantly adorned, the University recognizes in him an eminence conferring a national position, which all lovers of learning may unite to applaud.

The University trusts that Sir George Gabriel Stokes may live long, full of health and years and honours, to be the ornament of his University and his Country.

G. W. KITCHIN, D.D.,
Warden of the University of Durham.

SOLAR PHYSICS COMMITTEE.

We, your Colleagues on the Solar Physics Committee, wish to avail ourselves of the opportunity afforded by the celebration of your jubilee as Lucasian Professor of Mathematics in the University of Cambridge, to express our admiration of your life-long labours in the cause of Science.

We feel proud to be associated with one whose name will go down to posterity as one of the founders of that Science of Spectrum Analysis which forms so large a part of the investigations directed by our Committee.

W. DE W. ABNEY.　　J. F. D. DONNELLY.
W. H. M. CHRISTIE.　　J. NORMAN LOCKYER.
G. H. DARWIN.　　RICHARD STRACHEY.

FRANK REDE FOWKE, *Secretary.*

UNIVERSITY OF LONDON.

Sir,—We the Chancellor and Vice-Chancellor of the University of London desire to offer to you, on behalf of the Senate of this University our hearty congratulations on the occasion of the Jubilee of your Professoriate.

The University of London has been foremost in recognizing the claims of Science as an avenue to University honours, and has spared no pains to secure and maintain the highest possible standard for its Scientific Degrees.

It is therefore appropriate that as representing that University we should express the deep sense which it entertains of the importance and magnitude of the services you have rendered to Science during a long and distinguished career, and join in the felicitations with which all interested in the progress of Science throughout the world now approach you on the completion of fifty years of active labour in the cause of Scientific Education.

KIMBERLEY, *Chancellor.*
HENRY E. ROSCOE, *Vice-Chancellor.*
June 1st, 1899.

THE CHEMICAL SOCIETY.

[Illuminated with a prism and a spectrum.]

We, the undersigned, President and Officers of the Chemical Society, beg on behalf of the Council and Fellows, to offer you our sincere and hearty felicitations on the occasion of your Jubilee as Lucasian Professor of Mathematics in the University of Cambridge.

The fifty years during which you have held a position so honourable in the ancient University which claims you as among the most distinguished of her sons have witnessed an extraordinary and unprecedented development in those branches of natural knowledge which are immediately connected with the work of your professorship. To this development your own labours as a teacher and an investigator have contributed in no small degree. You have profoundly influenced the teaching of physics wherever this subject is taught, and this influence has of necessity extended into those departments of chemical science which are directly dependent upon or are associated with physics. We gratefully acknowledge the services you have rendered to chemical physics by your researches on hydrodynamics, by your contributions to the theory and practice of spectrum analysis, and by your optical investigations. Your memorable contribution to our own Journal, "On the application of the optical properties of bodies to the detection and discrimination of organic substances," made more than a third of a century ago, has borne fruit an hundredfold, and to-day the refractometer and the spectroscope are as indispensable to the chemist as is his balance. We are glad to recognise that your interest in our special field of enquiry springing from that catholicity which is your characteristic has in no wise abated, for there is hardly a chemical subject bordering upon those branches of physical research which you have made more especially your own that has not been elucidated and benefited by your kindly criticism and advice. There are many workers in our Society who thankfully acknowledge the ready help which you have rendered to them by your counsel and suggestions.

That you may long continue to enjoy, in health and prosperity, the esteem and respect with which you are universally held by those who labour for the advancement of knowledge and the spread of learning, is the heartfelt wish of every member of that body on whose behalf we now address you.

T. E. THORPE, *President.* WILLIAM A. TILDEN, *Treasurer.* WYNDHAM R. DUNSTAN, ALEXANDER SCOTT, *Secretaries.* RAPHAEL MELDOLA, *Foreign Secretary.*

BURLINGTON HOUSE, LONDON, 1st *June*, 1899.

ROYAL ACADEMY OF SCIENCE OF THE NETHERLANDS.

MOST HONOURED COLLEAGUE.—On the occasion of the 50th anniversary of your election to the Lucasian Professorship of Mathematics the Royal Academy of Science of the Netherlands feels bound to offer you its hearty congratulations and to give expression to its feelings of high regard and admiration for what you have achieved. With unremitting zeal you have used your mathematical talent and experimental skill for the promotion of modern science, taking a part as well in the early development of spectrum analysis as in the recent study of new invisible radiations, and always setting the example of scientific inquiry, carried on in an honest, truth-loving, humble spirit, as you once so beautifully advocated it.

It is thus that your investigations in hydrodynamics, by which so many difficult points have been cleared up, and your researches on fluorescence and in theoretical optics, have secured you a high place in the ranks of those Natural Philosophers of whom the Country of Newton may justly feel proud. Nor has it been by your original work only that you have deserved the gratitude of men of science. Many of England's most distinguished physicists, among them the universally regretted Clerk Maxwell, were once your pupils. And in many ways, especially as Secretary and

President of the Royal Society, you have shown how warmly you love Science in general, and not only those branches in whose development you have had so prominent a share.

May Providence bless the evening of your life, and spare you for long years to the benefit of science and of your country.

For the Royal Academy of Science

H. G. v.d. S. BAKHUYSEN, *President.*

J. D. v.d. WAALS, *Secretary.*

AMSTERDAM, *June* 1, 1899.

QUEEN'S COLLEGE, BELFAST.

DEAR SIR,—On behalf of Queen's College, Belfast, we tender to you our very hearty congratulations on the completion of the fiftieth year of your Professoriate.

Your experimental investigations, especially those on the Change in the Refrangibility of Light, and those on the Extension of the Solar Spectrum beyond the Violet Rays, would alone entitle you to be enrolled amongst the great scientific discoverers of the century.

But your claim to our admiration rests upon higher foundations. We cannot but regard you as the Nestor and master-mind of the illustrious school of English Mathematicians, who have been so persevering and successful in applying the methods of exact mathematical analysis to the manifold problems which the rapidly developing Physical Sciences have presented for solution.

In a word, we cordially join in the universal verdict, that the great traditions of the chair of Newton have been, in your tenure, most worthily maintained.

It is naturally a matter of pride to us that we can claim you as a countryman, and that your good lady too hails from Ireland, where the name of her father, the late distinguished Astronomer of Armagh, is still a household word.

That God may grant you many more years of health and vigour in which to enlighten us by your insight into His laws is our earnest prayer.

Signed on behalf of the Council of Queen's College, Belfast,

T. HAMILTON, *President.*   J. PURSER, *Registrar.*

June, 1899.

QUEEN'S COLLEGE, GALWAY.

We, the Council of Queen's College, Galway, beg to tender our warmest congratulations on your having completed the fiftieth year of your Professorship.

In this Western Province of Ireland we feel that we have a peculiar claim to write our modest but hearty good wishes with the acclamations of the scientific world, as it is a proud boast of Connaught to number you among her most distinguished sons.

May you long be spared to look back on the great and good work you have done for Science, and to enjoy the universal esteem you have won in the half-century during which you have so worthily occupied your Chair.

ALEX. ANDERSON, *President.* EDWARD TOWNSEND, *Registrar.*

17*th May,* 1899.

THE ROYAL COLLEGE OF SCIENCE AND ROYAL SCHOOL OF MINES, LONDON.

ROYAL COLLEGE OF SCIENCE,
1*st June,* 1899.

The Council, Staff, and Students of the Royal College of Science, London, desire to congratulate you most warmly on the attainment of the fiftieth year of your tenure of the Lucasian Professorship.

At this moment when men of science are celebrating the lustre which for half a century your name has shed upon the University of Cambridge, we desire to recall to your recollection that you were once numbered among the teachers of the Institution which has since developed into the Royal College of Science and School of Mines.

Your successors are proud of the fact that they are reaping where you helped to sow, and the grain of the harvest of knowledge is richer in London as in Cambridge because you scattered the seed in days when such labour was less appreciated and more difficult than now.

Signed on behalf of the Council and Staff JOHN W. JUDD.

Signed on behalf of the Students FRANK MOULD.

OWENS COLLEGE, MANCHESTER.

The Senate of the Owens College, Manchester, offer to you their heartiest congratulations on the occasion of the Jubilee of your tenure of the Lucasian Professorship.

To many members of the Senate, the profound researches by which that tenure has been distinguished have long been a source of instruction and inspiration. Others also recall the courtesy and dignity which marked your long association with the great Scientific Society to which they owe allegiance. And all feel honoured by being allowed to join in the universal expression of admiration and goodwill which the present celebration has evoked.

Signed on behalf of the Senate

ALFRED HOPKINSON, *Principal.* ARTHUR SCHUSTER.
OSBORNE REYNOLDS. HORACE LAMB.

OWENS COLLEGE, MANCHESTER, *May,* 1899.

UNIVERSITY OF BOMBAY.

SIR,—On behalf of the Chancellor, Vice-Chancellor and Fellows of the University of Bombay, whom it is my privilege to represent to-day, I tender to you their cordial congratulations on the approaching completion of the period of fifty years which began on the date of your election to the Lucasian Professorship of Mathematics.

It would scarcely be becoming in the Delegate of a University which was not in existence till nearly fourteen years after that date to use the language of praise in reference to the labours by which, as Lucasian Professor, as Secretary and President of the Royal Society, as Lecturer in the School of Mines and elsewhere, and as the Author of numerous treatises on Mathematical and Physical Subjects, you have promoted the advancement of Physical Science in our time. But the University of Bombay can, with propriety, share with older Institutions their respectful admiration of your pre-eminent attainments and distinguished educational services. It can join in this tribute the more freely because, since its establishment, in the year 1857, it has always looked to Cambridge for light and guidance in regulating the methods and procedure of its Examinations in Mathematics for degrees in Arts.

In response to the invitation with which they have been honoured by the Senate of the University of Cambridge, the Chancellor, Vice-Chancellor and Fellows of the University of Bombay take their part in this memorable celebration of your Jubilee with the sincerest satisfaction; and I deem myself happy in the opportunity which has been given me, on this most auspicious occasion, of conveying to you this deferential expression of their hearty good-will, esteem and homage.

I have the honour to be, Sir, Your most obedient Servant,

H. BIRDWOOD,
Delegate, and late Vice-Chancellor.

CAMBRIDGE, *2nd June,* 1899.

IMPERIAL UNIVERSITY OF TOKIO.

[In Japanese: accompanied by a translation made by the Japanese Legation in London, as follows.]

IMPERIAL UNIVERSITY OF TOKIO,
17 *April*, 1899.

SIR,—It is with sincere gratification that I learn the news of the celebration, in the early part of June next, of the Fiftieth anniversary of your professorship at the University of Cambridge. Your meritorious services rendered for the cause of Science during the last half a century are well known in all parts of the globe. It is no wonder that the whole world heartily welcome this approaching celebration. I now have the honour to present this congratulation in the name of the Imperial University of Tokio, Japan, and beg to express my wish that your age will last long together with the University of Cambridge itself.

I have the honour to be, Sir, your obedient servant,

DAIROKU KIKUCHI,
President of the University of Tokio.

LONDON MATHEMATICAL SOCIETY.

SIR,—The Council of the London Mathematical Society have much pleasure in tendering to you their sincere congratulations on the approaching completion of the fiftieth year of your tenure of the Lucasian Professorship. They recognize that many of the memoirs published in the Proceedings of the Society bear witness to the great influence which your writings have had in opening new paths of discovery in Mathematics and in pointing the way to novel applications in the domain of Natural Philosophy, and they desire to express a hope that you may live long to represent in Great Britain the science which you have done so much to promote.

KELVIN, *President.*
ROBERT TUCKER, A. E. H. LOVE, *Secretaries.*

SOCIÉTÉ FRANÇAISE DE PHYSIQUE.

La Société Française de Physique, qui a le grand honneur de vous compter parmi les dix Physiciens auxquels elle a décerné le titre de Membre Honoraire, ne saurait rester indifférente à la manifestation organisée en votre honneur par le monde scientifique tout entier.

Elle a chargé l'un de ses anciens Présidents de vous porter

l'expression de ses sentiments d'admiration et de reconnaissance pour les travaux que vous avez accomplis, et de vous offrir les vœux qu'elle forme pour la longue continuation de votre brillante carrière.

A l'exemple du plus illustre de vos prédécesseurs, dans cette chaire que vous occupez avec tant d'éclat, l'immortel Sir Isaac Newton, vous avez su faire profiter de la puissance de votre génie créateur les diverses parties de la Science ; les Mathématiques, l'Astronomie, la Physique vous doivent des découvertes admirables.

Comment les Physiciens pourraient-ils jamais oublier, parmi tant d'autres travaux dont ils vous sont redevables, vos études relatives aux phénomènes les plus délicats de l'Optique ? Les résultats que votre esprit pénétrant a pu obtenir sur la diffraction, sur la réflexion, sur l'absorption demeureront à jamais classiques, et vos recherches sur la fluorescence suffiraient à immortaliser votre nom ; elles vous ont conduit à une loi dont l'importance au point de vue des conséquences pratiques, et au point de vue de la Philosophie Naturelle, grandit tous les jours.

Mais il ne saurait être question d'énumérer ici tous les titres que vous avez à l'universelle reconnaissance des Physiciens, tous les Membres de la Société Française de Physique les connaissent, aussi ont-ils unanimement décidé qu'ils vous prieraient de recevoir l'hommage de leurs sincères sentiments d'admiration.

Les Membres du Bureau de la Société Française de Physique,

GAL. BASSOT, *de l'Académie des Sciences, Président.*

L. POINCARÉ, *Secrétaire général.*

H. DESLANDRES.

UNIVERSITY OF NEW ZEALAND.

Quod nos comiter invitatione prosecuti estis, ut quendam legaremus, qui laetitiae vestrae testis et particeps adesset, idcirco maximas gratias et agimus et habemus. Vobis autem et viro multis nominibus laudando, Georgio Gabriel Stokes, professori vestro, de magna raraque inter homines felicitate gratulamur; huic quia per L annos publico munere in sua eademque vestra Academia praeditus scientiam rerum occultarum indagavit et adhuc indagat, et quasi facem philosophiae cum aequalibus tum posteris praelucet; vobis vero quod contigit tantae tanti viri gloriae communione illustrari.

Etenim nos, quamquam penitus a vobis toto orbe divisi sumus, tamen in vestro gaudio partem nostram vindicamus. Fuerunt quidem olim philosophi qui unam quandam omnium qui ubique sunt sapientium civitatem adumbrarent, quod somnium nostra aetas verum reddidit et repraesentavit. Nulla enim res usquam

a doctis hominibus potest reperiri, quin protinus eius notitia ad
sola terrarum ultima perveniat. Itaque de toto orbe terrarum
summi poetae verba, sententia immutata, usurpare possumus:
"mens agitat molem."

Et nos quidquid in vestra Academia geritur, vel proxime
tangit, non solum quod una nobis est vobiscum patria, et eadem
stirpe orti sumus, sed quia artiore vinculo inter nos copulati
sumus, ut alumni nostri hinc profecti non tamquam hospites in
Academia Cantabrigiensi versentur, sed praecipuo quodam societatis iure numero vestro adscribantur. Ergo laetamur ut qui
maxime, et nobilissimo professori vestro, qui in quaestionibus
naturalibus et mathematicis tam diu primas agit, qui tot rebus
obscuris, qui ipsi lucis naturae lumen attulit, laudem honorem
reverentiam, ut par est, deferimus, et dum festum eius diem
celebramus, omnia fausta eius causa precamur; simulque exoptamus ut ille ordo amplissimus virorum Cantabrigiensium egregia
cum laude in rerum natura investiganda exercitatorum, iam inde
a Newtono continuatus, etiam in omnium saeculorum posteritatem
propagetur.

DURHAM COLLEGE OF SCIENCE.

The Council, Professors, and Students of the Durham College
of Science, Newcastle-upon-Tyne, desire to offer their hearty
congratulations to Professor Sir George Gabriel Stokes, Bart.,
M.A., Hon. LL.D., Hon. Sc.D., F.R.S., on the attainment of his
Jubilee as Lucasian Professor of Mathematics in the University
of Cambridge.

They recognize, with admiration, the valuable work done by
Professor Stokes during the last fifty years, which has been by no
means restricted to his own University. His originality of thought,
alertness of mind, and soundness of judgment, have achieved
magnificent results, and greatly extended our knowledge of
Natural Phenomena. To his wonderful combination of mathematical power, resourcefulness, and experimental skill, as well as
to his stimulating influence, the world largely owes one of its most
brilliant schools of Natural Philosophers. He has proved himself
a worthy successor of the great master, "qui genus humanum
ingenio superavit." May he long be spared to adorn the Chair
whose highest traditions he has so nobly sustained.

Signed on behalf of the Council, Professors, and Students of
the Durham College of Science, Newcastle-upon-Tyne.

HENRY PALIN GURNEY, *Principal.*

UNIVERSITY OF ADELAIDE.

The Council of the University of Adelaide tender you their most sincere congratulations on the occasion of your Jubilee as Lucasian Professor.

They feel greatly honoured in being permitted to have a share in the manifestations of respect and admiration which this event calls forth on all hands; and they are pleased to think that their participation may serve in some degree as a symbol that the fame of your achievements is cherished, and the influence of your labours is felt, in whatever region of the earth science is taught or pursued.

It is the earnest prayer of the Council that abilities so pre-eminent and so faithfully used may long be spared to the service of your own University, and of the world-wide fraternity of seekers after Natural Knowledge.

UNIVERSITY COLLEGE OF WALES.

SIR,—The Council of the University College of Wales, Aberystwyth, regarding with feelings of high respect and admiration the brilliant services to Mathematical and Physical Science by which you have conferred an added lustre on your University and your Country, offer you their most cordial congratulations and good wishes upon the attainment of the Fiftieth Year of your tenure of the Lucasian Professorship of Mathematics in the University of Cambridge.

RENDEL, *President.* T. F. ROBERTS, *Principal.*

ABERYSTWYTH, *June 1st,* 1899.

YORKSHIRE COLLEGE, LEEDS.

SIR,—The Council and Senate of The Yorkshire College, Leeds, offer you their heartiest congratulations on the approaching completion of the Fiftieth Year of your tenure of the Lucasian Professorship of Mathematics in the University of Cambridge, and beg your acceptance of this Address, as a token of the esteem in which they hold your great achievements in the world of Mathematical and Physical Science, and of the gratitude which they feel for your splendid contributions to the advancement of human knowledge.

Signed by order and on behalf of the Council and Senate of the Yorkshire College.

ARTHUR G. LUPTON, *Chairman of Council.*

N. BODINGTON, *Principal, and Chairman of Senate.*

LEEDS, *June 2nd,* 1899.

ADDRESSES OF CONGRATULATION 465

PHYSICAL SOCIETY OF LONDON.

Sir,—I have been deputed by the Council of the Physical
Society of London to represent the Society at the celebration of
the Fiftieth Anniversary of your appointment to the Lucasian
Professorship of Mathematics in the University of Cambridge.

It is the desire of every Fellow of the Physical Society to offer
to you his hearty congratulations upon so memorable an occasion,
and at the same time to express the hope that you may long be
spared to continue the labours for the advancement of knowledge
that have rendered you justly famous and have been so fruitful in
the past.

The Fellows of the Physical Society have a keen sense of the
debt that is owed to you by the whole world for the great increase
that you have made in our knowledge of Physical Science, and
the whole Society is especially glad of the present opportunity of
thanking you for your work.

I am, Sir, Your obedient servant,

OLIVER J. LODGE, *President.*

26th May, 1899

UNIVERSITY COLLEGE, SHEFFIELD.

The Council and Senate of University College Sheffield
present their hearty congratulations to you and to the University
of Cambridge on the completion of your 50 years tenure of the
Lucasian Chair of Mathematics in that University.

The immense advance of physical knowledge during this period
has been in no small degree due to the results and methods
developed by yourself, to the brilliant series of Mathematicians
who owe inspiration to your works and teaching, and to the
assistance which you have ungrudgingly given to workers in
Physical Science.

Since your first paper in 1842 (on fluid motion) down to your
last (also on fluid motion) we owe to you amongst other results a
series of researches into the problems opened up by Fresnel's
establishment a quarter of a century earlier of the true theory of
light, and into the problems of hydrodynamics few of which before
the publication of the methods with which your name will always
be associated had been successfully attacked.

We trust that you may live long to see those studies to which
you have been devoted advance still further into the unknown;
that in those advances the younger provincial Colleges such as
our own, which owe their establishment to the quickened sense of
the value of Scientific Knowledge, may take a not unworthy part.

NORFOLK, *President.*

FREDERICK MAPPIN, HENRY STEPHENSON⎫
HENRY C. SORBY, WILLIAM DYSON ⎬ *Vice-Presidents.*

UNIVERSITY COLLEGE, DUNDEE.

The Governors, Council, Professors, and Students of University College, Dundee, desire to offer you their congratulations on the completion of the fiftieth year of your tenure of the Lucasian Chair of Mathematics in the University of Cambridge. During the whole of this long period you have devoted yourself to the advancement of your Science, and the record of your worth is one of which your fellow-countrymen are justly proud. The results which you have achieved will never be forgotten, but those who now address you would express the hope for yourself that you may be permitted for many years to enjoy the honours which your great labours have so signally merited and so abundantly called forth.

JOHN YULE MACKAY, *Principal.*

June 2nd, 1899.

VICTORIA UNIVERSITY.

The Court and Convocation, together with the Professors, Lecturers, and Scholars of the Victoria University, beg to offer you their most cordial congratulations on the occasion of the Jubilee of your tenure of the Lucasian Professorship.

They share to the full the universal admiration of the solid and enduring contributions to the fabric of Science by which that tenure has been marked, and of the lofty and sincere character which has added fresh dignity to a Chair already so illustrious, and they recall with especial gratification the fact that the University has been permitted to inscribe your name on its roll of honorary graduates.

The Victoria University trusts that you may long be spared to enjoy the position which you have attained in the respect and affection of your own University and of the Scientific world.

Signed on behalf of the University,

N. BODINGTON, *Vice-Chancellor.*

1st June, 1899.

ROYAL UNIVERSITY OF IRELAND.

SIR,—We the Members of the Senate of the Royal University of Ireland desire to offer you our hearty congratulations on your completion of a period of fifty years as Lucasian Professor of Mathematics in the University of Cambridge and we have commissioned one of our body to convey to you this expression of our high admiration and regard.

Your eminent services to the Mathematical and Physical Sciences are known and appreciated here as they are wherever these Sciences are cultivated. These services have been recognised not only by the honours conferred on you by the Universities and learned Societies of the British Isles but also by the illustrious Order which you received from the Sovereign of a Foreign State.

We remember with pride and pleasure that you are a member of an Irish family of high intellectual distinction and that a portion of your early education was conducted in this city.

We trust that you will enjoy health and happiness during the closing period of your honourable career.

Given under the Common Seal of the University this 11th day of May, 1899.

THOMAS MOFFETT, *pro-Chancellor.*

JAMES CREED MEREDITH, JOSEPH M^CGRATH, *Secretaries.*

ROYAL COLLEGE OF SCIENCE, DUBLIN.

We, the Dean and Council of the Royal College of Science for Ireland, beg to offer our most sincere and hearty congratulations on the attainment of your Jubilee as Lucasian Professor of Mathematics in the University of Cambridge, and desire to join in offering homage to one who has added still more fame to the Chair once filled by the illustrious Newton.

Your splendid series of researches have enlarged the boundaries of knowledge and enriched the sciences of Mathematics, Physics, and Chemistry; and your labours and influence as Secretary and afterwards as President of the Royal Society have been of inestimable service in stimulating and guiding the zeal of scientific workers not only in the University but throughout the nation and the whole civilized world.

We sincerely hope that your vigour of mind and body may be preserved and your brilliant career prolonged for many years.

Signed on behalf of the Council

WILLIAM McF. ORR, *Dean of the Faculty.*

26th May, 1899.

UNIVERSITY COLLEGE, LIVERPOOL.

I am desired by the Council and Senate of University College, Liverpool to offer you their sincere congratulations on the celebration of your Jubilee as Lucasian Professor in the University of Cambridge.

The sole survivor of that trio of Senior Wranglers who have made Cambridge Mathematics famous in this century, the most distinguished of the successors of Newton, you have held his

Chair for fifty years, and throughout that time have been recognized as a leader among those who have improved Natural Knowledge.

As students we are grateful to you for your work; some of us can claim to be your pupils, and we recognize how much Science owes to you, and we dare to express the hope that, even now, more of the treasures you have stored may be made available for others.

It is not for us to describe your work in Hydrodynamics, in Optics, and in Acoustics; your writings form the foundation of a large portion of our knowledge of these subjects; therefore we value this opportunity afforded us to-day of recognizing your great merits, and of congratulating you on the long period of your service to this Ancient University.

R. T. GLAZEBROOK.

1 *June* 1899.

UNIVERSITY COLLEGE OF NORTH WALES.

The Senate of the University College of North Wales offers its congratulations to the University of Cambridge on the Jubilee of Sir George Stokes as Lucasian Professor of Mathematics. The last fifty years has been a period of rapid and sustained progress of discovery in Physical Science, and, more than any other similar period in scientific history, has been remarkable for the number and importance of the applications of Science to Industry and the Arts. In the studies of the University of Cambridge the Mathematical Principles of Natural Philosophy have, ever since the Lucasian Chair was adorned by the great discoveries of Newton, held an important, if not a predominating place, and the work that the eminent men trained in its Mathematical discipline have done has contributed, above any other cause, to advance the science of our time.

The influence of the work of Sir George Stokes in this Department of Science has been wide-reaching and profound. His discoveries in Hydrodynamics and Physical Optics, and in the great mathematical theories which deal with physical subjects, rank with the productions of the illustrious mathematicians of the preceding half-century, and have done much to inspire the work of other discoverers who were his friends and pupils. His theories, especially those in Wave Propagation and Optics, have found unexpected but important application in the explanation of some of the recondite phenomena that have been discovered by experiments suggested by the great generalization which we owe to another Cambridge Professor, the Electromagnetic Theory of Light.

But above all, the Senate desires to recognise the example which Sir George Stokes has set to scientific workers. Ever slow to publish results until they have been amply confirmed by further investigation, he has pursued his work for the sake of truth alone, and has sought only that appreciation which is shown by the success of researches to which he had pointed the way.

The Senate heartily wishes Sir George Stokes many more years of health and strength for the continuance of labours which have been of so much advantage to the best interests of mankind.

H. R. REICHEL, *Principal*. JOHN EDWARD LLOYD, *Registrar*.

May 18*th*, 1899.

ROYAL INDIAN ENGINEERING COLLEGE.

DEAR SIR,—The Members of the Royal Indian Engineering College, Coopers Hill, desire to join in the congratulations which you are receiving from all parts of the civilized world on the memorable occasion of your completion of fifty years' occupation of the Lucasian Chair of Mathematics at Cambridge.

Within those fifty years, Applied Mathematics has experienced a development and acquired a power for the investigation of almost all natural phenomena which are, we venture to think, beyond all precedent in the history of human knowledge. To such an extent has this been the case that the world has been almost as much impressed by the achievements of the Mathematician in the domain of Physics as by those of the Inventor.

This glorious result has been produced in England mainly by the genius of a few men—Faraday, Kelvin, Clerk Maxwell, and yourself; and the University of Cambridge attests to-day, by the splendid work of her alumni, the power and efficacy of the teaching which she has received, during your fifty years of tenure, from her illustrious instructors. There is no need for us to particularise your contribution to this result. The great masters of Modern Science have repeatedly paid tribute to your work, both as an experimentalist and as a mathematician. Your name remains for ever impressed, both by experimental discovery and by all-important theorems, on Physical Optics and on Hydrodynamics.

For these things, Sir, and for something which is outside them—your invariable urbanity and courtesy to all who have worked in your domain and sought your assistance and advice—we beg to assure you of our profound admiration, and our desire that you may be still long with us as one of the brightest lights of British Science.

JOHN PENNYCUICK, *President*.

31*st May*, 1899.

VICTORIA INSTITUTE.

We, the Vice-Presidents, Council, Members and Associates of the Victoria Institute desire to offer our most sincere congratulations to you, our President, on the occasion of the Jubilee of your tenure of the Lucasian Professorship of Mathematics in the University of Cambridge.

We shall not attempt to enumerate your varied and valuable achievements in different branches of Science, but will confine ourselves to expressing our deep appreciation of your labours in the field of Natural Religion. It has strengthened the faith of many to find that one so well able to judge as yourself has not found Religion and Science antagonistic, but that you have in many instances—notably in your Gifford Lectures—either given able and convincing expositions of their harmony, or have clearly demonstrated the wisdom of a suspense of judgment till the advance of human knowledge may enable solutions to be discovered for problems as yet unsolved.

We trust that the Victoria Institute and its high objects may long have the advantage of your Presidency, and that every blessing may be vouchsafed to you, to Lady Stokes, and to your family.

HALSBURY.

F. W. H. PETRIE, *Capt., Hon. Sec.*

VICTORIA INSTITUTE, *June 19th*, 1899.

PROFESSOR D. MENDELEEFF.

I am very sorry not to be able to attend the celebration of the anniversary of your fifty years' glorious activity as Professor of Mathematics at the University of Cambridge; I beg you to accept my sincerest congratulations and best wishes to this rare event, which reminds me of my student days, when I first was delighted in reading your ingenious investigations in Natural Philosophy.

ST PETERSBOURG, 16/28 v. 1899.

INDEX TO VOL. I.

Printed in the United States
By Bookmasters

Printed in the United States
by Baker & Taylor Publisher Services